"*Ravenous* tells the story of an ex... ...ry work that sustained it. . . . [An] ...l-written book."   —Thoma... ...*pplement*

"The research that [Otto] Warburg is best known for today, and the work that forms the backbone of *Ravenous*, is his discovery that cancer cells behave differently from healthy cells in two very specific ways: They consume massive amounts of glucose—[Sam] Apple compares them to ravenous shipwrecked sailors—and they eschew aerobic respiration in favor of fermentation. . . . Apple covers everything from Hitler's obsessive preoccupation with cancer to how the German Empire's transformation into an industrial powerhouse led to a Romanticism-fueled movement that emphasized both environmental and racial purity. The fact that Apple can make these stories . . . feel so immediate is a testament to his canny knack for choosing apposite details."
          —Seth Mnookin, *New York Times Book Review*

"[Apple] weaves together this complex narrative in a way that makes arcane science accessible and fascinating. The book is also thought-provoking for anyone interested in avoiding cancer—and who isn't?"
          —Marie McCullough, *Philadelphia Inquirer*

"[A] fascinating new book about the link between diet and cancer . . . [*Ravenous*] has received positive reviews from both cancer researchers and general readers interested in how the way we eat in Western societies (specifically the inordinate amounts of sugar we consume) makes us vulnerable to cancer."   —Renee Ghert-Zand, *Times of Israel*

"*Ravenous* is a page-turner, and much of its success is due to Apple's fluid, approachable writing. . . . A joy to read and an utterly fascinating tale."   —Juli Berwald, Jewish Book Council

"A fascinating account of Warburg."
—Sylvia R. Karasu MD, *Psychology Today*

"[A] spellbinding new work of reporting by science journalist Sam Apple. Rarely has such an array of medical troubles, historical events, bench science and political intrigue come together as they do in *Ravenous*." —Paul John Scott, *Post Bulletin*

"Apple skillfully blends science writing with biography to present the story of this quirky, arrogant, and brilliant scientist, who revolutionized research on cancer and photosynthesis (how organisms use energy to make glucose). . . . An illuminating account that makes Warburg (the man and the scientist) accessible to general readers." —*Library Journal*

"A long-overdue biography of German biologist Otto Warburg (1883–1970), who won the Nobel Prize for his work on cell respiration and metabolism, especially as related to cancer. . . . A welcome addition to the library on the disease and one of its most successful enemies." —*Kirkus Reviews*

"Otto Warburg's decades-old science is central to a revolution in thinking about cancer as a metabolic disease. Sam Apple's riveting book, *Ravenous*, reveals Warburg in all his brilliant, bizarre complexity and is a must-read for anyone interested in the science behind low-carbohydrate/high-fat and ketogenic eating."
—Gary Taubes, science journalist, author of *The Case Against Sugar*

"Sam Apple's *Ravenous* is biography at its best. Otto Warburg is an uncommonly good subject—a cell biologist who could not stand his fellow humans but devoted himself to saving them from the scourge of cancer. The author's understanding of Warburg's life and scientific legacy is perceptive and subtle, his biology lessons are a joy to read, and his history of the connections between Hitler and Germany's early cancer research is a small masterpiece."

—Patricia O'Toole, author of *The Five of Hearts*, *When Trumpets Call*, and *The Moralist*

"A remarkable book that just might make you rethink your diet. It's well known—or should be—that poor nutrition can disrupt metabolism and lead to obesity, diabetes, and heart disease. In *Ravenous*, Sam Apple reveals that many of the most deadly cancers are connected to this very same diet-driven disease process."

—Mark Hyman, MD, author of *The Pegan Diet* and head of Strategy and Innovation at the Cleveland Clinic Center for Functional Medicine

"A brilliant weave of history and science, *Ravenous* tells the riveting story of how Otto Warburg, a Nobel Prize–winning biochemist and a gay man of Jewish descent, survived the Third Reich in a posh Berlin suburb, and how his theories of metabolic cancer cells may yet hold the key to finding a cure for the defining disease of our time."

—Helmut Walser Smith, professor of German studies at Vanderbilt University and author of *The Butcher's Tale* and *Germany: A Nation in Its Time*

"*Ravenous* reads like a cancer mystery with the larger-than-life Warburg in the role of the determined detective. By learning of the scientific struggles of the past, you'll gain a new appreciation for the modern focus on hormones, such as insulin, in the development of cancer."

—Benjamin Bikman, associate professor at Brigham Young University and author of *Why We Get Sick*

"A fantastic read. If you're interested in history or science—or just need inspiration to eat less sugar—this is the book for you."
—Nina Teicholz, science journalist and best-selling author of *The Big Fat Surprise*

"While tobacco-induced cancer deaths continue to decline, the second major cause of cancer—obesity—moves to center stage. Few realize its profound importance in causing cancer. Sam Apple has written an endlessly interesting and carefully researched book."
—Robert A. Weinberg, founding member of the Whitehead Institute and professor of biology at MIT

"A gripping and smart page-turner, Sam Apple's *Ravenous* tells two fascinating interwoven stories: that of the pioneer of cancer metabolism research Otto Warburg, who in the twenty-first century finally has been proven right, and that of Hitler's fear of cancer, both as the disease that had killed his mother and as a political metaphor."
—Thomas Weber, professor of history and international affairs at the University of Aberdeen and author of *Becoming Hitler*

RAVENOUS

RAVENOUS

Otto Warburg, the Nazis, and the
Search for the Cancer–Diet Connection

SAM APPLE

LIVERIGHT PUBLISHING CORPORATION
A Division of W. W. Norton & Company
Independent Publishers Since 1923

Copyright © 2021 by Sam Apple

All rights reserved
Printed in the United States of America
First published as a Liveright paperback 2022

For information about permission to reproduce selections from this book, write to
Permissions, Liveright Publishing Corporation, a division of W. W. Norton & Company,
Inc., 500 Fifth Avenue, New York, NY 10110

For information about special discounts for bulk purchases, please contact
W. W. Norton Special Sales at specialsales@wwnorton.com or 800-233-4830

Manufacturing by Lakeside Book Company
Production manager: Beth Steidle

Library of Congress Cataloging-in-Publication Data

Names: Apple, Sam, author.
Title: Ravenous : Otto Warburg, the Nazis, and the search for the cancer-diet
connection / Sam Apple.
Description: First edition. | New York : Liveright Publishing Corporation, a division of
W.W. Norton & Company, [2021] | Includes bibliographical references and index.
Identifiers: LCCN 2021005472 | ISBN 9781631493157 (hardcover) |
ISBN 9781631493164 (epub)
Subjects: LCSH: Warburg, Otto Heinrich, 1883–1970. | Biochemists—Germany—
Biography. | Cancer—Nutritional aspects. | Cancer cells—Growth. | Cancer—Research—
Germany—History—20th century. | Science and state—Germany—History—
20th century. | National socialism and medicine. | National socialism and science.
Classification: LCC QP511.8.W37 A67 2021 | DDC 616.99/40092 [B—dc23
LC record available at https://lccn.loc.gov/2021005472

ISBN 978-1-324-09201-8 pbk.

Liveright Publishing Corporation, 500 Fifth Avenue, New York, N.Y. 10110
www.wwnorton.com

W. W. Norton & Company Ltd., 15 Carlisle Street, London W1D 3BS

1 2 3 4 5 6 7 8 9 0

For everyone who helped—
you know who you are.

Contents

Introduction   ix

PART I

A "Disease of Civilization"
(Late Nineteenth Century–1918)

1 "A Chemical Laboratory of the Most Amazing Kind"   3

2 "The Great Unsolved Problem"   16

3 Magic Bullets   33

4 Glucose, Cancer, and the Crown Prince   43

5 "Slaves of the Light"   54

PART II

"The Meaning of Life Disappears"
(1919–1945)

6 The Warburg Effect   71

7 The Emperor of Dahlem   90

8 "The Eternal Jew"   113

9 "The Herb Garden" of Dachau   137

10 The Age of Koch   152

11 "I Refused to Intervene"   165

PART III
The Seed and the Soil
(Postwar)

12   Coming to America   181

13   Two Engines   195

14   "Strange New Creatures of Our Own Making"   207

15   The Prime Cause of Cancer   220

16   Cancer and Diet   230

PART IV
Pure, White, and Deadly
(The Twenty-First Century)

17   Lost and Found   243

18   The Metabolism Revival   260

19   Diabetes and Cancer   273

20   The Insulin Hypothesis   288

21   Sugar   302

22   The Evil Twin   314

Postscript   331

Acknowledgments   333

Notes   337

Suggestions for Further Reading   369

Illustration Credits   371

Index   373

Introduction

ON JANUARY 16, 1934, a Nazi customs official arrived at the door of the Kaiser Wilhelm Institute for Cell Physiology with a stack of papers. A recently issued Nazi decree required scientific institutes to obtain a special license for the purchase of ethanol. The Institute for Cell Physiology had sent in its application six days earlier but had failed to return all of the necessary documents. Noticeably missing was a "declaration of Aryan descent" from its director, Otto Warburg. The Nazi official had brought along a blank form so that this oversight could be corrected.[1]

Like most of the other scientific institutes of the Kaiser Wilhelm Society, the Institute for Cell Physiology was located in Dahlem, a quiet neighborhood of elegant villas in the southwest corner of Berlin. Though connected to the city by train, Dahlem, sometimes called the Oxford of Germany, was a small world unto itself, home to a significant portion of the greatest scientists alive. In the years before the Nazis came to power, it would have been hard to walk from one end of Dahlem to the other without bumping into a Nobel laureate. Otto Warburg himself had won the Nobel Prize only three years earlier. He was considered by many to be the greatest biochemist of his generation.

Standing on the doorstep of Warburg's institute, the Nazi customs official, Mr. Tesch, was likely wearing the uniform of his office, with a green, swastika-adorned sash visible on the sleeve of his long wool coat.

As Tesch would have known, it was no accident that the "declaration of Aryan descent" was missing from Warburg's application for an ethanol license. Warburg may have been the most famous Jewish name in all of Germany. Though best known as the family behind the legendary Hamburg-based M.M. Warburg & Co. bank, the Warburg influence extended beyond finance. Members of the family were leading scholars, artists, and philanthropists, a veritable German-Jewish aristocracy in a country that, long before Hitler, had severely restricted the place of Jews in public life.

Otto Warburg's mother was not Jewish, but it only took one Jewish grandparent to make him a "non-Aryan" according to the Nazi regulations of 1933. Warburg had two Jewish grandparents, and his famous Warburg cousins in finance were frequent targets of Nazi propaganda, the very symbols of Jewish capitalist greed. Given that Warburg also lived with a male servant rumored to be his homosexual partner, he had as much reason to fear the Nazis as almost anyone in Germany.

The door to the institute was opened by an employee, probably one of Warburg's research assistants, who told the Nazi official that Warburg was unavailable. Tesch left the blank Aryan descent form with the employee, making clear that he needed it returned within 48 hours. Three days later, still awaiting the filled-out form, Tesch called the institute. A secretary answered and passed the handset to Warburg. "I served in the military and was an officer," Warburg blared into the phone. "It is out of the question that I will sign this form."

Though Tesch already knew that Otto Warburg was not of Aryan descent, there was something else about Warburg that he likely did not know when he placed that call. Whether Warburg was the greatest biochemist of the era was debatable, but he was almost certainly the most self-important biochemist who ever lived. As a colleague put it, measuring arrogance on a scale from 1 to 10, "Warburg rated 20." Warburg was so enamored with himself that he once refused even to be photographed with a group of scientists he deemed beneath him—

and a number of the scientists in the group were Nobel Prize winners. For someone entirely convinced of his own greatness, the idea of Nazi lowlifes telling him which chemicals he could and could not order was almost unthinkable. As Warburg once put it to his sister: "*Ich war vor Hitler da*"—"I was here before Hitler."[2]

Still fuming days after the phone call, Warburg ordered a secretary to call the customs office that had sent Tesch to his institute. "Professor Warburg does not wish to see the customs official who delivered the forms again," the secretary announced, "and, if necessary, would have him removed from the building."

The Nazi official on the other end of the line was stunned. But Warburg's secretary wasn't finished. Asked to explain why Warburg was behaving so rudely to a government official, the secretary responded that Tesch had arrived at the institute "unshaven" and had "spread unpleasant odors around him." The odors, he clarified, had "presumably originated from an unclean body." The Institute for Cell Physiology, the secretary explained, "had to ensure a meticulous level of cleanliness."

To the hygiene-obsessed Nazis, few insults could have been more offensive. While the archival documents suggest the secretary was a "she," it was likely Warburg's male partner, Jacob Heiss, who told off the customs officials. Heiss, who would later become the administrator of the institute, was rarely far from Warburg's side, and he was in the habit of shouting people down on his behalf. A glassblower who worked at the institute after the Second World War remembered Heiss screaming at unwanted visitors in exactly the manner of the secretary on the phone with the Nazis.[3]

Warburg's message did not get through to the customs office. Tesch reappeared at his institute that same day demanding the completed Aryan descent form. An employee led him to a laboratory room with an open door. Standing before the entrance to the room, the Nazi official came face-to-face with several researchers, including Warburg.

Warburg was a handsome and compact man. He kept his hair short, neatly parted to one side and swept back over an always clean-shaven face. On the day that Tesch returned to the institute, Warburg may have been wearing a white lab coat over one of the cardigans or tailored English sports coats he favored.[4]

Few things annoyed Warburg more than interruptions of his work. To avoid unwanted visitors, he installed a brass plaque by the door of the institute indicating that visiting hours began at 6:30 p.m. In photos, Warburg's heavy-lidded blue eyes have a dreamy quality, leaving the impression of a man lost in deep thought. But if Tesch directed his gaze to Warburg at this moment, he almost certainly would have been met with eyes that could fill with "spark-shooting rage," as one biochemist recalled.[5]

Tesch had never met Warburg and did not recognize him as he stood before the laboratory door. As Warburg approached, Tesch raised his right arm to the ceiling and stiffened it at 45 degrees in a rigid Nazi salute. Warburg was expected to salute back. Instead, he walked past Tesch and into the hallway without speaking. Tesch was flabbergasted. It was an "outrageous disregard of a civil servant who is a representative of the National Socialist State," he wrote in a report on the incident.

With Warburg now standing behind him in the hallway, Tesch demanded his name. According to Tesch's testimony, Warburg turned halfway around, announced who he was, and pointed down the hall. "There is the door," Warburg said. "Leave the building!"

Warburg would have to repeat the demand several more times before Tesch finally departed. Warburg immediately filed a formal complaint with the customs office, stating that the institute no longer had possession of the form Tesch wanted and that it no longer needed any ethanol.

For all his bravado, Warburg was likely nervous. After sending off his complaint, a secretary (presumably Heiss) called the customs office and asked what would happen if Warburg were never to provide evidence of his Aryan descent. The matter, the secretary was told, would be addressed with the Kaiser Wilhelm Society, the parent organization of Warburg's institute. At this point, according to the report of the customs office, Warburg's secretary announced in "a tone of derision" that Warburg's institute was privately funded and did not need to follow instructions from the Kaiser Wilhelm Society.

The Nazi customs officials had heard enough. The office sent a report of the incident to Max Planck, the president of the Kaiser Wilhelm Society and the effective head of German science. In it, Tesch said that Warburg had "grossly insulted" him and demonstrated a "disregard of the German salute" that was, in his opinion, "characteristic of Prof. Warburg's attitude toward the current State." Unless Warburg apologized, the report stated, further steps could be taken against him.

Planck recognized the seriousness of the threat. On February 13, 1934, he sent Warburg a telegram: "Honored Professor. The President asks you to appear for a consultation on Friday, the 16th of this month, 12 o'clock at noon, here at the palace. Heil Hitler!" While there is no record of what was said at that meeting, Planck let the customs office know that he had spoken to Warburg about his behavior and that future requests for ethanol would come from the administration of the Kaiser Wilhelm Society, rather than directly from Warburg's institute.

Warburg, having seemingly avoided even the slightest consequence for his actions, might have left the matter there. Instead, five days later, he sent a note to Planck stating that the new Nazi restrictions made for impossible working conditions. Something needed to be done, and Warburg had a suggestion: He wanted Planck to contact the Reich Ministry of Finance with instructions on how to adjust the existing racial decrees. Warburg even offered specific language for the revised legislation, so that it would be clear to all that non-Aryan institute directors were "to be treated like Aryan directors."

In 1934, at a moment when Hitler had already begun sending Germans to concentration camps, Otto Warburg, a gay man of Jewish descent, wanted Nazi laws rewritten according to his personal needs.

THIS BOOK IS an attempt to unravel two related mysteries. The first is the mystery of Otto Warburg's life and science. The episode with the Nazi customs office was only one of many clashes between Warburg and the Nazis. And yet, remarkably, Warburg would prevail again and again. As the Nazis goose-stepped their way across Europe, rounding up and murdering Jews, Otto Warburg woke up every morning in an elegant, antique-filled home and rode horses with his male partner. He spent summer vacations at his private residence on a windswept island in the Baltic Sea. When Allied bombing made it too dangerous to remain in Dahlem, the Nazis relocated Warburg's institute to a countryside estate that had once been a favorite hunting spot of Kaiser Wilhelm II.

Warburg proved a tricky case for the Nazis in the 1930s. He was a Nobel Prize winner, and his institute had been paid for by the Rockefeller Foundation. His persecution would not have gone unnoticed in other countries. But once the war began, international opinion mattered little. In the spring of 1941, the Science Office of the Reich Ministry of Education evicted Warburg from his institute. It appeared to be the end for Warburg.[6]

In a desperate bid to hold on to his career, and, perhaps, his life, Warburg rallied his most influential colleagues to his defense. Their efforts to save Warburg led to a meeting on June 21, 1941, at the New Reich Chancellery, the seat of the Nazi government in Berlin. At the meeting, Warburg learned he would be allowed to continue working at his institute so long as he focused on cancer research. Though neither Hitler nor Heinrich Himmler is known to have been present at the meeting, they may well have been aware of Warburg's presence in the

Chancellery. Himmler's appointment book reveals that he had a meeting to discuss Warburg's fate that very afternoon.[7]

That the Nazi leadership was focused on Warburg during the war is perplexing in itself. But June 21, 1941, was not merely one more day in the conflict. Only hours after Warburg left the building, Germany would invade the Soviet Union. It was the largest military operation in history and the most critical moment of Hitler's political life. Otto Warburg should have been the last thing on the minds of the Nazis at that moment.

WHILE THE NAZIS often gravitated toward quackery, their interest in Otto Warburg was rooted in the most respectable science of the time. In the early twentieth century, Warburg created several new tools to measure the metabolism of cells. When he used those tools to study cancer cells, what he found amazed him. The cancer cells weren't eating like other cells. They were swallowing up huge portions of glucose, or blood sugar—as much as 10 times that of healthy cells in the same tissues. The cancer cells were eating like shipwrecked sailors. They were ravenous.

And that wasn't even the strangest part of what Warburg found. Instead of burning almost all of their food with oxygen, as healthy cells typically do, the cancer cells were chopping glucose molecules in half and spitting the fragments right back out of the cell. Cancer cells, Warburg realized, were fermenting glucose just as simple organisms like yeast and bacteria do. It was the same cellular process that gives us beer and wine and bread and so many other common foods.

Warburg knew that human cells could ferment glucose, but fermentation, which does not require oxygen, was supposed to function like a backup generator, only kicking in when there was not enough oxygen to sustain energy production in the cell's power stations, the mitochondria. Cancer cells were starting up their backup generators even when they had all the oxygen a cell could want. It made no sense, and with each passing decade, Warburg grew increasingly convinced that the phenomenon he had discovered—the cancer cell's reliance on fermentation—was the key to understanding the disease. "From the standpoint of the physics and chemistry of life," Warburg said, the "difference between normal and cancer cells is so great that one can scarcely picture a greater difference."[8]

Warburg eventually came up with an explanation for the phenomenon. Burning food with oxygen is a much more efficient way to generate energy. If cancer cells were running their backup generators even when there was plenty of oxygen for the power stations, it could only be because those power stations were somehow damaged.

It mattered little that Warburg had only scant evidence for this hypothesis. When Otto Warburg had an idea, other scientists took it seriously. He was widely regarded as a genius, a singular figure in his field. "It's like having God around," the future Nobel laureate Arthur Kornberg wrote when Warburg came to work at his institute as a visiting scientist.[9]

Warburg's influence would spread far beyond the scientific community. He is the first cancer scientist discussed in Rachel Carson's *Silent Spring*, the best-selling book often seen as a foundation of today's environmental movement. In Carson's telling, artificial chemicals were likely causing cancer exactly as Warburg said, by interfering with a cell's use of oxygen and forcing it to ferment glucose.

That cancer cells typically eat enormous amounts of glucose and ferment much of it was confirmed by other researchers in the decades after Warburg made his discovery. However, in the 1950s, some scientists pushed back against Warburg's explanation of the phenomenon. They could find no signs of the damage to cellular breathing, or respiration, that Warburg insisted was the true cause of cancer. Warburg defended his position in his usual vociferous way, but even as the battle raged in the pages of scientific journals, the larger war was rapidly coming to an end. In 1953, James Watson and Francis Crick—relying on the research of Rosalind Franklin—deciphered the structure of DNA and launched the new era of molecular biology. Other researchers soon set off in search of the specific genes, the so-called oncogenes, that cause cancer. By the mid-1970s, they were beginning to find them.

Though the DNA-focused molecular biologists were aware of the metabolic reactions Warburg studied, by the final decades of the twentieth century, Warburg's work was no longer of any interest. The metabolic proteins responsible for how cells eat would become known as "housekeeping enzymes"—necessary to deliver fuel to growing cells when called upon, but irrelevant to the deeper mysteries of cancer. Warburg's insistence that cancer should be thought of as a problem of energy came to sound as absurd as the notion that a computer is best understood by studying the power supply to the machine.

Thomas Seyfried, a biologist at Boston College, described the rush away from metabolism as a "stampede." "Warburg was dropped like a hot potato," Seyfried told me. In 2006, the first edition of a widely hailed 800-plus-page cancer textbook did not mention Warburg's once-famous discovery; nor does Warburg's name appear in *The Emperor of All Maladies*, Siddhartha Mukherjee's brilliant, Pulitzer Prize–winning history of cancer, from 2010.[10]

If Warburg's story had ended there, it's unlikely that I, or anyone else, would be writing about him today. But in the last years of the twentieth century, something entirely unexpected happened: cancer science began to come full circle. A small group of researchers hunting for cancer-causing genes conducted experiments that led them directly back to Otto Warburg and the science of how cancer cells eat and fuel their growth. It turned out that a number of the same genetic mutations that drive a cell to divide without restraint also drive a cell to eat without restraint.

This shift in our understanding of cancer-causing genes is more than an addition of nuance. It changes our very conception of what cancer is and how it begins. There is good reason to think that a cell's uncontrolled eating is the most fundamental step on the path to cancer. If a cell gains the ability to divide on its own but hasn't yet gained the ability to eat on its own, it is more likely to die than to give rise to a life-threatening cancer. The attempt to grow before securing a supply of nutrients creates a "catastrophe" for the cell, as Craig Thompson, president and CEO of the world-renowned Memorial Sloan Kettering Cancer Center, put it to me.[11]

By 2010, the rediscovery of Warburg's metabolic approach to cancer had led to a full-fledged revival, with new scientific conferences, new drugs designed to starve cancers of the nutrients they need to grow, and thousands of journal articles. A cell turning to fermentation when oxygen is available is now known as aerobic glycolysis, or "the Warburg effect." It's estimated to occur in at least 70 percent of all cancers. In recent decades, the positron-emission tomography (PET) scan has emerged as one of the most important tools for staging and diagnosing cancer. It identifies the location of cancers by revealing the places in the body where cells are consuming extra glucose. The more glucose being consumed by the cancer cells, the worse a patient's prognosis is likely to be.

In the course of researching this book, I have spoken with dozens of the scientists responsible for the revival of cancer metabolism research. A number of them are now the leaders of America's top cancer hospitals. Most of them disagree with Warburg's explanation for why cancer cells eat so differently from other cells. But they increasingly embrace a position Warburg always maintained: The change in the way a cell takes up and uses its food is not merely one of a series of events on the path to cancer. The unrestrained eating sits at the very origins of the disease, awakening the cell to the possibility of growth and driving the transformation to cancer that follows.

Cancer, these scientists argue, is a genetic disease, but the genetic transformation cannot be understood separately from the metabolic transformation. As Seyfried said of the rediscovery of Warburg's thinking, "We found out that the son of a bitch was right."[12]

RUNNING THROUGH THE mystery of Otto Warburg and his research on cancer cells is another mystery: why cancer became so much more common across the Western world starting in the nineteenth century. Cancer is an ancient disease and is found across the animal kingdom. But until the second half of the nineteenth century, it was not especially common. Over the next 100 years, cancer rates in many countries grew at an astounding rate, decade by decade. "Nearly every medical man" was asked the same question again and again, a British public health official observed in 1899: "Why has cancer increased so much?"[13]

Though cancer authorities were alarmed by the rising number of cancer deaths, they would have struggled to imagine the vast human toll cancer takes today. In 1915, the cancer statistician Frederick Hoffman warned of a terrifying future in which as many as 100,000 Americans would die of cancer every year. Today, approximately 600,000 Americans die of cancer every year, an increase that vastly outstrips population growth. Cancer is now the defining illness of our time, a disease that, in its refusal to succumb to our most ingenious schemes to defeat it, has come to symbolize the limits of medical progress.[14]

Cancer is closely linked to aging, and some of the increasing incidence of cancer can be explained by longer life spans alone. Newer, more precise methods of diagnosing cancer and more rigorous record-keeping also account for a portion of the increase. And yet, an immense body of evidence suggests there is more to the story. While cancer was on the march in many countries at the turn of the twentieth century, in others it remained largely absent. The medical literature of the period is full of accounts of Western doctors relocating to less developed regions only to be amazed by the rarity of cancer among the locals. Often people in these indigenous populations lived as long as their European and

American counterparts. Western physicians were struck by the scarcity of cancer precisely because they regularly found malignant growths on people of similar ages in their own countries.

A number of these doctors noticed something else as well: once indigenous populations began to transition to more modern Western lifestyles, cancer soon followed. Cancer appeared to be intimately linked to modernity itself. In the racist terminology of the nineteenth century, it became known as a "disease of civilization." The expression was still in use in the decades after World War II, a period when it was widely agreed that 70 percent or more of all cancers were caused by something in our environments—meaning something other than inherited or random genetic mutations, something other than bad luck.

How modern lifestyles make cancer more common has been among the more important questions in medicine for nearly 200 years. Because the types of carefully controlled experiments that might definitively show cause and effect are nearly impossible to carry out for a disease that develops over many years, it is a maddeningly difficult question to answer. But in the late 1990s, an important new clue was emerging. At the same time that molecular biologists were rediscovering Otto Warburg's research on the unusual appetites of cancer cells, a different set of researchers was arriving at another striking finding: many cancers are closely tied to being overweight.

The suspicion that "overnutrition," as the phenomenon was often referred to, contributed to cancer had been considered since at least the nineteenth century, but it took the American obesity epidemic of the late twentieth century to clarify the picture. While it is difficult to tease out clear links between specific foods and cancer, the connections between obesity and cancer are not ambiguous. At least 13 different cancers have now been strongly linked to excess body fat, including many of the most common and deadly cancers of all: pancreatic, thyroid, ovarian, uterine, colon, and (in postmenopausal women) breast cancer. (Less strong evidence points to a link between obesity and still more cancers, including prostate cancer, which afflicts one in every nine American men.)[15]

Obesity is a consequence of our cells ending up with more food than they should. Warburg discovered that cancer cells overeat. That the two phenomena must somehow be connected seems almost inescapable. And yet because the researchers who study cancer rates occupy a different silo of the medical world than those who explore the metabolism of cancer cells, the overlap between these two fields is often overlooked.

Drawing connections between diet, obesity, and cancer remains a controversial enterprise. Highly respected researchers continue to disagree about almost every aspect of the science. In this book, I've attempted to explore the science from the beginning, using Warburg's life as a frame for a narrative that unfolds across three different centuries and includes the strange story of how the rise of cancer influenced Adolf Hitler's own life and thought. In the final chapters of the book, I explore one particular hypothesis about cancer and diet that is increasingly accepted by mainstream cancer authorities and that may go a long way toward explaining why cancer is now such a common disease.

THOUGH CANCER WAS commonly referred to as a "disease of civilization" in the nineteenth and twentieth centuries, the term took on special resonance in Nazi Germany. The increase in cancer that would invariably accompany a country's industrial development fit all too neatly with the Nazi view that modern, urbanized life was profoundly corrupt. The term also captures the terrible irony at the center of this story. While cancer did grow more common in modern industrial societies starting in the 1800s, modernity's worst plague was an ideological rather than a physical sickness. The true "disease of civilization" was Nazism itself.

Otto Warburg was a victim of that Nazi disease, but he was not among the more sympathetic ones. And while this book is, in part, the story of Warburg's scientific redemption, it is not meant to be a complete vindication. Warburg was on the wrong side of a number of scientific debates. His biography can be read as the story of a visionary whose ideas about cancer were overlooked for too long. It can also be read as a parable of the dangers of certainty.

Because the book is in part an endorsement of Warburg's metabolic understanding of cancer, I suspect he might have appreciated some of the arguments in the pages that follow. I am more confident that he would have found a thousand things to hate.

Otto Warburg, around 1931.

PART I

A "Disease of Civilization" (Late Nineteenth Century–1918)

Faustus:
Be a physician, Faustus. Heap up gold,
And be eternized for some wondrous cure. . . .
The end of physic is our bodies' health.
Why, Faustus, hast thou not attained that end?
Is not thy common talk found aphorisms?
Are not thy bills hung up as monuments,
Whereby whole cities have escaped the plague
And thousand desp'rate maladies been eased?
Yet art thou still but Faustus, and a man.

—CHRISTOPHER MARLOWE,
DOCTOR FAUSTUS

CHAPTER ONE

"A Chemical Laboratory of the Most Amazing Kind"

THE STORY OF modern cancer research begins, improbably, with the sea urchin. At the start of the twentieth century, the German scientist Theodor Boveri turned to sea urchin eggs to answer one of the central questions in biology: how are the instructions for making a new organism passed from one generation to the next? It was clear that a fertilized sea urchin egg knew how to grow from a single cell into a spiny round creature that could inch its way along the ocean floor. And yet, despite the countless hours Boveri and his colleagues spent looking, the instructions themselves were nowhere to be found.

Boveri was a kind, reserved man, serious and meticulous. He kept his research bench at the University of Würzburg pristine. When lecturing, he stood immobile before the chalkboard and spoke quietly, "his eyes boring into" his audience, a student recalled.[1] Much of Boveri's working life was spent hunched over his laboratory table, gazing into his light microscope. The rest of it was spent drawing the cells he observed. Even when photographic reproduction became possible, Boveri, an amateur painter, preferred to rely on his own hand.

The sketching of the cells wasn't merely an act of reproduction. Boveri's former student and biographer Fritz Baltzer wrote that when drawing, Boveri carried on "an undisturbed and intimate dialogue with nature." For Boveri, drawing was a way of seeing, and he saw in a way that few others could: "One had the feeling," Baltzer wrote, that Boveri "saw more in a few minutes than the students had perceived in hours or days."[2]

Though he lived and taught in Würzburg, Boveri was happiest during his regular stays at the Naples Zoological Station. Opened in 1873 by Anton Dohrn, a wealthy German biologist, the zoological station became a major research site for Europe's leading developmental biologists. The building, as Boveri once described it, was a "beautiful white edifice" flanked by "dark green oaks." Standing in the second-floor corridor beneath the red arches, Boveri could look out onto the calm waters of the Gulf of Naples, where, hidden beneath the surface, lay the true treasures: the simple sea creatures that had become the preferred experimental subjects of physiologists and developmental biologists of the era.[3]

In the 1880s, Boveri was studying chromosomes, the little threads that had recently been discovered in the nucleus of the cell. At the time, scientists thought that chromosomes appeared anew in every cell. Through his careful examination and repeated drawings, Boveri saw that this was wrong. Cells grow by division: one cell becomes two; two become four; four become eight; and so on. Shortly before a cell divides, each chromosome is copied and pinned to its twin at the center. As a cell begins to split into two new cells, a quiet drama unfolds: the twins, pulled apart from their centers, fold inward at either end, appearing as if they're reaching out for one another, as if they're not quite ready to begin a new life alone.

That chromosomes are copied and passed from a cell to its progeny was a critical clue for Boveri, but it proved nothing in itself. He still needed a way to test whether chromosomes influence the development of a sea urchin. In 1901, while working in Naples, he found it, thanks to an experimental innovation that allowed him to create sea urchin eggs that lacked a complete set of chromosomes.

Some of the sea urchin embryos without the proper chromosomes would form hollow balls; others would grow half-formed parts and collapse in on themselves. Boveri had turned developmental biology into a theater of the grotesque, but the lesson was clear: each chromosome held a different part of the instructions for a new life. Without the right chromosomes, a fully formed and properly functioning sea urchin would never arise, just as a complete building can't be built from a blueprint that is missing critical sections.

If Boveri had stopped there, his 1902 paper summarizing his findings would still be among the most significant in the history of biology. We now know that chromosomes are made of tightly coiled strings of DNA that contain the instructions for how to build an organism. Boveri's mangled sea urchin embryos made possible the extraordinary genetic discoveries of the twentieth century.

Even though he was focused on how an organism develops normally, Boveri couldn't resist ending his paper with one additional observation: some sea urchin embryos with the wrong number of chromosomes develop "tumorlike" growths. Since it was already known at the time that cancer cells often had abnormal chromosomes, Boveri couldn't help but wonder if, in searching for how chromosomes shape the development of life, he had also found how chromosomes shape the development of cancer. Perhaps, he reasoned, cancer arises when the instructions for proper growth are jumbled.[4]

Boveri's cancer theory received little notice during his lifetime. But by the end of the twentieth century, he would be celebrated as the first scientist to recognize that cancer is a disease of bad information, a disease of damaged DNA providing cells with the wrong instructions.

As IT TURNS OUT, another scientist, Otto Warburg, was studying the growth of sea urchin eggs at the Naples Zoological Station at approximately the same time as Boveri. Though still a medical student, Warburg had already announced his desire to cure cancer. Because eggs, like tumors, grow by dividing again and again, they made for a logical place for Warburg to begin his experiments.[5]

Beyond a shared interest in sea urchin development and what it might reveal about cancer, Warburg and Boveri had little in common. Warburg didn't need to *see*. As a chemist, he was at home in the world of the invisible, with what he could infer by measurement and calculation alone. And so while Warburg was also studying sea urchin eggs in Naples and also interested in what the eggs might teach him about cancer, he approached his research with an entirely different set of questions. Boveri wanted to understand how changes to chromosomes provide cancer cells with the instructions to grow. Warburg wanted to understand how changes to breathing patterns provide cancer cells with the energy to grow. If a cell's path to cancer can be thought of as a building project gone awry, Boveri focused on what went wrong with the blueprint; Warburg on what went wrong with the power stations needed to run all of the construction equipment.

Warburg and Boveri weren't only different types of scientists. They were also very different types of people. Boveri was a man of doubt. After suggesting in his 1902 paper that cancer could be explained by chromosomes, he waited over a decade before expanding on the idea for fear of how it would be received. Warburg, who spent his entire career feuding with other researchers, at times seemed constitutionally incapable of doubt. When Warburg, an avid equestrian, fell during one of his morning rides and fractured his pelvis, he blamed his horse. A glassblower who worked closely with Warburg for eight years couldn't remember him ever admitting he was wrong on a single matter.[6]

Boveri once compared himself to a chandelier that had "ceased to give light" but "nevertheless decorated the room." Warburg, in his own mind, illuminated every room he stepped into. And he expected those rooms to be well furnished. He collected antique British furniture and rugs and owned silver from one of the oldest noble families in Europe. When his sister sent him a clock while he was away in medical school, he thanked her in a letter but confessed that he didn't like it. "It seems to be putting on airs," Warburg wrote. "It wants to be a bronze clock on a marble stand," but is made of "gilded wood that has been lacquered like a Berlin tenement."[7]

There might be no better example of Warburg's famed arrogance than his response after learning he had won the 1931 Nobel Prize: "It's about time." Eight years earlier, he'd been so sure he would win the Nobel Prize that he traveled to London to have a suit tailored for the ceremony. Warburg wasn't delusional. He had good reason to expect the Nobel Prize at some point in his life. He was raised to win it. "I was so much encouraged to become a scientist," Warburg once said, "that I deeply pitied every other occupation of men."[8]

Warburg's father Emil, himself nominated for a Nobel Prize in 1929, was among the leading physicists of his generation. (Warburg's mother, Elisabeth, managed the household and cared for Warburg and his three sisters.) In 1895, Emil Warburg was named professor of physics at the University of Berlin, placing him at the pinnacle of German science at a time when someone of Jewish heritage rarely managed to attain such heights. The family moved from Freiburg to Berlin, and Otto Warburg's childhood home became a gathering place for a circle of extraordinary German scientists, among them Max Planck, Fritz Haber, Walther Nernst, and Albert Einstein.

Given his parentage, it was little surprise that Otto Warburg would develop an almost pathological dedication to his science. According to the Nobel Prize–winning physicist James Franck, Emil Warburg's proposition to his students was straightforward: if you aren't prepared to dedicate "every bit" of yourself to science, it is best to keep your "hands off it."[9]

When studying under Emil Warburg, Franck and his fellow students took few breaks. Emil would sometimes turn up at the university lab at midnight ready to discuss their experiments as though it were perfectly normal to be debating the finer points of gas laws in the middle of the night. And God help the student who wasn't prepared for his unannounced arrivals. As Franck recalled, when a student couldn't answer questions about his research, Emil would sometimes simply walk away without a word so that "the unfortunate sinner" might absorb "the full measure of his ignorance."[10]

Otto Warburg may well have surpassed his father's attachment to science. The biochemist David Nachmansohn compared Warburg's passion for his work to "the religious fervor of some historical figures, whose whole life was almost exclusively devoted to their relationship with God." Warburg himself liked to say that a scientist should be "ready to die for the truth," and he expected everyone who worked in the lab to participate in his martyrdom.[11]

Still, it's one thing to be a person who lives only for science and another to be the child of such a person. Warburg's sister Lotte kept a diary, and Emil Warburg's choice of work over family is among the dominant themes in it. (Einstein once condescendingly suggested that she had a "daddy complex.") In her diary, Lotte describes Emil as "objective to the point of numbness." He could be "selfless and friendly" with his students, but when it came to family, he was "loveless." In another entry, Lotte remarks that her father wouldn't have been able to find the bed he supposedly shared with her mother and that it would have been better had he never married. Emil's thoughts, Lotte writes, led an independent life and "forced him to give up his own life."[12]

So as not to interrupt Emil's thoughts, the Warburg home ran according to strict dictates. The rules weren't merely implied; Emil had written them out by hand. They included restrictions on tipping one's chair onto its back legs while seated and putting one's elbows down on the table. In fairness, his son may have required a bit of heavy-handed parenting. Though little is known about Otto Warburg as a young boy, when he was 12 his parents received a letter from the elite Berlin *gymnasium* he attended. While the letter notes that Warburg demonstrated "considerable talents and accomplishments," there was bad news as well: "The pupil Warburg" the letter states, has "repeatedly taken part in gross misconduct and he has encouraged fellow pupils to join in."[13]

Whether Warburg was disappointed by his father's sternness or absence from family life is difficult to know. It is clear that Warburg saw his father as a competitor. Warburg had been a good violinist but gave it up when he saw that he would never be able to play music as well as Emil, a pianist who sometimes played chamber music with Einstein. The competitiveness might have been driven by spite. Emil was known for his frugality and resented his son's opulent tastes. And he appears to have never quite fallen out of the habit of criticizing his son. In an exchange of letters in 1912—Warburg was 29 at the time and already one of the most promising scientists of his generation—Emil expressed genuine interest in Otto's work, but also couldn't resist pointing out that he found it "increasingly difficult to decipher" his son's handwriting.[14]

Nevertheless, with his approach to science and insistence on repeated and meticulous measurements, Otto was practicing the ideals of his father, who once wrote that learning to measure accurately leads to the development of a "a serious, manly scientific character."[15]

If Warburg strayed from the path set before him as a child, it was by journeying beyond the pure science of physics and chemistry into the realm of the living. Emil Warburg had found new ways to understand the laws of the universe. For his brilliant son, that would not be enough.

AFTER TWO YEARS studying chemistry at the University of Freiburg, Warburg moved back to Berlin in 1903 to train in organic chemistry under Emil Fischer, one of the luminaries in his father's circle. Fischer was then the most distinguished organic chemist in the world. Tall and stern, he would stand perfectly straight in the lab in his formal black hat and pince-nez. The moment a student turned a conversation in a more personal direction, Fischer turned it right back to science. Warburg recalled that even after he had crystallized a molecule numerous times, a decidedly unimpressed Fischer would say, "Now go ahead twenty-five times more."[16]

Fischer had won the Nobel Prize a year earlier for his work on the chemical structures of sugars and purines. When Warburg arrived at his lab, he was studying proteins. Most researchers at the time believed that proteins were little more than unstructured clumps of smaller molecules. Fischer knew better. Proteins, he recognized, were built from simple molecules, amino acids, but they weren't the amorphous clumps other researchers described. They were intricately structured micromachines, each one designed to perform a unique job inside of the cell based on its unique form.

To decipher the structure and function of a protein was a Herculean task in its own right, but Fischer wasn't interested in merely understanding how proteins work. He hoped to one day make proteins from scratch. It was, he knew, an ambitious goal, but by 1903, he had already managed the very first step: synthesizing amino acids into short strings. Warburg wrote his dissertation on the process, and had he devoted the rest of his life to studying only the structures of organic molecules, he might still have become one of the most important scientists of his generation. But after receiving his degree, Warburg realized he didn't want to stop at organic chemistry.

The problem wasn't that Warburg's research under Fischer was too dull. On the contrary, the science felt revolutionary. Science, in Emil Fischer's words, was coming to seem like "the true land of unlimited possibilities." If a strand of amino acids could be synthesized in the lab, how long would it be until scientists could piece together an entire protein? And how long after that until organic molecules might be "tricked," as Fischer once put it, into doing the scientist's bidding so that the cell itself could function as "a chemical laboratory of the most amazing kind"?[17]

In the early twentieth century, Fischer anticipated a future, if only "half in a dream," when scientists would reinvent "the living world" itself. Warburg did not even need to dream to see that future. In 1906, just as Warburg was completing his training in organic chemistry, the German American physiologist Jacques Loeb published two popular monographs in German that made Fischer's grandiose visions seem almost modest.[18]

Though largely forgotten today, in the first decades of the last century, Loeb may have been the most famous scientist in America. Regularly featured in newspapers and magazines, he was the inspiration for the character of Max Gottlieb in Sinclair Lewis's novel *Arrowsmith*. Loeb, too, spent time at the Naples Zoological Station, where he performed a series of Frankenstein-esque experiments on the sea creatures that lived in the shallow waters of the gulf. He discovered that if he took tubularias, plant-like animals that spring from the ocean floor, and grew them upside down, he could "produce at desire a head in place of a foot." By cutting away a piece of the animal and suspending it in an aquarium, Loeb could even grow a two-headed tubularian. In another experiment, Loeb suspended a starfish in water by attaching its "arms" to small pieces of cork. A starfish, if turned upside down, will flip back over. But with no surface to orient itself, the suspended starfish could no longer distinguish up from down and would flip again and again, the suspended sea creature itself a metaphor for the dizzying state of the biology of the age.[19]

Loeb left Germany in 1891, convinced that his Jewish background would make a university appointment impossible. He first settled in Zurich, finding work as an ophthalmologist. It didn't go well. The philosophically minded Loeb was almost laughably unsuited to caring for patients. One night, while walking along a wooded hill on the outskirts of the city, he confessed to his American wife, Anne, that he was miserable. He had so many big questions that he couldn't explore while rubbing yellow ointment into people's eyes all day. By that point, Anne had likely already come to terms with her husband's obsessions. The couple had honeymooned in Naples the year before so that Loeb could immediately return to work at the zoological station. The Loebs immigrated to the United States, where Loeb found a position at Bryn Mawr. Less than a decade later, he would carry out the experiment that would make him famous around the world.[20]

Once again it was the sea urchin that played the starring role. Loeb found that if he placed a sea urchin egg in a jar of water mixed with minerals and then returned the egg to seawater, something utterly strange happened: the egg began to grow and develop as though it had been fertilized by a sperm. It was a simple experiment, but through Loeb's visionary gaze, it would come to seem like the start of something much bigger and more profound. On November 19, 1899, the *Chicago Sunday Tribune* announced Loeb's discovery with a detailed sketch of a sea urchin and a blaring headline across the top of page one: "Science Nears the Secret of Life." Loeb could already see the day, he told another newspaper, when a scientist might mix chemicals in a test tube and end up with a "substance" that would "live and move and reproduce itself."[21]

As a young researcher, Loeb had argued that scientists didn't necessarily need to understand every last mechanism of life so long as they succeeded in engineering living organisms to meet humanity's needs. He absorbed this belief from German plant physiologists of the era who were intent on securing Germany's food supply by creating better crops. But for all his claims about pragmatism, Loeb hungered for deeper knowledge. With each new discovery, he would find himself in need of another scientific explanation. What Loeb really wanted, he would eventually state, was to explain all "life phenomena" in terms of the "motions of electrons, atoms, or molecules."[22]

At the height of his fame, the *New York Times* compared Loeb to Copernicus, calling him "undoubtedly one of the greatest experimental geniuses of whom we have any knowledge." It might have helped that Loeb spoke in a thick German accent and looked the part of the eccentric genius—or that he was conducting experiments that seemed straight out of science fiction: "I wanted to go to the bottom of things," Loeb once said of his research program. "I wanted to take life in my hands and play with it . . . to handle it in my laboratory as I would any other chemical reaction—to start it, stop it, vary it, study it under every condition, to direct it at my will!"[23]

To his critics, including many of Germany's leading biologists, Loeb's quest to understand life at the level of the chemical reaction seemed hopelessly naive. Perfect knowledge of any given cellular function, according to this view, would never provide real insight on the organism as a whole. A dismissive piece in the *New York Times* in 1905 questioned the significance of Loeb's research and suggested that biologists see him as "a man of lively imagination" as opposed to a flawless "investigator of natural phenomena." But for his admirers, including Mark Twain, who responded to the attack with an essay in defense of Loeb, the bold futuristic claims were irresistible.[24]

Loeb made it seem as if science would soon achieve full mastery over life, and few scientists of the era were hungrier for mastery than the young Otto Warburg, who already regarded himself as something of a deity. But if Warburg was going to take life into his own hands, he would first have to learn how life worked. In 1906, Warburg moved on from Fischer's Berlin lab to the University of Heidelberg to study medicine. Two years later, on break from his medical classes, he traveled to the Naples Zoological Station and measured the breathing rates of sea urchin eggs. When his paper was published, he sent it to Loeb, who had already taken an interest in the topic. Loeb responded by thanking Warburg and inviting him to America. The two would continue to exchange letters until Loeb's sudden death in 1924. Though the 24-

year-old Warburg had initiated the correspondence, Loeb would fall into the role of fawning admirer, praising Warburg's "stunning" work again and again. In one instance, Loeb apologized profusely out of fear he had offended Warburg by suggesting a mistake in his work.[25]

Loeb's deference to Warburg was likely rooted in his fear of criticism and his sense that Warburg was an ally. His fame and success notwithstanding, Loeb took each new attack on his science as a personal affront. Anne Loeb once recalled that her husband had spent the first months of their marriage pacing the Gulf of Naples in fits of frustration over what he perceived to be unfair criticisms. Given that nothing bothered Loeb more than the failure of German physiologists to appreciate his physiochemical approach to the study of life, nothing would have pleased him more than Warburg's emergence as a supporter of his work. Who better to demonstrate the superiority of his approach to biology than the son of a world-class physicist, a man who, furthermore, had studied under the world's most distinguished organic chemist? Had Loeb ventured to create a new champion for his science in a test tube, he could hardly have done better than Warburg.[26]

If the relationship meant more to the mentor than the mentee, Loeb, nevertheless, had an immense influence on Warburg's career. Near the end of his life, Warburg said that the goal of his work had always been, and still remained, "to find out to what extent the processes in living organisms can be resolved in terms of physics and chemistry." And so while Warburg's early work on sea urchin eggs is typically explained as a first effort to understand cancer, he was also exploring questions Loeb had raised years earlier, questions aimed at even deeper mysteries.[27]

As Loeb makes clear in the introduction to *The Mechanistic Conception of Life*, his best-known book, he was interested in how sea urchin eggs begin to breathe because he saw it as a path to answering one of biology's greatest puzzles: how life itself begins. Loeb was in search of the on switch for life. Warburg's earliest studies of sea urchin eggs in Naples were designed to help him find it.

IN THE YEARS AHEAD, Warburg, like Boveri, would build on what he had learned from studying sea urchin eggs at Naples to arrive at an entirely new understanding of cancer. Boveri would become the father of the genetic approach to cancer; Warburg, the metabolic approach. Modern cancer science itself would, in a sense, develop from the eggs of the sea urchin.

But at the end of the first decade of the twentieth century, Warburg's cancer breakthrough was still more than a decade away. In the meantime, he had other great scientific problems to solve. Fischer and Loeb had taught Warburg that science could reveal every secret of life, and Warburg wanted to know them all. He planned to figure out not only how much oxygen a cell breathed but which hidden molecules inside of a cell made breathing possible. It was an extraordinarily ambitious goal, and while he was at it, Warburg was going to unravel the mysteries of photosynthesis as well.

Though journalists sometimes referred to Loeb as a modern Faust intent on pushing science to once unthinkable places, Loeb, for all his scientific hubris, was a humble man. If there was a true Faust in the first decade of the twentieth century, it was not Loeb but his mentee.

Warburg was even following the same academic path as the original Faust. The character is thought to be based on the sixteenth-century German physician and alchemist Georgius of Helmstadt, who called himself Doctor Faustus and who, like Warburg, studied at the University of Heidelberg. Helmstadt is believed to have died violently in the mid-1500s. By the end of the century, his legend was already the subject of a hugely popular book of tales and had inspired Christopher Marlowe's play *Doctor Faustus*, the first great literary treatment of the story.

As the classic version of the legend goes, a brilliant scholar, longing for deeper knowledge of the world, arrives at a deal with the devil. In exchange for greater scientific powers, he will give up his soul. Warburg, in his mid-20s, was already ravenous for knowledge and power. The devil's offer was still to come.

CHAPTER TWO

"The Great Unsolved Problem"

A NEW UNDERSTANDING of how cells use oxygen was not the only thing Otto Warburg searched for while in medical school at Heidelberg. He also hoped to find a wife, and as Warburg was happy to point out, he had a number of options. "I can tell just by looking at her that she wants to marry me," Warburg wrote to his sister, Lotte, about one young woman. The woman in question once asked Warburg where hysteria came from. If she had truly wanted to marry him, his response—"Great passion is always hysterical"—might have given her second thoughts.[1]

It's not hard to understand why women may have fallen for Warburg. As a young man, he was dashingly handsome. His blue eyes, if capable of ferocity, often idled in dreaminess. And Warburg was always impeccably dressed. A self-proclaimed Anglophile—he said he liked the English because they tolerated eccentrics like him—Warburg traveled to England twice a year to purchase his tailored suits and riding gloves. Upon meeting him at his institute for the first time, one colleague, observing Warburg's gray tweed trousers and elegant waistcoat, described him as "the picture of a nobleman of the British School."[2]

His extraordinary self-involvement notwithstanding, Warburg could also be charming. He "had contempt for most of his fellow humans," his cousin Eric Warburg observed, "but he had a good sense of humor." When a journalist once asked Warburg if it was true that he was a great scientist but a rotten human being, Warburg said that he was glad it was put that way, as the reverse would be worse. On another occasion, a reporter arrived at Warburg's institute unannounced to ask about his cancer research. Warburg answered the door himself and told the journalist, who didn't recognize him, that Otto Warburg had died.[3]

The Nobel laureate Hans Krebs worked at Warburg's institute as a young man and remained a friend and admirer of Warburg throughout his life. He knew Warburg as well as almost anyone. And yet, even Krebs found it hard to understand how Warburg could seem so fierce in one moment and so relaxed in the next. Krebs, remembering a sweet letter and gift Warburg had sent to his wife—Warburg addressed her in English as "Lady Krebs"—noted that Warburg could be particularly kind and considerate with women. Warburg had a number of female friends, and near the end of his life, he selected a woman, the University of Chicago biochemist Birgit Vennesland, to succeed him at his institute.[4]

But when Warburg was angry, which was often, women were not spared his ire. On Warburg's request, Vennesland had once translated a German document about his long-standing dispute over the number of photons required to power photosynthesis. Though the original document referred to the debate as a "war," Vennesland had used the word "argument." Upon reading the translation, Warburg was pleased and praised Vennesland. "Since he was never prodigal with compliments, I felt pretty good," Vennesland recalled.

The good feeling was short-lived. As Vennesland discovered, there was another side of Warburg, "a Warburg No. 2," as she put it, "who acted as a censor." This other Warburg was a man "of violent emotions." When Warburg looked over the manuscript a second time,

"Warburg No. 2" was reading it, and this Warburg noticed the change from "war" to "argument." "I won't use any of it," he shouted.[5]

The young women who courted Warburg during his Heidelberg years seem to have been aware of "Warburg No. 2." In one letter to Warburg, a law student confessed that she had half loved and half hated him. But now she was "begging for his love," for any sign that he did not "despise" her any longer. Whether she ever received this sign from Warburg is unknown. When talk of marriage grew serious, Warburg told her that his financial situation made it impossible. And though Warburg did occasionally have financial disputes with his father, it was clearly an excuse. On another occasion, Warburg told Lotte that he wasn't planning to marry and have children because his scientific work didn't leave enough time to properly care for a family.

Warburg did meet at least one young woman he genuinely liked. After growing close with the woman over a four-week period, he proposed to her in a letter. The woman—her identity is unknown—struggled with the decision for three days, only to reject the proposal via a letter of her own. She feared, she wrote to Warburg, that it wasn't "in her power" to make him happy—an intuition that was almost certainly correct.[6]

Shortly after receiving the rejection, Warburg wrote to Lotte that he had moved on. It was a typically Warburgian show of pride, but at least during this period of his life, Warburg was capable of vulnerability as well. In a 1912 letter to his fellow student, the future Nobel laureate Otto Meyerhof, Warburg seems to have said that he was worn down from work and struggling with his emotions. Though Warburg's letter has not survived, Meyerhof responded with a recommendation for a psychotherapist who treated compulsive thoughts and mild cases of hysteria. He also suggested that Warburg try gymnastics.[7]

It was at about this time that one of Warburg's medical professors suggested that he give lectures on his research to students, then a required step in the path to becoming a professor in a German university. Though Warburg would later give many public talks, as a young man the thought of lecturing was more than he could bear. In Warburg's own telling of the story, he comes across as almost Bartleby-esque in his inexplicable refusals:

> When I was young, my professor said, "The time has
> come. You have to give some lectures."
> I said, "I won't."
> He said, "You must."
> I said, "I won't."
> He said, "Oh come, just a few, not many, you can
> easily do it."
> I said, "I won't."

Warburg, laughing as he recounted the tale to a colleague, said that the medical faculty eventually held a meeting during which they concluded that Warburg suffered from paranoia and that nothing could be done about it.[8]

In his brief biography of Warburg, Hans Krebs wrote that Warburg's attachment to science was "the dominant emotion of his adult life, virtually subjugating all other emotions." But subjugation is never foolproof. "He has chosen to wear the mask of the unshakable, unflinching, not-to-be-moved man before the world," Lotte wrote in her diary. "But there is something behind the mask, I am firmly convinced of that."[9]

Warburg's most revealing comment about his internal life slipped out during a conversation with a colleague who was telling him about a mutual acquaintance dealing with emotional struggles. Warburg's advice: "Tell him not to think about anything but science—think about absolutely nothing else—only science."

The man wasn't even a scientist.[10]

THAT WARBURG SAW science as everything, as the solution even for emotional distress, is in part the story of a boy whose father was a celebrated physicist. But it is also the story of *fin de siècle* Germany. Warburg grew up during a period in which German science was conquering the world. He might have been more attached to science than almost anyone else, but his belief in science was a national phenomenon.

In the early nineteenth century, many of the various German-speaking states that would come to form the German Empire were poor and underdeveloped. As England and France raced into the industrial era, the German states lingered in a preindustrial world. And then, almost overnight, everything changed. The Germans, aware they had fallen behind, were anxious to catch up. And if the German states (unification would only come in 1871) didn't have the natural resources or extractive colonial regimes of England and France, they had something else: a growing, well-educated population that believed in the transformative power of science. Germany, known at the start of the century as the "land of poets and thinkers," became the land of steel production and synthetic dye manufacturing. By 1900, Germany had the largest economy in Europe. The pace of change could be dizzying. When Mark Twain visited Berlin in 1892, he anticipated "a dingy city in a marsh." Instead he found a sparkling new metropolis that looked "as if it had been built last week."[11]

As it industrialized, Germany also transformed itself into the world's leading scientific nation. Between 1901 and the start of World War I in 1914, Germans won a third of the Nobel Prizes given out to scientists. Emil Fischer revolutionized organic chemistry. Max Planck stumbled into the quantum nature of energy. German bacteriologists identified the microbes responsible for one deadly infectious disease after another. Germany's scientific dominance was so pronounced that German became the *lingua franca* of many scientific fields.

Perhaps the more surprising part of Warburg's story isn't that he was consumed by science but that he was consumed by cancer in particular. Photosynthesis and respiration, Warburg's other fields of interest, were much more obvious choices. In studying how cells breathe and how plants harness the energy of the sun, Warburg remained in the realm of foundational science. Cancer brought Warburg into a world of patients and disease. Respiration and photosynthesis were measurement and math. Cancer was messy and human.

On one occasion, Warburg himself wavered in his commitment to studying cancer. He asked Walther Nernst, the Nobel laureate with whom he had briefly studied as a young scientist, whether he should focus on photosynthesis or cancer. "Do cancer research," Nernst said. "Photosynthesis is working fine."[12]

But if Warburg's focus on cancer appears misplaced in the context of his scientific predilections, it makes sense in the context of modern Germany. For all the astonishing scientific and technological triumphs of the time, cancer was one challenge Germany could not overcome. Despite devoting more and more resources to cancer research in the late nineteenth century, more and more Germans were succumbing to the disease. In 1881, 65 out of every 100,000 Berlin residents died of cancer. By 1901, 65 had increased to 115. Ten years later, it was 138 per 100,000. Records from Munich, Hamburg, Stuttgart, and other areas followed the same pattern. Overall, Germany witnessed a 287 percent increase in the rate of cancer deaths from 1876 to 1910. Cancer was making a quiet mockery of the extraordinary march of German science.[13]

That, at least, is how it appeared to many German medical experts at the time. Others found it difficult to believe that the increase in cancer was real. Perhaps, one argument went, cancer deaths were rising not despite Germany's mastery of science and medicine but because of it. The success against infectious diseases had made it possible for people to live longer, and given that cancer is strongly associated with aging, an older population inevitably meant more cancer. Better recordkeeping and more sophisticated means of diagnosing cancer, in turn, had made it easier to identify and count each new victim of the disease.

Even today such arguments sound like simple explanations of the increasing number of cancer deaths. And yet, a closer examination of the evidence makes the skeptical position seem much less obvious. The first reports that deaths from cancer were on the rise in Europe date to the 1840s, before the average life expectancy began to shoot up. And though life expectancy did begin to increase rapidly in the second half of that century, the numbers can be misleading. The dramatically longer life expectancies didn't mean that Germany had a large population of elderly citizens for the first time. The increases in life expectancy were mostly due to a single factor: far more children were surviving childhood. Old age itself was not new to Europe and had not been for centuries. Between 1816 and 1849, the average life expectancy in the German states was 31 years, but someone who lived to age 20 could expect to live to 55, and many lived much longer.[14]

The percentage of the German population that was 65 or older did increase from 4.6 percent in 1871 to 5.8 percent in 1925, but cancer experts of the time understood that it was misleading to compare the number of cancer deaths in one year to the number of cancer deaths in another if there were more elderly people in one group. To account for this problem, they made statistical adjustments. And the "age-adjusted" or "standardized" cancer rates they produced did not make the increase in cancer deaths go away.[15]

Those who doubted that the rise of cancer was real could turn to another, more persuasive argument. It was possible that the number of deaths caused by cancer had not increased, but rather that the number of deaths attributed to cancer had. And more careful examinations and better tracking of cancer deaths are very likely a part of the story of the increasing prevalence of cancer. But this argument, too, had its weaknesses. For centuries, growths of all kinds—polyps, fibroid tumors, lesions caused by tuberculosis—had been considered cancer. As microscopic analysis of cancers became more common in the late nineteenth century, these conditions were recognized for what they were, meaning that more sophisticated diagnostic techniques could push the cancer rates down as well as up.

Moreover, the cancer statistics in dispute measured the number of people dying of cancer, not the number diagnosed with cancer. Doctors of the era had a difficult time detecting cancer at the early stages of the disease but a much less difficult time detecting the advanced cancers that kill people. As one indignant cancer expert put it in 1915, to suggest that the increase in cancer could be explained by a failure on the part of Germany's rigorously trained physicians to identify cancer at the time of death was "equivalent to a charge of gross malpractice."[16]

The best answer to skeptics could only be offered decades later. If the rising rates of cancer in Germany was only a result of older populations or more sophisticated diagnostic and recordkeeping practices, the sharp increase in cancer rates could have been expected to slow as the rise in life expectancy leveled off and physician and hospital practices grew more uniform. Instead, the cancer death rates only went up and up. In 1932, 87,000 Germans succumbed to cancer, making it the second-leading cause of death in Germany.[17]

Cancer rates were also rising across Europe and the United States. But no other country would respond to the emerging cancer crisis with the focus and determination of Germany. Toward the end of the nineteenth century, German researchers launched what amounted to the original war on cancer, becoming the first to identify one carcinogen after another, from sun exposure, to various industrial chemicals, to secondhand smoke. In 1900, Germany opened the world's first state-sponsored cancer agency, becoming the first country to attempt to register all cases of cancer, as opposed to only cancer deaths.[18]

Otto Warburg once said that a scientist "must have the courage to attack the great unsolved problems of his time." His model was Pasteur, and when Warburg opened his own institute, he commissioned an oil portrait of Pasteur and hung it in the library across from where he sat to write each day. Warburg turned to cancer research, ultimately,

Otto Warburg, 1931.

because it was the disease of his time and so the most direct route to a Pasteur-like acclaim. Science might have helped Warburg manage his psychological turmoil as a young man, but his research was never dispassionate. Cancer, for Warburg, was a path to glory.[19]

THE RISE OF CANCER in Germany also explains the story of Frederick Hoffman, a slim, energetic German who moved to America in 1884. After a series of odd jobs, Hoffman found work as an actuary at the Prudential Insurance Company, where he noticed something peculiar in the firm's files: the words "malignant neoplasm" were showing up more and more often in the records of recently deceased policyholders. Hoffman, likely already aware of medical reports pointing to a rise in cancer, decided to investigate and produce a report on the matter.[20]

By the time Hoffman completed his cancer report in 1915, it had turned into a book of over 800 pages filled with tables showing cancer death statistics from every corner of the world. The book, *The Mortal-*

ity from Cancer throughout the World, is believed to be the most comprehensive statistical accounting of a single disease ever assembled by that point in history. The mountain of scientific data that Hoffman collected led to a conclusion he found inescapable: the story of cancer in Germany was not unique to Germany. Cancer deaths were growing more common everywhere, "practically from year to year and from decade to decade." Cancer, Hoffman wrote, was "increasing at a more or less alarming rate throughout the entire civilized world."[21]

It was a sensational claim that reignited the debate as to whether the increase in cancer was real or merely a statistical mirage. But that cancer was rising in so-called "civilized" industrial societies was not even the most provocative argument Hoffman made. Even more striking than all the places where he found more cancer were all the places where he could find almost no cancer at all. Among the native populations Hoffman studied, cancer was either rare or unheard of.

Hoffman first rose to prominence in 1896 after publishing a grossly racist monograph in which he concluded that the health problems faced by the "American Negro," such as high rates of tuberculosis, were due to a "constitutional weakness." And though he publicly reversed this position years later, he was hardly inclined to see nonwhite populations as more robust. He believed that cancer was absent in indigenous populations because reports of the phenomenon had been accumulating for more than half a century.[22]

The first to issue such a report may have been Stanislas Tanchou, a French physician and veteran of Napoleon's army. Tanchou was studying death registries in Europe in search of cancer trends when he learned of new data that appeared to fit with no trend at all. French colonial doctors in northern Africa could find almost no trace of the disease. One doctor who had established a hospital "in the midst of the Arabs" had failed to find "a genuine cancer" among the 10,000 patients he treated. Based on these reports and his own research, Tanchou arrived at a conclusion that he presented to the French Academy of Science in 1843. Cancer, he said, was like "insanity." It "seems to increase with the progress of civilization."[23]

Tanchou's name would probably have been lost to history had he been the only one to notice that cancer was far more common in some populations than in others. But Tanchou was far from alone. In the following decades, frontier and missionary doctors from every part of the world—from Brazil to Fiji to Borneo—would report the same finding: cancer was either rare or nonexistent in places that hadn't adopted Western lifestyles.[24]

These reports were typically compiled by physicians who had spent years, sometimes decades, working far from Western nations. After spending six and a half years treating a Bantu population of some 14,000 people in South Africa, the surgeon F. P. Fouché noted, in 1923, that he had not diagnosed a single cancer. What made the absence of cancer so odd, he reflected, was that cancer was "frequently seen among the white or European population" in the same region.[25]

This phenomenon would be observed again and again. In 1908, the prominent physical anthropologist Aleš Hrdlička authored a 460-page Smithsonian report on Native Americans of the southwestern United States and northern Mexico. Hrdlička, who held a medical degree, consulted with resident physicians. He heard of "tumors" and identified a few "of the fibroid variety," but he did not see "a clear case" of cancer. Hrdlička also examined the remains of Native Americans but was unable to find "unequivocal signs of a malignant growth on an Indian bone." If cancer existed at all among the population, Hrdlička wrote, it "must be extremely rare." Six years later, Hoffman's own survey of physicians at the Bureau of Indian Affairs turned up only two reports of cancer deaths among 63,000 Native Americans from different tribes.[26]

The virtual absence of cancer in the Arctic was yet another data point. In 1884, a whaling ship captain who was also a highly regarded amateur surgeon was asked by his brother, a doctor, to look for cancer among the "Eskimos" he encountered in the course of his travels through northern Alaska, Canada, and Siberia. The captain, George B. Leavitt, knew how to identify cancer and would sometimes diagnose members of his crew. But over a 15-year period during which he is estimated to have annually encountered 50,000 "natives," he could not find a trace of the disease. It wasn't that Leavitt never had the chance to examine the Inuit closely. On the contrary, as he told the Harvard-trained anthropologist and explorer Vilhjalmur Stefansson— who, in 1960, wrote an entire book on the absence of cancer among non-Western populations—the Inuit actually liked to consult doctors.

Leavitt eventually gave up his search for cancer among these populations, convinced that he would never find a malignant growth on a native "Eskimo." He would not be the last to look. After 11 years among the Inuit of northern Canada, the physician Samuel Hutton wrote that he had "not seen or heard of a case of malignant growth in an Eskimo." As late as 1952, there were still no reported cases of cancer among Inuit following a traditional lifestyle.[27]

Perhaps the most famous physician to note the absence of cancer in a non-Western population was Albert Schweitzer, the Nobel Peace Prize laureate who studied medicine in Germany at the same time as Warburg, before moving to Gabon. The Gabonese, happy to have a German doctor among them, lined up to see Schweitzer. He estimated that he examined 2,000 patients during his first nine months in the country. "On my arrival in Gabon, in 1913, I was astonished to encounter no case of cancer," Schweitzer would later write. "I can not, of course, say positively that there was no cancer at all. But, like other frontier doctors, I can only say that, if any cases existed, they must have been quite rare."[28]

Nicholas Senn, a prominent Chicago surgeon and onetime president of the American Medical Association, was the rare medical expert who could attest to the absence of cancer in two very different parts of the globe. In 1905, Senn traveled to the Arctic with one of Robert

Peary's famous polar expeditions. The next year he traveled across Africa. "After closely observing the conditions of health" in both regions, Senn became "convinced that cancer is purely a disease of civilization."[29]

Even as they accumulated in number, such reports became more difficult for many Western doctors to accept. The most common objection—that everyone in native populations died in their 30s, before cancer was typically diagnosed—was easy to refute. Hrdlička's Smithsonian report, which found no cancer among Native American populations, also noted that the Native Americans included in the survey lived as long as the local whites (who would often get cancer).

The Irish physician and nutrition researcher Sir Robert McCarrison joined Britain's Indian Medical Service in the first years of the twentieth century. He testified that he saw no cancer after nine years and more than 3,600 surgeries "in a remote part of the Himalayas, among isolated races far removed from the refinements of civilization." According to McCarrison, who spent the last part of his career as the director of postgraduate medical education at Oxford, the problem wasn't that the locals were dying young: "Certain of these races are of magnificent physique, preserving until late in life the characters of youth; they are unusually fertile and long-lived, and endowed with nervous systems of notable stability."[30]

McCarrison may have romanticized the indigenous populations he treated, but his claims are consistent with contemporary research. A 2018 review of 12 hunter-gatherer and subsistence farming populations found that it was typical for those who survived into adulthood to live into their 60s and 70s or later. Researchers now believe that humans have been living into their 50s and beyond for at least 30,000 years, and the best evidence suggests that prior to the nineteenth century, cancer was a very rare disease. A 2018 study of 1,087 ancient Egyptian skeletons buried between 1,500 and 3,000 years ago found only six with cancer. A study of remains from a London crypt used between 1729 and 1857 found only a single cancer among 623 individuals.[31]

Cancer is also almost completely absent in written sources from antiquity. A disease believed to be breast cancer was described in an Egyptian papyrus that is thought to be some 4,500 years old, but as far as we know, cancer was not written about again for another 2,000 years. "Other diseases cycled violently through the globe, leaving behind their cryptic footprints in legends and documents," Siddhartha Mukherjee writes in *The Emperor of All Maladies*. Cancer, by contrast, "virtually disappeared from ancient medical history."[32]

The evidence points overwhelmingly in one direction: though cancer is an ancient disease that can be found throughout the animal kingdom, it remained fairly uncommon among humans until the nineteenth century. Cancer as we know it today, a disease that eventually afflicts one in every two American men and one in every three American women, might be partially explained by longer lives, but the more powerful explanation is not in how long we live but rather in how we live.

The doctors who worked in nonindustrialized regions over a century ago found no cancer when they arrived to treat native populations. They also saw that cancer would gradually arise as soon as local populations began to follow a more Western lifestyle. Albert Schweitzer would live in Africa for half a century, and over that period, some of the locals did get cancer. As Schweitzer saw it, local people were developing the disease as a consequence of "living more and more after the manner of the whites." The Chicago surgeon Senn made the same observation. "The pure native never has cancer," he said. "Only by gradual intermingling with the whites, are they susceptible."[33]

But what within the Western lifestyle had set off the cancer explosion? As the British cancer authority Charles Powell White pointed out in 1908, it is possible that "no one factor in civilization" was responsible for the rise of cancer, but that it was "the condition as a whole that is at fault." Cancer was likely caused by a combination of "unnatural and excessive food, unhealthy surroundings, indoor and sedentary occupations, and the mental anxiety and worry which are inseparable from civilized life."[34]

White's conclusion, that cancer is, essentially, inextricable from modern Western life was reasonable, given the evidence available to him. But in 1908, scientists still had little understanding of the cancer cell itself and so could not convincingly explain how anything causes the disease. When Warburg studied the breathing of sea urchin eggs that year, it was his first step toward bringing clarity to this confusing state, toward identifying a central mechanism that might explain all the different causes of cancer in a single stroke, just as Pasteur had explained all infectious diseases as a problem of microbes. Warburg didn't yet have answers, but his path to greatness, he knew, was hidden inside a cancer cell.

THE AUSTRIANS across the border from Germany were recording the same increases in cancer deaths as other Western populations. In 1907, as Warburg was setting out to study the disease, a woman in Linz felt a sharp pain in her chest. The woman, Klara Hitler, summoned the family's Jewish doctor, Eduard Bloch, who saw right away that she had breast cancer. Bloch remembered Klara as a "modest, kindly woman" with "brownish hair which she kept neatly plaited, and a long, oval face with beautifully expressive gray-blue eyes." Not wanting to upset Klara, Bloch reported the news to her children.[35]

One of those children, Adolf, was 17 when Bloch delivered the news. Bloch remembered him as a "frail looking" boy who "lived within himself." Hitler, a terrible student, had dropped out of school two years earlier. Though he was not working, he had been unwilling to help his widowed mother around the house. To the extent that Hitler had any ambition, it was to be a great artist; he stayed up all hours working on sketches of buildings and bridges.

By his teenage years, Hitler was already showing signs of the man he was to become. He was prone to fits of rage and quick to blame others for his failures. Hitler could be "like a volcano erupting," August Kubizek, his one close friend from the time, remembered. "It was as though something strange, other-worldly, was bursting out of him."[36]

When Bloch told him about his mother's cancer, Hitler wept openly, his "long, sallow face" twisting in agony. "In almost forty years of practice, I have never seen a young man so utterly filled with pain and grief," Bloch recalled years later.[37]

Though the prospect of losing his mother would have been devastating regardless of the cause, at the time cancer was an especially awful diagnosis. Because it was often diagnosed at an advanced stage, as was the case with Klara Hitler, the disease was frequently accompanied by infections and rotting flesh. A 1908 German book on caring for people with incurable cancer warned that contact with the "terrible-smelling secretions is revolting and also harbors a danger for those nearby." A cancer diagnosis was a source of great shame, a secret to be kept from one's neighbors.[38]

Klara Hitler underwent a double mastectomy. In the following months her condition improved. In September, Adolf left his mother behind in Linz and traveled to Vienna to take an entrance exam for the Academy of Fine Arts. He made it beyond the first round of elimination but was not among the 28 applicants who advanced to the third round.

Crushed by his rejection, Hitler returned to Linz in October of 1907 and soon received even worse news from Dr. Bloch. His mother's cancer was back, and incurable. Decades later Kubizek would still remember Hitler's reaction to the news.

> *His eyes blazed, his temper flared up. "Incurable—*
> *what do they mean by that?" he screamed. "Not that*
> *the malady is incurable, but that the doctors aren't*
> *capable of curing it. My mother isn't even old. Forty-*

seven isn't an age where you give up hope. But as soon as the doctors can't do anything, they call it incurable."

I was familiar with my friend's habit of turning everything he came across into a problem. But never had he spoken with such bitterness, with such passion as now. Suddenly it seemed to me as though Adolf, pale, excited, shaken to the core, stood there arguing and bargaining with Death, who remorselessly claimed its victim.[39]

Hitler pleaded with Dr. Bloch to do whatever possible to save his mother. Bloch turned to an experimental method, reopening Klara Hitler's wounds and applying iodoform, a treatment thought to kill bacteria. Iodoform contains iodine. The treatments would have left Klara Hitler writhing in pain. Bloch returned to the Hitler home for 46 consecutive days, applying the iodoform again and again as young Adolf looked on.

Klara Hitler died during the night of December 21, 1907. When Bloch arrived in the morning, Hitler was sitting by his dead mother's side. The mastery and control of life that Jacques Loeb and Otto Warburg sought through science, Hitler would later seek by other means.

CHAPTER THREE

Magic Bullets

IN 1906, in yet another sign of the country's dominant position in cancer science, Germany hosted the world's first international congress of cancer research. The opening speaker, Professor Ernst von Leyden, provided an overview of the previous century's cancer research, insisting that the statistics demonstrated a clear increase in cancer rates.

Much of the conference took place at the University of Heidelberg, where Warburg was then beginning his medical studies. While there is no record of Warburg's attendance at the conference, he was almost certainly aware of it. Among the luminaries taking part in the event was Paul Ehrlich, a man surpassed perhaps only by Pasteur in Warburg's hierarchy of scientific greatness. While Pasteur's portrait would later occupy the central spot in Warburg's library, a portrait of Ehrlich would hang only a few feet to its right.

Born in 1854, Ehrlich had been studying medicine in Strasbourg when he came across a monograph by Emil Heubel, of the University of Kiev, that would forever change his life. Though Heubel had once studied frogs thought to be in a state of hypnosis, it was his canine work that captivated Ehrlich. Along with their daily servings of a half-pound each of meat and bread, Heubel's dogs received a considerable helping of pure lead. After a few months, the dogs would invariably die, at which point Heubel would cut them open and remove their organs to see how much lead each one had soaked up. What struck Heubel—and later Ehrlich—was how unevenly the lead would be distributed inside of the dogs. Some organs—the liver, kidneys, and brain—were quick to take up lead, but other organs were left virtually untouched. The heart, Heubel noticed, could not be poisoned by lead.

To confirm his findings, Heubel took some of the canine organs and placed them directly in a solution of lead only to see the same pattern repeat itself. Some seemed to soak up lead like a sponge; others could sit in a tub of lead forever and remain lead-free.

It wasn't lead poisoning itself that fascinated Ehrlich but the underlying principle: lead solutions could only poison tissues where lead was able to form a chemical bond to specific types of cells. If Ehrlich could only figure out the secret of chemical attraction, why one molecule in the body will grab hold of another and never let go, medicine would no longer be "shrouded in darkness."

Heubel had been curious about poisons, but Ehrlich was interested in cures. If a poison could attach to one type of cell yet not another, then so, too, could a medicine. The goal, Ehrlich said, was "to take aim, in a chemical sense," to find a drug that would hit only its specific target. "It was a revelation to me," Ehrlich later wrote, "and a sort of destiny."[1]

Ehrlich, still a medical student, had arrived at the idea of chemotherapy—a term he himself would later coin. Though chemotherapy is now thought of as a general toxin that kills healthy cells along with cancer cells, leaving patients sick and often hairless, it began with Ehrlich's dream of precision targeting. After reading Heubel's monograph, Ehrlich dropped all other projects and spent his days gazing at slides of lead-poisoned brain tissues. The professor who was sup-

posed to be Ehrlich's examiner in chemistry at the time told another
colleague that he couldn't recall having seen Ehrlich at a single lecture.

Ehrlich's medical training turned into a "disaster," he recalled. He
was ignoring his other work, yet making no progress in understanding why lead went into some tissues and not others. It was clear that
lead bonded to something in the brain, but the brain was full of different types of cells, and any one of them might have been responsible for forming the link. Had he been a more typical student, Ehrlich
might have given up on chemotherapy and returned to his studies. But
Ehrlich had never been typical. As part of his high school exit exam, he
had been assigned to write an essay in response to the prompt "Life—
A Dream." Ehrlich took the opportunity to argue that dreaming was
merely a chemical process in the brain. (He almost failed to graduate.)
Later in life, he carried pencils at all times so that he could quickly
write down a thought or sketch out a molecule before it slipped his
mind. If there was no paper nearby, Ehrlich would write on the nearest
surface he could find: doors, tablecloths, and on at least one occasion,
the cuff of a presumably stunned colleague.[2]

So, instead of retreating to his planned course of study, Ehrlich
searched for a better way to investigate how cells form chemical bonds.
The answer, he soon realized, had been staring him in the face the
entire time. Like all medical students of the era, Ehrlich had learned
how to stain tissue samples with dyes. (The little lines Boveri spent
his life gazing at and drawing were called chromosomes, or "colored
bodies," because they would take up the dyes.) But why, Ehrlich now
wondered, should the use of dyes stop with the staining of cells on
microscope slides? Why not inject the dyes directly into an animal
and then look to see where they end up—to which types of cells they
bond—by observing which parts of the animal changed color?

Ehrlich's idea was to redo Heubel's experiments with dyes of every
color. If he could determine the underlying principle—why a chemical will bond with one type of cell yet not another—he would be able

to take aim at the targets he most wanted to hit: the living organisms that invade our bodies and cause diseases. By the late nineteenth century, Ehrlich had plenty of such organisms to choose among in the form of bacteria and parasites that had recently been linked to infectious diseases.

The first step was to identify chemicals that would attach to the invaders. But Ehrlich's goal was not to make the unwanted organisms more colorful. He needed chemicals that could both bond to the invaders and kill them. What he needed, Ehrlich would say years later, were "magic bullets."

The term "magic bullets" had been popularized by the nineteenth-century German opera *The Marksman*. In it, Max, a forester, is in love with Agathe. But before he can marry her, he must first beat his competition for her hand in a shooting contest. With so much at stake, Max is unable to resist when he is offered bullets that will hit whatever target he chooses. Max doesn't realize that the magic bullets come at a steep price. Six of the bullets will hit whatever mark the shooter wants. The seventh bullet is guided by the devil.

If Ehrlich had alighted on his idea of targeted chemotherapies a decade earlier, he might have made little progress. But by 1878, the year he completed his medical studies, he had an array of different bullets to test. Germany was awash in new synthetic dyes. The dazzling new colors had emerged from the least likely of places: coal tar, the foul-smelling black gook that was left behind in gaslight lamps and gathering by the barrelful throughout Europe. William Henry Perkins, an English chemist, launched the synthetic dye industry when, at age 18, he discovered that a compound in coal tar could give rise to a pale purple solution. Yet it was a new generation of German technical school graduates who mastered the art of making chemical dyes.

Ehrlich became so lost in his dye research that the other students began to joke about him. A classmate recalled that Ehrlich "always went around with blue, yellow, red and green fingers." His laboratory bench was said to "gleam in all colors of the rainbow." Even Ehrlich's face would sometimes show traces of the colors.[3]

Ehrlich's initial plan had been to use the dyes to understand how chemical bonds form in living tissues. But in some cases, the dyes themselves appeared to be magic bullets. In the early 1890s, Ehrlich injected malaria patients with a variant of a dye known as methylene blue. The injections turned the whites of the patients' eyes blue, but some also noticed that their fevers were improving. For Ehrlich the methylene-blue-derived malaria treatment was an early success—it would remain a malaria drug until the middle of the next century—but not the precision weapon he had in mind.

Ehrlich needed more powerful and more precise bullets, and he knew where to look for them. The immune system, Ehrlich had come to understand, is equipped with molecules that single out and execute invading germs while leaving the body's own cells unharmed. His turn to immunology would lead Ehrlich to a new theory of how the immune system works and a series of breakthrough treatments for infectious diseases.

By 1899, the dreamy medical student with colored fingers had his own beautiful institute in Frankfurt and an entire team of scientists working under him. He had become the thing that Otto Warburg most wanted to become: a celebrated German and one of the most highly respected medical researchers in the world. Warburg could only read about Pasteur's legendary feats of a generation past. Paul Ehrlich was Warburg's living model.

GIVEN THAT EHRLICH WAS a famous medical researcher in Germany in the early twentieth century, it was inevitable that he would take aim at cancer. But while Ehrlich is celebrated by cancer scientists as the inventor of chemotherapy, it did not take him long to appreciate that treating cancer was a different challenge than treating an infectious disease. The entire point of chemotherapy, from the beginning, had been specificity, eliminating unwanted cells and only unwanted cells. But cancers form from our own cells. Borrowing another term from German folklore, Ehrlich called cancer cells "hostile brothers." Whatever killed cancer cells, Ehrlich saw, would also kill the body in which the cells resided. Firing a bullet at cancer was like turning a gun on yourself.

Ehrlich needed a different approach for cancer, a way to undermine the hostile cells without poisoning their nonhostile siblings. He now turned to a new line of research pioneered by Leo Loeb, the younger brother of Warburg's mentor Jacques Loeb. In 1897, after earning his medical degree from the University of Zurich, Leo Loeb moved to Chicago, where Jacques was then teaching. It was not a hard decision. He was repulsed by rising German nationalism, and there was no one left for him back in Germany: the Loeb boys had been orphaned when their father died from tuberculosis (Jacques was 16 at the time, Leo only 6).

Without an academic appointment, Leo Loeb started a private practice, but like Jacques, he lacked the temperament for the work. Beneath his quiet exterior, Leo, too, had a passion for bold experimentation, for pushing biology beyond its known limits. While in medical school, Loeb had experimented with transferring skin from one part of an animal to another. In Chicago, he rented a small room behind a drugstore and filled it with mice and guinea pigs purchased with his own money. Years later, Loeb was delighted to see his landlord, the owner of the drugstore, in the audience at one of his lectures. But at the time, the landlord likely couldn't have imagined that the quiet young immigrant working behind his pharmacy would soon be among the most highly regarded cancer researchers in the world.[4]

In America, Loeb moved from transplanting skin to transplanting tumors. (His interest in cancer and cellular growth had been sparked, at least in part, by his familiarity with the studies done on sea urchin embryos at Naples.) Though others had worked on tumor transplants, it was Loeb who perfected the technique. In a paper published in 1901, he described transferring 360 fragments of thyroid cancers from a single white rat into some 150 host animals. Among Loeb's many striking observations, one stood out: When the implants took hold and a tumor grew, the cells of the host animal did not become cancerous. Only the cells from the transplanted tumor multiplied. And if Loeb kept transferring these cancer cells from one animal to the next, they could live on and on, seemingly forever.[5]

It was one of nature's cruel jokes: the possibility of immortality discovered not only by someone who had been orphaned at age 6, but in the very cells that kill us. Yet Loeb's work on transplants and the immortality of cancer cells helped turn the study of cancer into a modern science. At the time, researchers had no reliable method for inducing cancer in animals. Thanks to Loeb, research labs—including Warburg's—would fill with cancer-stricken animals, and countless new experiments would become possible.

Still, if Loeb's transplantation studies were key to the rise of modern cancer science, his successful transplants weren't the entire story, maybe not even the most important part of the story. What intrigued Loeb most were the transplants that failed, those that were immediately destroyed by their new hosts. Was it something in the host animal's hereditary makeup, Loeb wondered, that caused it to accept or reject a cancer? Did the host's immune system attack the tumor?

Loeb believed that if he could only figure out how a body eliminated a transplanted cancer, it might lead to a cure or, at the least, a new understanding of cancer prevention. But Loeb had more questions than answers. He soon moved on to chemotherapy studies, building on Paul Ehrlich's work with chemical dyes. Ehrlich, meanwhile, had grown interested in Loeb's work on transplanted tumors. He, too, tried to determine why many of the transplanted tumors never took hold. In some instances, he could implant a tumor into a rodent, and it would grow in the usual fashion. But when he tried to transfer additional cells from the very same tumor into the very same rodent, the second transplant would often fail. It was as if his experimental animals had a one-transplant limit.[6]

Ehrlich, among the world's experts on the immune system, might have assumed that the body's own defenses were eliminating the second transplants. But that assumption was far from obvious at the time. If the animal's immune system had already accepted the tumor once, Ehrlich wondered, why should it reject it the second time? This puzzle led Ehrlich back to an outdated idea. Pasteur had once hypothesized that when a body eliminated an infection, it would destroy not only the invading organism itself but also whatever nutrients or chemicals the invading organism had relied on to grow. By the time Ehrlich took up the question, it was clear that Pasteur's idea was wrong. "Yet," Ehrlich wrote, "this old theory seems to me to contain a nucleus of truth, as is so often the case."

As Ehrlich saw it, the "nucleus of truth" in Pasteur's idea was that microbes will not grow without the right nutrients. He had seen that phenomenon in his own laboratory dishes. Some of his bacilli would not grow until he added hemoglobin to the cell culture. Perhaps, Ehrlich reasoned, the same principle held true for cancer cells inside of an animal body. Perhaps the second transplant couldn't grow because the first transplant was using all of the nutrients it needed.

Ehrlich did not have every detail right, but he had arrived at a "nucleus of truth" of his own. In an extraordinary 1907 lecture in which he described his experiments, he came closer than any scientist of his time to anticipating today's thinking on nutrition and growth factors in cancer. "[E]very proliferation," Ehrlich said, "depends in the first place on the avidity of the cells for the nutritive substances."[7]

Ehrlich, the man who brought the phrase "magic bullets" to medicine, had a knack for language. But the name he chose for his new metabolism-based theory didn't help it catch on. He called the starving of transplanted tumors "athreptic immunity." The term "athrepsia" comes from an ancient Greek word for malnutrition. At the time, it referred to a medical condition that caused newborn infants to waste away and die. The eccentric Ehrlich had looked at vanishing tumors and seen dying babies.

OTTO WARBURG MET Paul Ehrlich in 1912 during a visit to Ehrlich's institute in Frankfurt. By then, Ehrlich had achieved international fame with his most successful magic bullet of all, the one capable of seeking out and destroying the organism responsible for syphilis. After their meeting, Warburg and Ehrlich rode together in a horse-drawn carriage to a nearby town. "It was a beautiful summer evening," Warburg recalled. When Warburg told Ehrlich about his discovery that sea urchin eggs will take up more oxygen upon being fertilized, Ehrlich was unable to corral his excitement. As Warburg recalled, Ehrlich shouted "Yes, quantitatively!" again and again, growing so lost in the idea that "he seemed to see nothing of the mountains and valleys that surrounded us."

The two men met only one other time, at an academic reception in Berlin, not long after their initial meeting. When Warburg spotted him, Ehrlich was standing alone and appeared to be lost in his thoughts. Ehrlich may have been depressed. His secretary claimed that Ehrlich had once confessed to her that when feeling down, he would sometimes stand before his cabinet full of dyes and say to himself, "These are my friends which will not desert me." But Warburg saw no melancholy in Ehrlich's isolation. What he saw, instead, was a scientific imagination hard at work. "Ehrlich lived in two worlds," Warburg wrote, "and by putting into practice in this world what he saw in that world of his own, he arrived at one of the greatest scientific achievements of all time."[8]

Ehrlich continued to pursue a cure for cancer even as his syphilis treatment was changing the course of modern medicine. In 1909, Carlo Moreschi, an Italian researcher then studying at Ehrlich's institute, added a new twist to the study of "athreptic immunity." Before carrying out the tumor transplants, he first put some of the mice on low-calorie diets. The results were striking: though it was sometimes possible to successfully transplant tumors into the underfed mice, the tumors wouldn't grow nearly as well as those transplanted into mice that were able to eat all they wanted.

While Moreschi was studying the quantity of food consumed by rodents, other researchers began to wonder whether the type of food mattered as well. A number of studies found that diets free of carbohydrates, which break down to glucose, could protect mice from cancer in much the same way as low-calorie diets. In 1913, two Cornell University cancer researchers published a review of these studies in the *Journal of Medical Research*, concluding that enough experiments had been done on enough animals to eliminate any doubts that carbohydrate-free diets made rodents "more resistant to tumor growth." The article made no claims about a cure, noting that the effect of the carbohydrate-free diet was observed not at the beginning of the tumor growth but on "its continued progress." Yet the contrast between carbohydrate-free diet and standard diet was clear: "When the diet includes carbohydrate," the authors noted, "the tumors grow luxuriantly."[9]

Cancer, it seemed, had an unusual appetite. The question was why. A decade later, Warburg would observe cancer cells swallowing up enormous amounts of glucose and become convinced that he had arrived at the answer.

CHAPTER FOUR

Glucose, Cancer, and the Crown Prince

WHILE IN MEDICAL SCHOOL in Heidelberg in his late 20s, Otto Warburg, as always, found much to complain about. In 1907, he sent a letter to his sister Lotte in which he noted that his recent experiments had been "ruined by two jackasses." "Medical doctors," Warburg added, "can really botch things up."[1]

Still, Warburg had plenty to be grateful for at Heidelberg. The well-known physician overseeing his studies, Ludolf Krehl, was an ideal mentor for him. Krehl believed that good medicine began with foundational science, and as a result, Warburg's medical training was almost entirely devoted to research.

Krehl, like Emil Fischer and Emil Warburg, saw no reason to take breaks from science—students were expected to be at work when Krehl stopped by the laboratories on Sunday evenings. When Warburg went on vacation, it was to the Naples Zoological Station for more research. Under the influence of Jacques Loeb—and with cancer seemingly already in mind—Warburg had begun his work at Naples with a basic question: how much oxygen did growing cells consume? But over time, Warburg shifted from the question of "how much" to the question of, simply, "how." For all the scientific advances of the era, one of the most important questions in biology—how food and oxygen combine inside us to sustain life—remained unanswered.[2]

That nutrients we eat are burned with something in the air was clear even to Galen, the Roman physician to the gladiators who remained the most influential medical thinker for much of the next two millennia. Galen was convinced that food and air came together in the heart, which burned with the heat of a flame. Some 1,600 years later, the great French chemist Antoine Lavoisier alternately placed a guinea pig and a burning piece of charcoal inside a chamber of ice and demonstrated that Galen had been at least partially right in conceiving of animal breathing as a burning fire. The ice around the animal and the ice around the charcoal would melt at different rates, but in both cases, the amount that melted (a measure of heat) was in direct proportion to the carbon dioxide given off. Respiration, wrote Lavoisier in 1790, is "similar in every way to that which takes place in a lamp or lighted candle."[3]

Lavoisier also grasped another critical fact about respiration: the carbon we breathe out in the form of carbon dioxide comes from our food. But Lavoisier, guillotined during the French Revolution, never had the opportunity to solve the rest of the puzzle. Our bodies might burn food in much the same way a lamp burns oil or a candle burns wax, but oxygen reacts with lamp oil or candle wax only when we raise the temperature with the heat of a flame. The question left for scientists after Lavoisier was how the same reaction could take place in the absence of that flame. Why does a slice of bread that won't burn when sitting on a table begin to burn once it is inside of us? Food and oxygen were clearly the fuels needed to sustain a cellular fire, but the match inside of our cells remained elusive.

In 1910, Warburg still had only a vague notion of how oxygen reacts with molecules from our food. He knew there had to be a molecule that made the reaction possible. Such molecules were already known as enzymes, though Warburg, distrustful of the enzyme science of the time, preferred the older term, "ferment." Since chemicals that damage cell membranes were known to slow respiration, he reasoned that the key molecule was likely attached to the cell's surface. But Warburg had little idea of how that molecule—he called it the "respiratory ferment"—worked or how it interacted with a membrane. As Warburg wrote to Loeb, the membrane's role in respiration remained "a total mystery."[4]

Three years later, Otto Warburg completed his medical studies in Heidelberg. Though he had not solved the mystery of cellular breathing, he had already written some 30 scientific papers and distinguished himself as one of Germany's best young scientists. He graduated magna cum laude, missing the rank of summa cum laude not because of the quality of his work, but rather, one adviser noted, because of the "dictatorial certainty" with which he had spoken during his dissertation defense.[5]

Warburg returned to Berlin. His timing was perfect. Germany might have been at the forefront of international science in the first decade of the new century, but the German scientific establishment was not resting easy. England and the United States were both threats to German hegemony. The individual scientist causing Germans the most anxiety was none other than Jacques Loeb. In 1910, the Rockefeller Institute had lured Loeb away from academia with the promise that he would be free to pursue his scientific interests without the obligation to teach. Loeb, already the most prominent figure in the world of cell biology, was poised to usher in a new scientific era in which Germany would be unable to compete.

What their nation needed, prominent German scientists argued, was a Rockefeller Institute of its own, a place where that nation's best researchers could work on whatever matters they chose without distraction. In 1911, the Kaiser Wilhelm Society, a joint venture of the German state and private donors, was founded. Emil Fischer, the famed scientist who taught Warburg organic chemistry, was one of the leading fundraisers and advocates for the new society. Fischer told one scientist he recruited to the Kaiser Wilhelm Society that he would have so much freedom he could choose to spend years simply walking in the woods, perhaps pondering "something beautiful."[6]

The first two institutes were devoted to chemistry. A third Kaiser Wilhelm institute, devoted to biology, was scheduled to open in 1914 with Theodor Boveri as its director. Boveri's most important task, at the outset, was to select the leaders for the various departments within the new institute. He soon found himself trapped in a thicket of political negotiations. Boveri was already in poor health, and the stress of staffing the institute only made things worse. "I am at the end of my forces," he told a friend in January of 1913. "For four months this inner struggle has torn me."[7]

Boveri finally gave up. He would never direct the new institute. But the department heads he chose would remain in place. To lead the cell physiology department, he had tapped the most promising German scientist in the field: Otto Warburg. In so doing, Boveri, then in the midst of writing the famed monograph that would lay the foundations of the genetic understanding of cancer, also helped further the metabolic view of cancer.

Not yet 30 when Boveri made his decision, Warburg now had among the most envious positions in all of science—his own lab in an elite scientific institute in the world's most advanced scientific country. Warburg was free to study photosynthesis or respiration or any other topic he cared to. Loeb had taught him that a scientist could remake the biological world. Thanks to Boveri, Warburg now had everything he needed to achieve his grand ambitions.

Because the Kaiser Wilhelm Society Institute for Biology had not yet been built in 1913, the one thing Warburg lacked, in the near term, was a place to conduct his experiments. But finding a laboratory bench would not prove difficult for the son of Emil Warburg, whose influence on German science had only grown since Otto had left home. In 1905, Emil became the president of Germany's Imperial Physical and Technical Institute, then one of the premier physics institutes in the world. The campus, located in the posh Charlottenburg district, included an elegant two-story villa for the president's family. Otto once said the home was "like a palace."[8]

Warburg continued working on cellular respiration in the laboratory of Walther Nernst, a future Nobel laureate and friend of the family. Whether Warburg was able to appreciate his good fortune is difficult to know. He may still have suffered from the distress he had alluded to in his 1912 letter to Otto Meyerhof, his medical school classmate. Meyerhof had sent a sympathetic response and recommended a doctor, but if Warburg was feeling better upon his return to Berlin, it was likely Meyerhof's science, rather than his kind words, that made the difference.[9]

Only months after Warburg completed his research on cellular respiration at Heidelberg, Meyerhof brought a surprising finding to his attention: certain acids could slow the breathing of growing sea urchin embryos. Since the same acids were known to bind to metals, Warburg wondered if metals played a role in allowing a cell to breathe by making oxygen more reactive. Specifically, he wondered if the respiratory ferment (enzyme) he was searching for worked because it contained a metal. Perhaps, like a dancer stepping in to swoop away someone else's partner, the acids were slowing respiration by grabbing on to the metal and preventing it from reacting with oxygen as it normally would.

Warburg wasn't the first to suspect that metals might play a role in respiration. That iron, when wet, would react with something in the air and turn to rust had been understood for thousands of years. And iron was already known to carry oxygen in the blood, where respiration was once thought to take place. Asking whether iron might be the key to breathing required little in the way of scientific brilliance. The hard part was providing conclusive evidence. Warburg had a hunch that iron was critical for respiration and thus life, but he could not be certain.

To learn more, Warburg sprinkled iron salts on breathing sea urchin eggs. More iron, he saw, increased the amount of oxygen taken up. Next Warburg wanted to know if iron compounds could trigger reactions in a model system—a system outside of living cells. To find out, he heated hemoglobin from blood, which he knew contained iron, until it turned into charcoal. He then added the charcoal to a solution containing amino acids and oxygen. The charcoal worked precisely as Warburg pictured a respiratory ferment working on the surface of a cell. Before the addition of the charcoal, the oxygen wouldn't react. The iron-rich charcoal was like a love potion. As soon as Warburg added it to the solution, the oxygen sprang to life and instantly attached to the surrounding amino acids.[10]

More revealing still, whether in living cells or in his model system, the iron compounds would lose their catalytic magic when cyanide was added to the solution. Cyanide was understood to be the world's deadliest poison due to its ability to interfere with how our bodies use oxygen. Now Warburg saw why. It reacted with the iron in his respiratory ferment that would otherwise be available to react with oxygen. Cyanide, Warburg realized, is like a pillow being pressed down on the face of a cell. It kills by suffocation.

Even with these findings, Warburg still didn't have anything close to proof that respiration is dependent on iron. Cyanide can react with other metals as well, and copper was also known to be present in cells. But the evidence was growing. Galen, it seemed, had turned to the wrong artifact of ancient life in trying to understand why we breathe. The most revealing clue hadn't been fire, but the rusting iron swords of the gladiators he had once treated.

IN TURNING TO BIOLOGY, Otto Warburg had moved beyond his father's physics. Now, having returned to Berlin, he would pause to look back. Less than a decade earlier, Einstein had pondered what happens when light hits a metal surface and arrived at one of the most startling insights in the history of science: Light wasn't simply a wave as every physicist believed. It could also behave like a particle. It was an idea so outrageous that hardly any physicists accepted it. Even Planck, whose own calculations had laid the groundwork for Einstein's breakthrough, dismissed the notion as a youthful mistake of an otherwise promising physicist.

But not everyone thought Einstein was wrong. In 1907, Emil Warburg conducted his own investigations of light's energy. In the following years, he would provide the first experimental evidence for Einstein's theory of how light particles interact with molecules. "You are making real," Einstein wrote to Emil Warburg, "that which for years I had only vaguely dreamed about."[11]

Otto Warburg, too, would become fascinated by the transfer of energy from light to matter. He began to study the phenomenon at Emil Warburg's institute. But rather than shining light at metal, he would aim light at living cells to understand how photosynthetic organisms use energy to make glucose. With that simple redirection, Warburg had arrived at the perfect bridge between his own research and the research of his father. In a letter that Emil Warburg sent to his son, he even mentions an experiment the two intended to work on together, a surprising plan given the tension in the relationship. While it appears that the joint experiment was never carried out, Otto Warburg would go on to revolutionize the field of photosynthesis over the next decade, becoming the first to truly grasp the process in the context of Einstein's new understanding of light.[12]

Otto Warburg won the Nobel Prize for his research on cellular respiration, and he is best known today for his cancer research; but for all of the many bitter scientific disputes he engaged in, none was more personal to him than the dispute over photosynthesis. Though accepted as scientific fact for several decades, Warburg's finding that only 4 photons were necessary to release one molecule of oxygen would be overturned in the second half of the twentieth century. The revised figure—8 to 12 photons—took little away from Warburg's critical contributions to the field, and yet Warburg was never able to acknowledge that he was wrong or let go of his indignation. His hope, he wrote of the photosynthesis researchers who disputed his findings, was that they would be "punished already in this world."[13]

Warburg's obsession with the number of photons required to power photosynthesis was a mystery to some of his colleagues. They failed to grasp that photosynthesis wasn't just another area of interest for Warburg; it was his inheritance.

They may have also failed to grasp that Warburg saw photosynthesis as a natural extension of his study of cellular breathing. Photosynthesis is essentially respiration in reverse, and Warburg understood the two processes as different aspects of the same underlying phenomenon. In his early work on photosynthesis, he picked up almost exactly where he had left off in his respiration studies by investigating the role of metals in sparking reactions. When another researcher complained that the study of animal physiology was overtaking the field of plant physiology, Warburg told his sister Lotte that he found the objection ridiculous. "This all belongs together," he said.[14]

Cancer, from Warburg's perspective, belonged with both photosynthesis and respiration. Warburg had not yet discovered that cancer cells ferment glucose, but his work on sea urchin eggs started with the assumption that cancer was a problem of growth and so must have something to do with the way a cell uses energy. The legendary twentieth-century biologist Theodosius Dobzhansky famously wrote that it was impossible to make sense of any aspect of biology without examining it through the lens of evolution. Warburg might have said the same of energy processes. Now back in Berlin, he was poised to establish his own scientific identity in the new Kaiser Wilhelm Institute, but even when studying cancer, he would remain Emil Warburg's son.

THAT GERMANY'S NEW scientific society had been named after Kaiser Wilhelm II was more than an obligatory show of respect. Wilhelm was an enthusiastic supporter of the society and its institutes and was said to have personally designed the organization's flag and court uniform. Though his interest was less in the substance of the science than in the prestige it conferred on him, Wilhelm did once attend a lecture Ehrlich gave on chemotherapy. As the story—which may be apocryphal—goes, Wilhelm later called Ehrlich to a special audience and asked him when he was going to cure cancer. When Ehrlich could not provide the answer he wanted to hear, the kaiser lost interest and turned away.[15]

Wilhelm had good reason to be concerned about cancer. By 1901, both of his parents had been killed by it. His father, Kaiser Friedrich III, was the first to succumb. In early 1887, Friedrich, then still the crown prince and heir to the throne, complained of a sore throat. It seemed he had only a cold. By the end of the month, Friedrich's throat was so swollen he could barely speak. Upon examining the crown prince, his doctor noticed a small growth and cauterized it with an electric wire. This didn't help. Over the course of the year, as Friedrich deteriorated, his treatment devolved into a dark comedy of errors as his British and German physicians bickered over whether to operate.

For the many Germans who loved Friedrich, his demise was painful to witness. He was a fundamentally decent man who detested war and favored democratic reforms—he once attended a synagogue service in his Prussian military uniform to show his solidarity with Berlin's Jews. By November 1887, Friedrich's doctors were desperate, which probably explains why they listened when a Bavarian duke told them about a new cancer therapy he had heard about from his sister, Empress Elizabeth of Austria. Elizabeth had supposedly learned of the new experimental therapy during a visit to a hospital in Vienna.[16]

The new therapy was the creation of Ernst Freund, a 23-year-old Jewish doctor in Vienna who had just earned his medical degree. Two years earlier, while still a student, Freund had published an obscure five-page paper describing the elevated glucose levels he had detected in the blood of patients with carcinomas (cancers that arise in the epithelial cells that line the surfaces of our organs). A full 62 of the 70 carcinoma patients Freund had examined appeared to have too much glucose in their blood. Since the link "can hardly be considered accidental in so many cases," Freund wrote, "I believe that the presence of an abnormal of amount of sugar or glycogen in the blood is necessary for the existence of the carcinoma."[17]

Newspaper reports from late 1887 offer conflicting reports of what happened after Friedrich's doctors took an interest in Freund's research. It appears that the crown prince had blood drawn from his neck and that his physicians confirmed that he did, in fact, have elevated levels of glucose. The physicians then introduced Freund's experimental glucose-lowering treatment, which involved both dietary changes and an unspecified medication. It may have been the first cancer ever treated with a modern metabolic therapy.

According to some accounts, the treatments helped. On December 23, 1887, the *New York Times* reported on the crown prince's therapy under the headline, "The Crown Prince's Malady: A new theory regarding the treatment of cancer." The very next day, the paper published a second story, citing a clearly perturbed cancer authority: "The theories of Dr. Freund of Vienna concerning the cause of the cancer in the German Crown Prince's throat are generally discredited by New-York medical men," the article stated. The article added that there was "no relation whatever between cancer and sugar in the blood." For readers of the *New York Times*, at least, the new metabolic understanding of cancer was dead within 24 hours of its arrival.[18]

It is possible that Freund's dietary regime made a difference. When Wilhelm I died in March 1888, Friedrich was well enough to assume the throne. But if the treatment helped, the effect was short-lived. Friedrich died that June after four months in power, leaving the throne—

and the German military—in the hands of his notoriously foolish and aggrieved 29-year-old son, Wilhelm II. Wilhelm read little outside of newspaper clippings about himself and would erupt into angry fits when portrayed in an unflattering light. "In order to get him to accept an idea you must act as if the idea were his," Wilhelm's closest friend once explained. Worst of all, he surrounded himself with warmongers who were skilled in the art of manipulating his rage. Wilhelm's hatred of England was so fierce, a *New York Times* reporter noted in 1908, that "his eyes snapped" when the subject came up.[19]

It is impossible to know how a living Kaiser Friedrich would have changed the course of the twentieth century, but Germany would almost certainly have followed a different path had he survived. While countless different social and economic factors played a role, the critical turning point for Germany was the start of the First World War, a war that left 20 million dead and set the stage for the Second World War. And the First World War would likely never have happened if not for the ascension of the recklessly militant and volatile Wilhelm. John Röhl, the celebrated British historian who spent nearly 30 years writing a three-volume biography of Wilhelm, concluded that even though he was not acting alone and was not the principal advocate of war in the summer of 1914, Wilhelm nevertheless had "perhaps the heaviest overall" responsibility "for having brought about Europe's great catastrophe."[20]

Friedrich was only 56 at the time of his death, and his own father had lived into his 90s. The cancer that killed Friedrich and put Kaiser Wilhelm II in charge of the German military might have been the most consequential in history, a mass of overeating cells in the throat of one man that led to a half century of hell.

CHAPTER FIVE

"Slaves of the Light"

KAISER WILHELM II put Germany on a path to destruction, but the fire of war requires a match. It was struck on June 28, 1914, when Gavrilo Princip, a young Serbian nationalist, raised a .32 caliber pistol in the direction of Franz Ferdinand, heir to the Austro-Hungarian throne, and pulled the trigger. "Where I aimed, I do not know," Princip later said.

That Princip did not know where he was shooting mattered little. The devil's bullet knew which way to go. Franz Ferdinand died; Austria-Hungary made plans for war, and Kaiser Wilhelm II promised Germany's unconditional support. On August 1, 1914, Kaiser Wilhelm II signed the mobilization order that launched World War I. That same day, Jacques Loeb, sitting at his desk on the other side of the Atlantic, wrote a letter to Otto Warburg. Loeb was responding to a letter Warburg had sent to him in June. Warburg, still waiting for construction to finish on the Kaiser Wilhelm Institute for Biology, wanted to spend the winter of 1914 doing research with Loeb at Stanford University's Hopkins Seaside Laboratory. "Would the sea urchins be fully developed at the time?" Warburg had asked. Though Loeb left open the possibility of Warburg's visit, he was not optimistic. He hoped, he told Warburg, that "the war clouds in Europe" would "be dispelled" by the time his letter arrived.[1]

A committed pacifist, Loeb was monitoring Europe's "terribly stupid aristocratic governments" with disgust. He believed the process through which war propaganda convinces people to sacrifice themselves could be understood scientifically. The idea stemmed from his own research on insects. Loeb was struck, in particular, by one observation: upon emerging from their nests, typically near the bottom of a plant, newly hatched caterpillars would invariably make their way up the stalk, where they would find their first meal in the form of the plant's leaves. How, Loeb wondered, did the newborn caterpillars always know which way to go? It was as if they were born already knowing where to find food.

There had to be a signal, Loeb reasoned, something in the environment that instructed a newborn caterpillar to make its ascent. There were plenty of possibilities. The caterpillars might have been responding to a specific visual cue or smell. But Loeb had another guess. He suspected that light itself was the signal. To test the idea, he brought caterpillars into his lab and placed them inside test tubes that had been arranged horizontally. Loeb put the insects' "favorite leaves" on one side of a tube and then shined a lamp on the other side. The effect was even stronger than Loeb had anticipated. The caterpillars always went toward the light and stayed near it. They were "slaves of the light," Loeb wrote. Their food, only inches away on the other side of the tube, might as well have been a continent away. The caterpillars would starve to death in the grip of the light rather than go in search of sustenance.

Though Loeb didn't think that people were as simple as insects, he did believe that people were strictly mechanical and that they could be controlled by manipulative language in much the same way that his caterpillars could be controlled by light. While the possibility that we are under the sway of hidden forces sounds chilling to modern ears,

Loeb saw his mechanistic understanding of mental life as a source of hope. Scientists studying hormones and behavior, he argued, would soon be able to explain how the process worked; Ivan Pavlov, through his experiments on salivating dogs, had already made great progress to that end. And once people recognized how their brains were being controlled, they would be set free from superstition and hatred—from nationalism and war.

Loeb eventually realized that his optimism was misplaced. He might have been right about the power of slogans to shape the human mind, but in 1914, there were no signs of anyone breaking free. The German masses were being lured toward their own destruction exactly like the starving caterpillars at the wrong end of the test tube.[2]

AFTER SIGNING THE mobilization order, Wilhelm returned to his Berlin palace and stepped out onto the balcony. Some 50,000 Germans were gathered below singing patriotic songs. The kaiser had stood on the same balcony the day before, telling the crowd that "envious people" were forcing "the sword" into Germany's hand. But with war now set to motion, the kaiser chose to stress the importance of unity at home. "Today," the kaiser said, "we are all German brothers and only German brothers."[3]

The kaiser's call for a "civil truce"—a message he would repeat in the following days—would give rise to feelings of togetherness in a country previously marked by intensely felt social and political divisions. German Jews, who would sign up to defend the fatherland by the tens of thousands, were particularly enthusiastic about the new sense of German unity. Jewish newspapers reprinted the kaiser's speech, declaring with pride that Jews could no longer be considered different from any other Germans. Some German Jewish Zionists who had emigrated to Palestine even returned to fight for Germany.[4]

If there was a note of desperation in the German Jewish response to the kaiser's small gesture, it was perhaps because German Jews were increasingly perplexed by their place in German life. In the eighteenth century, the vast majority of Germany's estimated 60,000 Jews lived on the fringes of society. Forbidden from most trades, they eked out livings as peddlers and pawnbrokers. Some roamed the countryside in groups of beggars or thieves. (Legend has it that the Jewish highwaymen of the time did not rob on the Sabbath.)

And then, decade by decade, throughout the nineteenth century, Enlightenment thinking reshaped German Jewish life. New laws granting Jews equal status spread from one German state to the next. By the second half of the century, the children of peddlers were studying to become lawyers and doctors, bankers and scientists. The new legal protections did not give Jews anywhere near full equality with Christians. Jews who hoped for success in the civil service or to reach the upper ranks of the military or academic institutions would often first have to convert to Christianity. Thousands of Jews would pay this steep price, including Emil Warburg. But German Jews were succeeding in ways that would have been unimaginable a century before.

As they assimilated, Germany's Jews came to think of themselves as fully German. Like a suitor whose advances have finally been accepted after a long period of rejection, they couldn't help but feel grateful for their gains. "Before Hitler rose to power, other Europeans often feared, admired, envied, and ridiculed the Germans," the Austrian-born Israeli author Amos Elon wrote. "Only Jews seemed actually to have loved them."[5]

And yet, rather than fading away as Jews entered the mainstream of German society, the animosity toward Jews only spread and turned more virulent. Some of the new anti-Semitism wasn't new at all. It was the same Christian bigotry that had plagued Europe for centuries. In 1891, riots broke out in the German town of Xanten after a Jewish butcher was accused of ritually murdering a 4-year-old Christian boy.

But there was something genuinely different about the anti-Semitism of the late nineteenth century, which gave rise to modern political parties devoted to combating the Jewish influence on German society. The rapid industrialization of Germany had created great wealth, but, for many, also anxieties about how they would fare in a new economic system. Romantic nationalists looked back to preindustrial times through a mythical lens in which ethnic Germans had supposedly lived in spiritual communion with nature. The Jews, as a separate people, could never be a part of this German *Volk*. It wasn't only that Jews had the wrong blood. Insofar as German Jews were associated with industrial and urban development, they were responsible for the destruction of an agrarian utopia, enemies of the soil itself.

The anti-Semitism of the Romantic nationalists blended easily with the pseudoscientific theories of the period that explained history as a struggle among races. That the German or Aryan race was superior was a given. Less certain was whether it would maintain its superiority. According to Joseph Arthur de Gobineau, the French writer who had helped popularize scientific racism in the middle of the nineteenth century, the downfall of society came about when a superior race mixed with an inferior race. For German scientific racists, the gravest threat posed by the Jews wasn't that they were manipulative or dirty or any of the other old anti-Semitic tropes. It was that they were assimilating and marrying non-Jewish Germans.

By World War I, German Jews had made extraordinary contributions to virtually every facet of German life. And yet the modern anti-Semites, with their mantra of "the Jews are our misfortune," seemed only more and more aggrieved. Like Boveri's sea urchin embryos with the wrong number of chromosomes, Enlightenment thinking, corrupted by the wrong information, was turning hollow and collapsing in on itself.[6]

AMONG THE MOST prominent anti-Semitic ideologues of the time was Houston Stewart Chamberlain, an Englishman who had relocated to Germany. Chamberlain, a friend of the kaiser and the son-in-law of the famously anti-Semitic composer Richard Wagner, saw the entire history of the West not merely as a struggle among races but specifically as a struggle between Aryans and Jews. In addition to their countless other crimes, Chamberlain maintained, Jews were responsible for the debasement of science. In 1912, Chamberlain singled out Jacques Loeb's "soulless mechanist technological Jewish science."[7]

The attack on Loeb might as easily have been made on his mentee. In 1914, Warburg would have sensed his vulnerability even as he was poised to assume one of the most prestigious positions in German science. Warburg might not have identified as Jewish, but Germany identified him as a Jew. He would have understood, as did most German Jews, that his commitment to his country needed to be made public. Not long after the kaiser's call for unity, Warburg was among the tens of thousands of Germans of Jewish heritage to sign up to fight.

The army probably had another, more distinctive appeal to Warburg. He might have been the rising star of Chamberlain's "soulless" science, but he was also an aristocrat and a snob. The kaiser's army was exactly the type of old-fashioned, hierarchical nineteenth-century institution to which he was drawn. Warburg joined an uhlan (cavalry) regiment made up of men who still carried titles, including "Count" and "Baron." Despite their pedigree, the Prussian uhlans had earned a reputation outside of Germany as ferocious warriors. Charles Darwin, upon encountering an uhlan for the first time, confessed that he thought they "belonged to a half-civilized tribe on the Eastern frontiers of Germany."[8]

By the First World War, the uhlans, who rode with 10-foot lances, were already an anachronism. Their elaborate uniforms included a field-gray tunic and a *czapka*, a square-topped helmet that gave the appearance of a graduation cap. In the dawning new era of trench warfare and deadly machine guns, the uhlans galloping on their horses with lances were almost comically unsuited for battle.

After the war began, Germany's military leadership quickly con-

cluded that cavalry regiments would be better suited to the Eastern Front, where the rougher terrain might make horses more useful. In the first years of the war, Warburg sometimes rode in patrols in advance of the front lines. He had spent almost all of his adult life, to that point, in a laboratory, and he could only keep the charade going for so long. He once acknowledged that his ranking officers had doubts about his future in the army, given his ineptness. But with time, Warburg seems to have gained the respect of his fellow soldiers. He rose to the rank of lieutenant, and after suffering a minor injury, received the Iron Cross, First Class.[9]

In 1917, Warburg's regiment was attached to an infantry division on the Eastern Front, for which Warburg served as an aide-de-camp, a role far more suited to his skills. In addition to acting as the regiment's physician, Warburg was named a "gas officer," meaning he was responsible for detecting gas attacks and ensuring that the division was prepared. The role of "gas officer" was typically assigned to anyone with even a modicum of knowledge of chemistry. The other men in his regiment might not have appreciated that Warburg may have been more skilled at measuring gas than any human alive.[10]

WHILE WARBURG WAS discovering the possibility of life without science on the Eastern Front, the world he had left behind was crumbling. The scientists at the new Kaiser Wilhelm Society put their research interests aside to make Germany's own chemical weapons. The Italian government seized the Naples Zoological Station. The steamboat so many scientists had used to gather sea creatures from the pristine waters of the gulf was converted into a warship.

As the fighting dragged on and German casualties mounted, the euphoria of 1914 turned to despair. Boveri, deteriorating from a mysterious medical condition, pondered whether "violent emotions" in the

Opposite: Otto Warburg in his military uniform, 1916.

face of so many unexpected deaths might be causing "latent processes to erupt" inside of him. "Dying of a broken heart," he wrote, "seems to me to belong in this category." Five months later, he was dead at 53.[11]

Ehrlich, too, was devastated by Germany's descent into war. Though he maintained that Germany had a clean conscience in its choice to fight, he saw from the start that there would be no happy ending. He died of a stroke in August of 1915, but not before he witnessed the chemical compounds in his dyes reassembled into explosives and poisonous gases.

After two years of war, German Jews saw that the kaiser's call for unity was a false hope. In response to anti-Semitic claims about Jews "shirking" their military duties, the Prussian war minister ordered a "Jew count" to tally the number of Jews serving on the front lines. The results of the humiliating exercise were never made public, but Germany's Jews felt betrayed. Some 80,000 of the 100,000 Jews who served in the kaiser's army did so at the front. By the time the war was over, 2 percent of all German Jews had died defending the fatherland.[12]

As the war dragged on, Warburg's mother, Elisabeth, was among the many Germans who had come to think of it as "a sin against humanity" and as "mass murder." She had already lost a brother in the fighting and was "consumed with worry and fear" that she would lose her only son. Emil Warburg shared her anxiety. In an effort to secure his son's release from the army, he wrote a series of letters to the Ministry of the Interior, arguing that Otto was important to German science and that his photosynthesis work might one day help Germany feed its population.[13] Likely after being prodded by Emil, the prominent German botanist Carl Correns also sent a letter to the Ministry of the Interior calling for Warburg's release. If Warburg's experiments prove successful, Correns wrote, it will be "of extraordinary importance for people's nutrition" and "open up a truly rich source of nutrients."[14]

The letter-writing campaign succeeded, but convincing the German military to allow Warburg to return to Berlin was only half of the challenge facing Warburg's parents. They also had to convince Warburg that leaving his regiment behind on the Eastern Front and returning to Berlin was the right decision. In 1918, they reached out to the one person they thought might be able to persuade their son to come home: Albert Einstein.

While Emil likely spoke to Einstein first, it was Elisabeth who sent a letter to Einstein on March 21, 1918:

> *My husband has always been so utterly delighted with*
> *his son. He once told me in confidence that he thought*
> *he was going to be one of the greats. Are our hopes for*
> *him to all be for naught? Is he to just throw away his*
> *destiny? . . .Why does it have to be his division that is*
> *put right up in front?*[15]

Einstein, who had left Germany for Switzerland as a teenager, in part to avoid being conscripted into the German army, needed no convincing that the war was a great and avoidable tragedy. "All our lauded technological progress—our very civilization—is like the axe in the hand of the pathological criminal," he wrote to a friend in 1917.[16]

Einstein composed a letter to Otto Warburg on March 23, 1918, two days after Elisabeth Warburg had written him. He began by acknowledging that he did not know Warburg well—but he knew him well enough to flatter him. "I gather that you are one of the most able and most promising younger physiologists in Germany and that the representation of your special subject here is rather mediocre." With that, Einstein then made an emotional plea:

> *I also gather that you are on active service in a very*
> *dangerous position so that your life continuously*
> *hangs on a thread: now for a moment please slip out*
> *of your skin and into that of another clear-eyed being*

and ask yourself: Is this not madness? Can your place
out there not be taken by any average man; is it not
important to prevent the loss of valuable men in that
bloody struggle? You know this well and must agree
with me.[17]

While Warburg's response hasn't survived, Einstein's letter appears to have been enough to convince Warburg to leave his regiment.

Warburg returned to Berlin several months before the armistice. It is possible that Albert Einstein saved his life and in so doing played his own small role in Warburg's cancer breakthrough that would come less than a decade later.

Though he left his unit prematurely, Warburg formed lasting friendships with a number of the other officers. "You can really only hold a proper conversation when in the company of officers from the Prussian Guard," he once told a scientific colleague, in what might have been his most obnoxious comment of all.[18]

Near the end of his life, Warburg would speak proudly of having worn "one of the finest uniforms of the old Prussian Army" and come to view his war years as formative. "I got to know the realities of life which had escaped me in the laboratory," as he put it. In addition to learning to "handle people" and to "obey" and "command," Warburg said, he learned that "one must be more than one appears to be."[19]

GERMAN JEWS WERE far from alone in their patriotic fervor in the lead-up to World War I. After his mother's death in 1907, Hitler again applied to Vienna's Academy of Fine Arts and again was rejected. In the aftermath of this second rejection, his friend and roommate, August Kubizek, noticed that he was growing increasingly "unbalanced." Hitler, Kubizek wrote, would "fly into a temper at the slightest thing."[20]

By late autumn of 1909, Hitler had reached bottom. He was sleeping on park benches or in Vienna's homeless shelters, several of which were supported by Jewish philanthropists. To earn spare change, he shoveled snow and carried people's luggage at the train station.

Hitler's only real income came from the hastily drawn postcards he peddled. By 1910, he had made enough money selling painted postcards to move into a men's home. He spent much of his time during his Vienna years in cafes, where he was exposed to the fanatical German nationalism and biological racism of the era. Though already a fervent German nationalist prior to the war, Hitler had not been overtly anti-Semitic. He was on friendly terms with a number of the Jewish residents at the men's home where he slept. One of those Jews, a one-eyed locksmith's assistant, periodically shared his disability allowance money with Hitler. What little money Hitler made himself often came from selling his paintings to Jewish merchants.

Upon turning 24 in April 1913, Hitler was able to claim his inheritance from his father, who had died when Hitler was 13. By then, Hitler was a Pan-German nationalist who longed to see the German part of Austria united with Imperial Germany. His inheritance made it possible for him to move to Munich and thus avoid being conscripted into the army of the Austro-Hungarian Empire he despised. Hitler later wrote that his plan had been to become an architectural draftsman. Instead, once in Munich, he returned to his prior line of work: selling hastily painted pictures around town, sometimes by approaching people in beer gardens relaxing after work.

If Hitler expected to find kinship with his fellow German speakers, the reality of his life in Munich probably came as bitter disappointment. He failed to make any close friends. His landlady at the time described him as a "hermit." Even Hitler's roommate moved away, possibly because he could no longer stand Hitler's diatribes.[21]

And so when the kaiser called for unity among Germans in August of 1914, it would lead to a darkly ironic moment in history: Hitler appears to have been excited by the kaiser's words in much the same way as Germany's Jews. The promise of war, both for German Jews and Hitler, came with the promise of a new beginning. In *Mein Kampf,* Hitler remembers his fanatical enthusiasm for the war as "a release from the painful feelings" of his youth. The day after the kaiser's call for unity, Hitler was among the thousands celebrating in the streets of Munich. Several weeks later, he enlisted in a Bavarian infantry regiment and soon found himself in northern France.[22]

On October 29, 1914, as Hitler's regiment engaged in heavy fighting, a British soldier raised his gun and fired in Hitler's direction. If Hitler's account can be trusted, the bullet tore through the sleeve of his shirt but failed to pierce his skin. "Miraculously," Hitler wrote in a letter to an acquaintance, "I remained without a scratch."[23]

Only three weeks later, a shell hit a regimental commander's hut and killed several of the staff inside. Hitler had been inside the tent only 5 minutes before. By that point, Hitler had already become a dispatch runner, responsible for delivering messages. Most historical accounts indicate that Hitler risked his life again and again. But documents recently discovered by the historian Thomas Weber show that Hitler spent almost the entire war out of harm's way. While some runners were required to travel to the front lines, Hitler, as a regimental dispatch runner, merely carried messages from one battalion headquarters to another. Other soldiers, resentful of the privileges of such runners, referred to them as "rear-area pigs."

"The front experience of Private Hitler consisted more in the consumption of artificial honey and tea than of the participation in any combat," a member of Hitler's regiment wrote in 1932. "Thousands of family fathers would have filled Hitler's little post behind the front just as well as him."[24]

In 1918, having already received an Iron Cross, Second Class, Hitler was awarded an Iron Cross, First Class. Such medals, ostensibly given out for bravery, were largely a reflection of a given soldier's connections at regimental headquarters. (Warburg probably earned his own Iron Cross for similar reasons.) Hitler received the honor on the recommendation of Hugo Gutmann, a lieutenant and the highest-ranking Jewish soldier in his regiment. (Gutmann would flee Nazi Germany in 1940.)

By war's end, Hitler, having seen thousands of soldiers die around him, had learned that "life is a cruel struggle." He was no longer the fragile young man who had tenderly cared for his cancer-stricken mother, nor the hapless, failed artist sleeping on public benches. Though far from a daring soldier, he had managed to carry out his duties and earn a modicum of respect. As Hitler biographer Volker Ullrich wrote, the "25-year-old loner finally thought he had found a way out of his disoriented, useless existence." Hitler, like Warburg, had learned to be more than he appeared.[25]

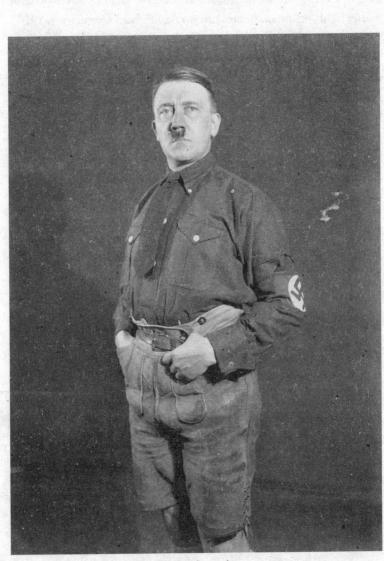

Adolf Hitler, date unknown.

PART II

"The Meaning of Life Disappears" (1919–1945)

Faustus:
Tell me, where is the place that men call hell?

Mephastophilis:
Under the heavens.

Faustus:
Ay, but whereabout?

Mephastophilis:
Within the bowels of these elements,
Where we are tortured and remain for ever:
Hell hath no limits, nor is circumscribed . . .

—CHRISTOPHER MARLOWE,
DOCTOR FAUSTUS

CHAPTER SIX

The Warburg Effect

THE BERLIN OTTO WARBURG returned to at the end of the war was not the Berlin he had left behind four years earlier. Up until the last days of the fighting, most Germans remained convinced that they would emerge victorious. It was more than self-delusion. As late as the spring of 1918, German forces were winning battles on the Western Front. When the momentum began to shift back to the Allies, the German propaganda operation successfully glossed over the bad news. The denial was so complete that some German troops returned home after their nation's defeat to celebratory parades.

Kaiser Wilhelm II was no more in touch with the reality of the battlefield than ordinary German citizens. Not until a meeting with his military advisers on November 9, 1918, did he realize that all was lost and his days as emperor were over. A German general recalled that Wilhelm's expression betrayed "amazement, then piteous appeal, and then—just a curious wondering vagueness." The general went on: "He said nothing, and we took him—just as if he were a little child—and sent him to Holland."[1]

By that point, Germans had been starving for years. The food shortages began early in the war after the British navy cut off supply routes to the Central Powers. As butter and meat grew increasingly scarce, food riots erupted throughout the country—more than 50 in 1916 alone. The government did its best to calm the public. The Germans were encouraged to eat turnips and even young crows, which, according to the Ministry of Agriculture, could make for a "suitably pleasant alternative" to traditional meat dishes. But whatever plan the government proposed, it was never enough. More than 400,000 Germans died from malnutrition or starvation during the war. One foreign visitor in Berlin witnessed a horse die in the street, only to have its flesh ripped off by a descending throng of German housewives.[2]

The naval blockade remained in place until June 1919, when the Germans signed the Treaty of Versailles. There is no record of Warburg ever going hungry, but he, too, had learned the value of a piece of meat. At one point during the war, Warburg exploited his military privilege to send half a sheep to Emil Fischer, the legendary University of Berlin chemist who had taught him to synthesize polypeptides. Fischer did not respond immediately, but several weeks later Warburg received a congratulatory telegram informing him that he had been made a "professor" at the University of Berlin, a title Warburg had no plausible claim on.[3]

If the food shortages did not affect Warburg's meals, they did influence his science. When Emil Warburg had argued that his son should be released from military duty because his photosynthesis research might help feed Germany, he had not been bluffing. Warburg resumed his photosynthesis research upon his return from the war, and the prospect of growing food more efficiently would remain one of his scientific aims for decades.

Warburg was now finally able to work at the Kaiser Wilhelm Society lab that had been promised to him shortly before he rode off to war with the uhlans. The Institute for Biology resembled a small bourgeois Berlin apartment building of the era. Warburg's lab occupied the top floor. The conditions were not what Warburg had envisioned before the war. Germany was bankrupt, and so, too, were many of the business titans who had promised to fund the new Kaiser Wilhelm institutes. In November 1920, a desperate Warburg sent a request for financial support to his cousin, Paul Warburg, a prominent American banker who had recently helped establish the US Federal Reserve System. Aware that Americans—even German Americans—might not be feeling especially charitable toward Germany after the war, Warburg mentioned that his own lab had played no role in producing gas used on the battlefield and that he didn't share the reactionary views common at German universities.[4]

Whatever money his famously wealthy Warburg relatives were able to provide, it wasn't enough. Warburg requested an additional 10,000 marks from the Kaiser Wilhelm Society, only to be told there was no money to give. He was encouraged to apply to the Emergency Association of German Science, but Warburg pointed out to the administrators that he didn't have a secretary. An official at the Kaiser Wilhelm Society sent a secretary to Warburg's lab. Warburg gave her a sheet of paper and instructed her to type a single sentence: "I require 10,000 marks." He then signed the paper and sent off what might be the most outrageous grant application in the history of science.[5]

Warburg received the money, but his financial situation remained challenging. "The decline of science in Germany is, as you can imagine, rapid, especially because young scientists do not earn enough to live from," Warburg wrote to Loeb in 1922. Warburg moved ahead as best he could. In keeping with the Kaiser Wilhelm Society's emphasis on total freedom for its scientists, Warburg's cell physiology division of the Institute for Biology was known simply as "the Warburg department." In control of his own lab for the first time, Warburg ran it like a unit of the kaiser's army. The employees were expected at the lab six days a week, from 8 a.m. to 6 p.m.—after which they would still have to find time to read and write scientific articles. As one member of

Warburg's lab remembered, there "were no reasonable grounds, apart from death, for not working." Warburg liked to point out that working hours had been much longer when he was a young scientist.[6]

At weekly lab meetings on Fridays, there was no discussion. Warburg would announce the latest findings from the institute then declare the meeting over. In his heart, one colleague said, Warburg always remained "a staff officer in plain clothes."[7] "I have been the anvil long enough," Warburg once told Hans Krebs, referencing a line from Goethe, "now I want to be the hammer."[8]

Though young scientists came to Warburg's lab for brief stints, most of the work was carried out by a small group of highly skilled instrument specialists, several of whom Warburg hired away from the Siemens Company. These technicians arrived with little, if any, training in chemistry, which was exactly how Warburg wanted it. While a number of them became world-class biochemists in their own right, Warburg liked that his technicians, unlike young scientists, had few academic interests of their own. To keep the technicians motivated, Warburg made them coauthors on his papers. He even shared some of his Nobel Prize money with Erwin Negelein, a technician turned biochemist who worked on many of Warburg's most important experiments.[9]

In the 1920s, at least, Warburg's behavior generated more wonder than bitterness. During their 30-minute lunch breaks, the young scientists would gather in a common room and gossip about Warburg's eccentricities over their boiled eggs and milk. Krebs, who joined the lab in the mid-1920s, sat directly across from Warburg's laboratory bench and remembered that Warburg would regularly insult whichever scientist he happened to be feuding with at the time. But Krebs saw the softer side of Warburg as well. When Krebs first arrived in Warburg's lab, he wasn't yet familiar with the complicated equipment. Warburg took the time to teach him what he needed to know and was ready to listen whenever Krebs had something important to discuss. He was an authoritarian, Krebs recalled, but his brilliant accomplishments and basic integrity had earned him the right to make demands of others.[10]

Given the relentless work schedule, there was little chance for anyone to maintain a relationship outside of the lab. And it was not obvious to German scientists of the time that love should be allowed to interfere with work. When Krebs, in a letter to his father, mentioned that he hoped to one day meet a girl, his father suggested that he read Rocco's "Love will not suffice" aria from *Fidelio*.[11]

Yet Warburg did now have an important new person in his life. Shortly after the war, he had hired a young man, Jacob Heiss, to serve as a full-time, live-in assistant at his home in the wealthy Berlin suburb of Lichterfelde. Heiss had served in the war and had been recommended to Warburg by his military colleagues. The role was akin to a butler or house boy. For the remaining five decades of Warburg's life, the two would rarely be apart.

While the precise nature of the relationship between Warburg and Heiss is impossible to know, every sign points to a loving homosexual partnership. Heiss "did everything he could to support and protect Warburg," Krebs wrote, "and to make his life as congenial as possible." After Warburg fell off his horse in 1924 and fractured his pelvis, he never rode again unless Heiss rode with him. They went to the opera together and vacationed together. On at least one occasion, Warburg agreed to an overseas invitation on the condition that Heiss be invited as well.[12]

Heiss, 15 years Warburg's junior, was the taller and stockier of the two. He was born in the town of Kirn and had a number of siblings, but little else is known about his background. He favored fedoras and checkered sports coats, apparently sharing Warburg's taste for fine things. By the 1950s, Heiss was bald and doing his best to manage a comb-over that would sometimes fly upward in a gust of wind. Like Warburg, Heiss was a lover of animals and raised his own chickens.

In various photos, he can be seen standing by Warburg's side, stiff and serious, in the manner of a bodyguard.[13]

Given the norms of the period, Warburg and Heiss were as open as was possible for a homosexual couple. A bachelor with a live-in male attendant wasn't a particularly common arrangement, and homosexuality could still be punished by imprisonment. In 1907, the revelation that the kaiser's closest friend was gay had turned into a national scandal.

Though Heiss was responsible for Warburg's home, over the years he would grow increasingly involved in Warburg's work life as well. After the war, Warburg made him the institute's official administrator, meaning that Heiss and Warburg spent nearly every hour together. Peter Ostendorf, a glassblower who worked at the institute, said that Heiss's role was to prevent anyone or anything from disrupting Warburg's research. Often that meant yelling angrily at the staff. Heiss once told Ostendorf that if the press showed up at an event honoring Warburg, Ostendorf should knock the cameras out of the photographers' hands. "I never saw Heiss laugh," Ostendorf recalled.[14]

On one occasion, Ostendorf remembered, Heiss appeared at the door of his workshop and shouted, "Are you trying to kill me, Mr. Ostendorf?" Ostendorf didn't understand. He soon discovered that his crime was failing to alert Heiss that an auditor had arrived at the lab. Heiss directed Ostendorf to chase the man off the property with one of the enormous Great Danes that he and Warburg kept as pets.

Warburg's cousin, the banker Eric Warburg, recalled that when he visited Heiss and Warburg for lunch in the late 1930s, Heiss— Eric Warburg referred to him as Warburg's "man-servant"—served the meal. When Otto asked his cousin if he should sell his shares in Deutsche Bank, Eric said that he might as well go ahead, given that the dividends had just been paid out. Otto, hesitant, suggested that they continue the discussion when Heiss was done serving them. After lunch, Heiss took a seat on a chair against the wall, leaving Eric Warburg with the impression that he wasn't permitted to sit at the table. Warburg now asked Heiss the same question about his Deutsche Bank shares that he had posed to his cousin over their meal, only to receive a more detailed analysis. "Heiss," Eric Warburg wrote, "was as well-informed in financial matters as our whole investment department."[15]

Heiss seems to have had a mind for chemistry as well. After the Second World War, he regularly exchanged letters with Dean Burk, the head of cell chemistry at the National Cancer Institute, who became Warburg's greatest champion in the 1950s. Despite possessing only a high school education, Heiss was able to discuss Warburg's research in considerable detail, a skill he likely needed as the institute's administrator. Krebs wrote that Heiss "did much in calming and appeasing Warburg's emotions when they were raised by controversy and resentment," and that may be so. But Heiss's letters give the impression of a true believer, a Sancho Panza always willing to affirm Warburg in his quixotic battles against the scientific establishment.[16]

Though Warburg's friends and relatives came to accept Heiss as a fixture in his life, often ending their letters with "regards to Heiss," in the early years they expressed their distaste. In a diary entry from 1933, Warburg's sister Lotte noted that many were saying that Warburg was "a homo and in the hands of Jacob." "Sometimes I almost believe it," Lotte continued, "because his nature has changed too much. He is spiteful against all, vicious and absolutely unreliable."[17]

Elisabeth Warburg struggled with her son's lifestyle up until her death in 1935. In a letter to Lotte in 1931, she expressed her concern that Warburg wasn't living according to "regular family relations" and would never pass on his genius to any offspring. On one occasion, after Elisabeth pressured him to find a wife, Otto said that he did not want to end up with a wealthy woman in a marriage motivated by reason rather than love. Elisabeth, sensing an opening, suggested that a "marriage based on reason" could, in fact, be an excellent idea. "Think about all that you can do with a girl who is that rich," she wrote. "And you can let go of all the rest—the lectures and exams, and you won't need to slave away like Papa did."

Warburg's sarcastic response might have been accompanied by an eye roll: "From then on it's just horseback riding and getting dolled up."[18]

THE NEW STABILITY in Warburg's personal life came against the backdrop of a nation in upheaval. After Germany failed to make reparations payments dictated by the terms of Versailles, French and Belgian forces crossed into the Ruhr Valley, the industrial region critical to the German economy. The local Germans went on strike with the support of their government, which now had to cover their wages the only way it could: by printing more and more money. The ensuing inflation left German money worthless. By November 1923, an American dollar was worth 4 billion marks. Some Germans took to hauling wheelbarrows of cash around to buy food. Others joined violent militias. Jacques Loeb, watching Germany crumble from afar, was horrified by "the blind passions of the reactionary forces." "I feel strongly," Loeb wrote to Warburg, "that our civilization is seriously threatened."[19]

The problems, Loeb grasped, were about far more than money. The reoccupation of the Ruhr Valley had reopened the psychic wounds from World War I that had only just begun to heal. More and more Germans would now treat those wounds with a salve of denial and scapegoating. Germany hadn't lost the war to other countries, according to this narrative. That was impossible. The German military was too powerful, the German people too racially superior. Germany could only have lost due to sabotage from within. The Germans, as the popular saying went, had been "stabbed in the back" by Jewish Bolshevik traitors.

If preoccupied by his work, Warburg was not unaware of what was happening outside his lab. In 1923, he encouraged Otto Meyerhof to accept an offer in America, telling him that he would be unlikely to find a decent position in Germany given the "prevailing anti-Semitism."

Warburg came to this conclusion even though Meyerhof had won the Nobel Prize only the year before. In a letter to Loeb later that year, Warburg wrote that Germany was still waiting for "a turning point" and that he wasn't sure it would arrive in time.[20]

Though Warburg had almost no contact with his famous cousins in banking at the time, he would have felt the weight of his last name. The Protocols of the Elders of Zion, the notorious Russian hoax that detailed a Jewish plan for global domination and influenced countless Nazis, including Hitler, first appeared in German translation in 1920. Among the capitalist "criminals" said to be responsible for the disastrous state of the nation was Otto Warburg's cousin, Max.

The new German government had urged Max, a famous banker, to take part in the reparations negotiations at Versailles, hoping the presence of a Warburg might inspire better terms from the Americans—Max's brothers, Paul and Felix, were both influential American bankers. Max resisted at first, aware that it could be disastrous for a Jew to be seen as responsible for the inevitably harsh terms of the treaty. Though Max eventually relented, he would soon regret it: "Whatever one chose, one chose hell," he later said of the negotiations. "It was simply a question, which hell promised to be of shorter duration."[21]

On November 16, 1923, Emil Warburg captured the mood in a letter to Einstein, who was abroad at the time. "You know how much it must mean to me that you stay with us, as you are the only colleague with whom I have a closer personal relationship and productive scientific exchanges," Emil wrote. Nevertheless, he felt it would be "too irresponsible" to encourage Einstein to return to Berlin, given the "increasingly dire" situation.

In the same letter, Emil told Einstein that he had recently played music with his daughter Käthe, who sang a Schubert song with the line "the world becomes more beautiful day by day." Added Emil Warburg, "That's over now."[22]

THE YEAR 1923 is significant in German history for another reason. It is the year that Otto Warburg published his first paper on the peculiar manner in which cancer cells eat. In 1908, at the Naples Zoological Station, Warburg had found that a sea urchin egg will consume six to seven times more oxygen after fertilization. The discovery made perfect sense to Warburg. A fertilized egg needs to grow, he reasoned, and so needs more oxygen to react with nutrients and generate energy.

Warburg anticipated that cancer cells would also take up more oxygen to fuel their growth, but there was no way to be certain of this with the tools available to him in 1908. At that time, Warburg had measured the breathing of sea urchin eggs only indirectly, through a chemical analysis of the oxygen in the water surrounding the eggs. He needed a more direct and reliable approach to advance his science, and he knew where to look. Researchers had begun to measure human breathing rates by strapping frightening-looking masks onto people's faces. The principle was simple. Every breath we take involves an exchange of gases: oxygen flows into the system, carbon dioxide flows out. The mask was connected by tubes to a rubber bag, and the rate of breathing could be measured by the time it took for carbon dioxide to fill the bag. It was an excellent device for measuring the metabolism of an entire body, but it left Warburg with a problem: How could he strap a mask onto the face of a microscopic cell?

Warburg might have given up and moved on to a more feasible experiment. But then he would not have been Otto Warburg. The same self-regard that made him so insufferable also made him a remarkable scientist. When Warburg encountered an obstacle in the lab, the question wasn't whether he would overcome it, but how.

A turning point came in March 1912, when, on a visit to the lab of an English colleague, Warburg spotted what he later described as a "beautiful blood gas apparatus." The device was nothing more than a U-shaped tube, known as a manometer. The tube was partially filled with fluid and connected to a small glass vessel containing blood cells.

When the oxygen bound to the blood cells was released, the pressure inside the tube would change and cause the fluid to rise on one side of the "U" and fall on the other. Warburg might have been the only person alive who could have looked at the device and seen beauty.[23]

It wasn't the first manometer Warburg had observed. Physicists had long used manometers to measure changing gas pressures for the study of inorganic processes. Emil Warburg had himself devised a special manometer for his study of ozone gas. And Otto had already realized that a manometer could be used to measure the breathing of cells. The beauty of the manometer he spotted in England was in the small technical modifications that made it more suitable for the study of life.

Warburg immediately began to use this new manometer in his research on respiration and photosynthesis, along the way introducing a series of his own technical innovations until he had arrived at a device that would become known around the world as the "Warburg manometer" or "Warburg apparatus." For the rest of his life, the manometer was Warburg's signature tool in the laboratory.

Even after perfecting the manometer, Warburg had another major hurdle to overcome before he could accurately measure the breathing of cancer cells. The manometer allowed him to put a mask over small slices of a tumor, but cutting these slices inevitably damaged the outer layers of cells. If the slice was too thin, a significant percentage of the cells in the sample might not be able to breath properly. If the slice was too thick, meanwhile, the oxygen would take too long to reach the inner layers of cells and this, too, could lead to errors. A tenth of a millimeter off in either direction and Warburg would not be able to trust his results.

To solve the problem, Warburg devised a complicated mathematical formula that could determine the precise width required for each specific tumor sample. Warburg's manometer and "tissue slice technique," as it was known, became critical tools for an entire generation of researchers. In the following years, these innovations made possible countless fundamental discoveries in biochemistry. But first they forever changed our understanding of cancer.

If growing cancer cells were like sea urchin embryos, they could be expected to take up oxygen at an unusually high rate. But when Warburg put the cancer tissue he had sliced from the tumor of a rat into a glass vessel and attached it to his manometer, the drama was not in what happened but in what failed to happen. The fluid in the tube did not rise or fall any faster than it would have with noncancerous cells. Despite their rapid growth, the cancer cells were not taking up more oxygen. Instead, they were gorging themselves on glucose and fermenting it.

Warburg's research had arrived at a peculiar place. Fermentation was the biochemical process behind beer and bread. It was not supposed to have anything to do with cancer. The development of agriculture, and thus modern civilization itself, can be traced, in part, to a passion for fermenting grains into alcohol. The ancient Sumerians credited fermentation to Ninkasi, goddess of beer, whose "lofty shovel" was said to sprout barley. Jews eat matzah on Passover as a reminder that the Israelites hurrying out of Egypt did not have time to let their bread rise—which happens because some types of fermentation give off carbon dioxide.[24]

In the nineteenth century, fermentation was among the most important and fiercely debated topics in science. By then it was clear that the process produces alcohol and carbon dioxide in some cases and lactic acid (as in the making of cheese, yogurt, or sauerkraut) in others. But little was known beyond that. Many of the leading chemists of the era were convinced that fermentation was merely a by-product of oxygen reacting with simple sugars, much as rust appears when oxygen reacts with iron. Others insisted that fermentation was the work of living organisms. According to this view, grape juice left to sit in a barrel turned alcoholic not because it was decaying but because tiny, invisible animals were feeding on the juice and leaving behind alcohol as a waste product.

Louis Pasteur proved that the second group was correct, but he did not stop at crediting microbes for fermentation. In 1860, he made the observation that would forever shape Warburg's thinking on cancer. Hunched over his microscope, gazing at a drop of fluid, Pasteur could see a tiny universe: rodlike specks moving about in a state of silent frenzy. Pasteur had seen similar scenes of microscopic life many times before, and yet, on that particular day, the bacteria weren't doing what they were supposed to do. Instead of gravitating toward the edge of the slide, where they could find more of the oxygen that every living organism needed to live and breathe, they were moving away from the oxygen, to the slide's center.

For Pasteur, seeing bacteria turn away from oxygen was akin to watching a group of people in the ocean choosing to drown themselves rather than swim to the shore. But the most shocking part wasn't that the bacteria were forgoing oxygen; it was that doing so did not seem to bother them in the least. Pasteur, having already proved that fermentation is the work of living organisms, had now found something even more surprising: fermenting organisms could use oxygen, but they didn't appear to need it. Overturning the most basic assumptions of biologists of the era, Pasteur declared that fermentation was "life without air."[25]

Determined to understand why an organism would ever choose fermentation over respiration, Pasteur carried out additional experiments. When he sent a stream of oxygen through the microbe-filled fluid, he saw that the production of alcohol or carbon dioxide would suddenly cease, indicating that the fermentation had stopped. Pasteur concluded that when an organism could use oxygen, it would do so. Fermentation, according to Pasteur's framework, was a backup process that activated and sustained the life of microbes when oxygen wasn't readily available. This fundamental relationship between respiration and fermentation is now known as the "Pasteur effect," and it was Otto Warburg who named it.

By the time Warburg began to study cancer, Otto Meyerhof had arrived at another breakthrough while studying muscle tissue. It wasn't only microbes that turned to fermentation. Animal cells could ferment glucose into lactic acid just like the invisible organisms that feed on the sugars in milk. It was a remarkable and, at the time, unsettling demonstration of the unity of nature. Humans and single-celled organisms weren't nearly as different as many wanted to believe.

For Warburg, there was an even more significant lesson in Meyerhof's discovery. Meyerhof had found that our cells turn to fermentation for the same reason that, according to Pasteur, microbes turn to fermentation. In both cases, fermentation was understood as an act of desperation, a means of generating energy if there is not enough oxygen available. This is what happens during vigorous exercise, when our heaving lungs can't deliver oxygen to our muscles quickly enough; the soreness we feel after intense exercise is a by-product of the lactic acid our fermenting cells produce.

Warburg's explanation for why cancer cells ferment even when oxygen is plentiful—the process that would later become known as "the Warburg effect"—was almost tautological: if a cell had access to oxygen and nevertheless chose fermentation, it could only be because something fundamental was broken and preventing the cell from using the oxygen properly. To test this theory, Warburg poisoned growing cells with chemicals that disrupted their use of oxygen. The result was exactly what Pasteur would have predicted: as respiration decreased, fermentation increased.

The more these cells fermented, Warburg reasoned, the more likely cancer became, because the energy of fermentation was "inferior," a cheap, inefficient substitute for the energy provided by respiration. Fermentation couldn't support the elaborate microscopic machinery of healthy tissues any more successfully than a small backup generator might keep the lights in a large building on for an extended period. Fermenting cells "have lost all their body functions and retain only the now useless property of growth," Warburg wrote. "Thus, when respiration disappears, life does not disappear, but the meaning of life disappears."[26]

Over the course of the 1920s, Warburg and other researchers would repeat his rat tumor experiments on other cancer cells again and again. They tested human cancers from the skin, throat, intestines, and penis and saw cancer cells overeating and fermenting glucose every time. Evidence of the fermentation was found not only in cancer cells in petri dishes but in the unusual levels of lactic acid in the blood of living animals with cancer. Everywhere Warburg looked, the result was the same: cancer cells were fermenting, sucking up glucose, and spitting out lactic acid—as much as 100 times the lactic acid normal tissue would produce, in some cases.

Fermentation was not merely less efficient than respiration. To the aristocratic Warburg, the process was also less noble, worthy only of "the lowest living forms." In cancer, Warburg once wrote, oxygen is "dethroned."[27]

WARBURG ONCE WARNED another scientist about the hazard of forming theories based on experimental findings. Such interpretations "are always wrong," Warburg said, insofar as they will inevitably change as new discoveries arise. As Warburg put it, "Good experiments are right forever." But for all the times he criticized others for arriving at conclusions that went far beyond their experimental findings, he could rarely resist the temptation himself.[28]

Warburg's initial studies had not shown that fermenting cancer cells were damaged and using significantly less oxygen than healthy cells. The studies showed only that, in addition to burning glucose with oxygen, the cancer cells were also taking up lots of extra glucose and fermenting it. There was no reason to assume that fermentation arose only in response to a cell's inability to power itself with oxygen, but Warburg couldn't set aside his reverence for Pasteur, who died believing that yeast ferment only when normal breathing with oxygen is insufficient. If the backup generator is on, Warburg concluded, then the power station must be damaged.

Had Warburg thought more broadly about respiration and fermentation, he might have wondered whether there was another explanation for his findings. In particular, he might have considered whether the type or quantity of the food available to the cells might be influencing the way they processed energy. The evidence for this possibility was there from the start. In his very first experiments on cancer cells in the 1920s, Warburg had detected an increase in fermentation only after adding extra glucose to the cell culture. And Warburg was likely aware of the latest cancer feeding studies, given that he knew Peyton Rous, a scientist who had picked up that line of investigation where Ehrlich's institute had left off.

Rous worked at the Rockefeller Institute, placing him within the same international circle of scientists as Loeb and Warburg. He is famous for his 1911 discovery that a particular form of cancer in chickens could be spread by a virus. Not long after wondering about the role of viruses in cancer, Rous moved on to another question: could giving an animal very little food slow a cancer's growth or prevent cancer altogether? Ehrlich and Moreschi had transplanted tumors into mice and then fed them different diets. Rous took their research one step further. It wasn't only transplanted tumors that were influenced by an animal's diet. Tumors that would typically arise spontaneously, Rous found, could be prevented or impaired by limiting the amount of food an animal ate.

Rous was underfeeding animals a full decade before Warburg discovered that cancer cells overeat and ferment glucose. In a letter to Warburg sent in December 1924, Rous practically begged him to investigate the metabolism of the chicken cancers, even offering to cover some of the expenses. Rous did not even give Warburg a chance to say no. "I am sending some tubes of the dried and powdered chicken sarcoma," he wrote. "If you do not care to have the material, someone else may desire it."[29]

Warburg accepted the powdered chicken cancer, but his interest was in how the cancer cells used oxygen, not in how much or how little food the birds might be eating. To most cancer scientists in the 1920s, the feeding studies were already old news. Though still convinced that cancer was a problem of metabolism, Ernst Freund, the Austrian scientist who detected glucose in the blood of cancer patients, had himself moved on to other interests.

WARBURG'S FIRST cancer discoveries occurred as political fires continued to sweep across Germany. These fires burned especially hot in Munich, where Hitler had returned at the end of the war. The information office of the German military, wary of uprisings, hired Hitler to join a group of instructors tasked with instilling a hatred of bolshevism in the troops. It would prove the perfect job for a man who knew nothing but resentment.

The historian Karl Alexander von Müller was in Munich at the time and was among the first to notice Hitler's talent for public speaking. Years later, he could still remember Hitler's "light blue, fanatically cold, gleaming eyes" and how his listeners had watched him, transfixed. "I had the strange feeling that he had got them excited" and that their interest had simultaneously "given him his voice." Hitler's power over his listeners, Müller wrote, was almost "like a magic trick."[30]

Hitler was aware of his newfound power. He once referred to himself as "the greatest actor in Europe." As a New York Times reporter observed in 1930, Hitler's followers seemed less interested in the specifics of his speeches than in his delivery. But Hitler did cling to a vaguely coherent set of beliefs until the end of his life. He began with a basic premise: Germany was sick. The signs of this illness, in Hitler's deranged view, could be found in every aspect of the nation's social and political life, from the weakness of its military, to its failed educational system, to the poverty of its cultural creations. Germany had become "a slowly rotting world" in which things "seemed to have passed the high point and to be hastening toward the abyss."[31]

The cure was to return to forceful and decisive action without fear of consequence. Nature, even its most brutal aspects—especially its most brutal aspects—was the highest ideal. Any departure from the ruthlessness of nature was, in turn, a form of decay. Brutality wasn't an unfortunate consequence of Hitler's other beliefs; it was the core idea. Only weak and corrupt Jews quibbled over ethical considerations, as he saw it. "Eternal hunger" and the drive to reproduce, Hitler wrote, "are the rulers of life."[32]

Hitler was never an original thinker. Philosophers from Hegel to Nietzsche had called for a world in which decisiveness and bold action were to be freed from moral concerns. And in the 1920s, Darwinian thinking was widespread even among those who did not use it to justify racism. Warburg, in fact, would use the same Darwinian language as Hitler on one occasion. Only, as Warburg expressed in a letter to Loeb in 1922, the villains subverting the natural order were not Jews but his fellow scientists at the Kaiser Wilhelm Society. The "well-fed schoolmarms," Warburg wrote, left the best, independent-thinking scientists to "perish" while they stuffed the last crumbs from their tables "into their own pockets."[33]

After the starvation of the war years and the success of the English naval blockade, Hitler was particularly fixated on Germany's longstanding quest to produce enough food to feed its population. The problem, Hitler concluded, was that Germany would never have enough fertile land to grow food for its large population. He recognized that it was possible to increase crop yields through scientific innovation—Warburg's plan to make photosynthesis more efficient was one possible approach—but by the 1920s, Hitler had already rejected all such ideas. "[F]or a certain time," the need for more food can be counterbalanced "by greater industriousness, more ingenious production methods or special thriftiness," he wrote, "but one day all of these means will prove inadequate."[34]

For Hitler, the deeper problem with growing more food through innovation was that it did not involve dominating and killing off other races. According to the laws of nature, the strong were supposed to eat and thrive, but the weak also had to starve and die. What Germany needed was not merely better agricultural methods, but the will to fight and conquer more *Lebensraum*, or "living space." "[F]rom the distress of war grows the bread of freedom," Hitler wrote. "The sword breaks the path for the plow."[35]

Hitler had a specific "living space" in mind. He dreamed of a new "German East" that would be settled by German soldiers. After doing away with the locals, they would plow the fertile Ukrainian grain fields and end German food insecurity forever. "I need the Ukraine," Hitler once said, "so that they cannot starve us out, as during the last war."[36]

The war Hitler longed for was coming. It would be, in part, an attempt to swallow up as much glucose as possible.

CHAPTER SEVEN

The Emperor of Dahlem

THOSE WHO ANTICIPATED Germany's collapse in 1923 were wrong. All hope was not lost—not yet. In the second half of the 1920s, the introduction of a new currency curbed the country's runaway inflation. In addition to an extraordinary blossoming of cultural and intellectual life in Berlin, the economic stability brought about a new political stability. The Nazis' fortunes at the polls began to suffer. Even sales of *Mein Kampf* would fall dramatically. "One scarcely heard of Hitler or the Nazis except as butts of jokes," an American journalist in Germany recalled.[1]

The Berlin of street battles and food riots turned into the Berlin of avant-garde art and bourgeois dinner parties. At a dinner party at Einstein's home in 1925, Warburg met a scientist who later recommended that he hire Hans Krebs. When the 25-year-old Krebs arrived at Warburg's lab for his interview that December, Warburg was characteristically blunt. Though he had been invited to dine with Einstein, Warburg described himself to Krebs, accurately, as an outsider who was disliked by scientists in academia. If Krebs hoped to have a career in academic medicine, Warburg told him, he would be better off attaching himself "to some old ass of a professor." Warburg was particularly outraged, even prepared to leave the Kaiser Wilhelm Society, when a man he disparagingly referred to as a "bee researcher" received a lifetime appointment to the Institute for Biology before the same offer had been made to him.[2]

Though the specifics of what transpired are not entirely clear, at one point in the 1920s, Warburg accepted an offer from an institute in Heidelberg. Warburg changed his mind, but he may have done so under pressure. Faculty members in Heidelberg—presumably ones who remembered Warburg from his medical school days—were rumored to be planning a boycott if Warburg took the position.

Warburg might have had legitimate grounds for his complaints about other scientists. In a letter to Jacques Loeb from the same period, Einstein noted, as an aside, that Warburg had "been quite nastily oppressed" by his colleagues. Whatever caused the friction, Warburg had few allies left in the Kaiser Wilhelm Society. Boveri, the man who had chosen him for his position, had been dead for years. Warburg's greatest supporter at the society, Emil Fischer, had committed suicide in 1919 after being diagnosed with terminal cancer.[3]

Warburg certainly had reason to be outraged in 1927. That year, he was one of two finalists for the 1926 Nobel Prize for Physiology or Medicine. (The 1926 prize was awarded retroactively in 1927 because no one had been deemed worthy of the award in 1926.) The other finalist, Danish researcher Johannes Fibiger, had discovered that he could induce cancer in rats by feeding them worm-infected cockroaches. Whereas Warburg saw cancer cells eating and multiplying like microbes when oxygen was scarce, Fibiger saw the microbes themselves as the cause of cancer.

Fibiger was far from alone in this respect. An entire generation of cancer researchers took up the hunt for the "cancer microbe," or "cancer bacillus," as it was commonly referred to. Warburg, to his credit, never believed that a cancer microbe would be found. As he put it in a talk before the German Chemical Society in the summer of 1926, there is "no cancer bacillus" just as there is "no diabetes or arteriosclerosis bacillus." Warburg's reasoning was difficult to dispute. It was already recognized that exposure to tar and radiation and various chemicals could cause cancer without any assistance from tiny organisms.[4]

That both finalists for the Nobel Prize in Physiology or Medicine were cancer researchers was evidence of the expanding place of cancer in science and in public life in the 1920s. But that two scientists with very different explanations for how the disease arises could both be nominated for the Nobel Prize was also evidence of the state of confusion within the field.

The deadlocked Nobel Committee eventually concluded that Warburg and Fibiger should share the honor. And yet when the announcement was made, it did not include Warburg's name. The Nobel Committee's last-minute decision to give the award to Fibiger alone remains unexplained. It would prove to be an embarrassing mistake. Fibiger had failed to run controlled experiments—an ironic turn of events, given that he was among the first researchers to run a randomized controlled trial. In addition to the worm-infected cockroaches, Fibiger's rats had eaten only bread and water. When researchers later repeated the experiments with rats eating well-rounded diets, they found no tumors. Still further research revealed that the tumors Fibiger had detected weren't even malignant.

If Warburg had good cause for complaint about the 1926 Nobel Prize, he also had reasons to feel grateful. In a time of relative political calm, Germany was quickly reestablishing itself as the world's leading scientific nation, and Warburg, hidden away in the top floor of the Kaiser Wilhelm Institute for Biology, was becoming something of a scientific legend. "Warburg already had a mystery about him," wrote the Nobel laureate Fritz Lipmann, who was then a young researcher working in Otto Meyerhof's lab in the same building. "We admired him boundlessly but saw little of him." If they were lucky, Lipmann recalled, young researchers might catch a glimpse of Warburg "when he descended into the lower regions."[5]

In addition to building on his cancer findings, Warburg now returned to his prewar investigations of cellular breathing. Warburg's aim remained the same: he wanted to understand how our cells make the nutrients we eat react with oxygen. In 1913, he had discovered that the enzyme required for cellular breathing—Warburg still called it "the respiratory ferment"—appeared to include iron. But Warburg had nothing close to proof. The next step was to crush yeast cells so that he could isolate the enzyme within the chemical soup that emerged. But given the limitations of his laboratory tools in the 1920s, Warburg had little hope of singling out a molecule that existed in only minuscule quantities inside the cell.

Warburg needed another approach, and he needed it quickly. By 1925, the Cambridge researcher David Keilin was close to solving the riddle of respiration. Keilin had identified a series of metal-infused proteins that were clearly involved in the breathing process, despite Warburg having dismissed them as "degenerate ferments." But none of the molecules Keilin had found would react directly with oxygen. The molecule Warburg sought was still hidden inside the cell, waiting to be discovered.

Warburg's breakthrough arrived during a serendipitous supper. At a Berlin dinner party in the winter of 1927, he found himself chatting pleasantly about carbon monoxide poisoning with the English physiologist A. V. Hill. Warburg knew that the gas is poisonous for the same reason that cyanide is poisonous: it binds to iron and prevents the meeting of iron and oxygen in our cells. It amounts to a chemical choking.

But that night, Hill mentioned something Warburg did not know. In 1896, a pair of Scottish researchers had made a peculiar observation: in the dark, carbon monoxide would react with the iron in hemoglobin, the molecule that carries oxygen in blood, but shining a simple beam of light on the hemoglobin could cause the gas to lose its deadly power. It was as if the light, like a security guard, were stepping in carbon monoxide's way before it could attach to the iron. As the story goes, upon hearing of this surprising finding, Warburg excused himself from the party and went directly to his lab.[6]

Warburg wasn't interested in preventing carbon monoxide poisoning. The moment Hill mentioned the strange discovery, Warburg understood that if light was preventing the carbon monoxide from binding to the iron in blood cells, it might have the same effect in every cell. And if the phenomenon did hold true for every cell, then light could provide Warburg with an entirely new way of identifying his invisible respiratory ferment.

Warburg's insight was based on a principle that had been appreciated for decades. Every molecule in a cell absorbs light differently according to its chemical properties. If Warburg could figure out exactly how much light, at each wavelength, would block the carbon monoxide from attacking iron, it would reveal his respiratory ferment's unique "absorption spectrum." If the light functioned as a security guard, the "absorption spectrum" was like a fingerprint the guard collected at the scene of the crime.

As in a criminal investigation, the value of a fingerprint for Warburg was that it could be compared with another fingerprint. Warburg's plan was to find the fingerprint of the molecule in yeast and then to repeat the same trick in his model system, in which charcoals containing iron functioned much like the respiratory ferment. If the two fingerprints turned out to be identical, it would be as close to proof as possible that Warburg had identified the most fundamental molecule of life, the chemical match in our cells that made it possible to burn food without fire.

The first task for Warburg was to confirm that what Hill had told him was true for blood cells was true for other cells, too. Warburg took a small glass vessel containing a solution with living yeast and oxygen and filled it with carbon monoxide. As expected, the yeast were soon suffocating. Warburg then inserted a lamp covered in a watertight jacket into the vessel. Though he would never have admitted to anything resembling giddiness, Warburg must have been filled with anticipation in that moment. The fate of the yeast in his vessel held enormous implications for his scientific future.

The young Hans Krebs stood by, sensing he was witnessing an important moment in the history of biochemistry. If the breathing rate of the yeast increased, it would be reflected by the rising fluid in Warburg's manometer. Gazing at the U-shaped tube, Warburg, like a doctor looking hopefully at a flat EKG, desperately needed a sign of life.

And then it happened: the fluid crept up the right arm of the tube. The yeast gradually began to breathe. Warburg immediately went to work shining lights of different wavelengths on the cells until he had his ferment's fingerprint. Next, he repeated the experiment in his model system with charcoals. The moment he saw that the two fingerprints were the same, he would have understood that his own life would no longer be the same.

On February 22, 1928, Warburg presented the new evidence for his respiratory ferment at a lecture held at the Berlin Palace, a sprawling fifteenth-century structure that had served as the residence of various Prussian kings over the centuries. David Nachmansohn attended the lecture. Decades later, he could still remember the electricity in the room. Those in the audience, he recalled, understood that Warburg hadn't merely demonstrated an important discovery. By linking cellular breathing to a specific part of a specific molecule, Warburg had launched a new era in science.[7]

Warburg once told Krebs that it was critical for a scientist to investigate a narrowly defined problem for many years and that such narrow pursuits would, paradoxically, provide answers to the most expansive questions in biology. Now, some 15 years after he had first sprinkled iron salts on crushed sea urchin eggs, Warburg had identified the mechanism of life itself.

As word of Warburg's discoveries spread, he was increasingly regarded as the leading figure in his field. In 1929, he traveled to the United States to present his latest work. After a lecture at Johns Hopkins University, Warburg met with officials from the Rockefeller Foundation, who were so impressed that they offered to support him with $20,000 per year for his research. Warburg returned to Dahlem in an unusually good mood, regaling Lotte with stories of the women who had pursued him during his journey home on a luxury ocean liner.

The Rockefeller money was an extraordinary gesture, given how recently America and Germany had been at war, but Warburg had something more in mind. In an exchange of letters with a foundation official, Warburg made the case for an institute of his own. His lab at the Kaiser Wilhelm Institute for Biology, he wrote, wasn't suitable for the "delicate physical measurements" he hoped to carry out. That was an exaggeration, if not an outright lie, but it mattered little. The Rockefeller officials believed in Warburg's genius and his importance to cancer science, in particular. On April 16, 1930, the foundation's board of trustees met in New York and passed a resolution granting the Kaiser Wilhelm Society a lump sum of $655,000 for the building of two new institutes: one for Warburg and one for the study of physics.[8]

Warburg suggested to the Rockefeller Foundation that his new building could be considered a Rockefeller institute—with the Rockefeller Foundation taking over all of the funding. The director of the Kaiser Wilhelm Society, Friedrich Glum, had not approved the idea and was dismayed when he learned of it. Glum wrote to the Rockefeller Foundation explaining that Warburg's suggestion that his own institute be given a special status within the Kaiser Wilhelm Society was impossible. Although Glum was, ostensibly, Warburg's boss and had every right to dismiss the proposal, he asked the Rockefeller Foundation officials not to reveal his intervention to Warburg.[9]

Additional money for Warburg's institute would come from the estate of a wealthy woman who was dying of cancer. The woman wanted the building designated a "cancer research institute," but Warburg refused, arguing that his research couldn't be reduced to a single medical application. It would be as "ridiculous and unscientific as if a physicist wanted to call himself a radio waves physicist," Warburg said.[10]

Work on the new institute in Dahlem began right away. The architect hired for the job drew up plans for a modern Bauhaus-style institute. Warburg glanced at the blueprint and was repulsed. It looked like a "factory," as he put it. Always fond of rococo, Warburg wanted the architect to model the institute on an eighteenth-century country manor he had seen while traveling with Heiss outside of Potsdam. Though there was no evidence it was true, Warburg liked to say that the manor had been built by one of Frederick the Great's generals.

The architect was horrified by Warburg's vision for the building, which would not look anything like the other Kaiser Wilhelm institutes, or any other modern scientific building, for that matter. But Warburg, now at the height of his influence, was not about to let someone else design his dream institute. He had secured the Rockefeller money, and he was going to have an eighteenth-century rococo manor. The windows were to be the same size and to have the identical shutters as the home he had seen near Potsdam; the roof was to be adorned with the identical slate shingles.

The new institute was built exactly to Warburg's specifications, which called for an imposing oakwood door and a small walkway flanked by rows of lime trees. The contrast between the baroque exterior and the sophisticated scientific equipment inside was a perfect mirror of Warburg himself. To ward off all distractions from his research, Warburg assigned each of his technicians a different managerial responsibility, from paying bills to ordering new glassware.

By the time the institute opened in 1931, Warburg and Heiss had already moved into a new home down the street. The house, too, was built according to Warburg's detailed instructions. He insisted on 13-foot ceilings, a stone-tiled hallway, and parquet floors. Behind the house, Warburg built a stable for his horses and created a large riding area. When Lotte visited in the spring of 1930, she found Warburg on the terrace drinking coffee with Heiss. A poodle had its head on Warburg's knee. Looking around at the mahogany furniture and the antique Persian rugs, Lotte was struck by her brother's talent for reconfiguring the world to his desires. Everything was unique, "as he would have it and no other." As Lotte said on another occasion, it was as though every last object was a "reflection of Otto's being."[11]

Neighbors who spotted Warburg atop his horse, galloping down the streets each morning in the boots and spurs he had worn with the uhlans, began calling him "the Emperor of Dahlem." Warburg was indeed shaping every last aspect of his life in a way that only an emperor could. His salary, 36,000 marks, was twice that of a typical science professor of his seniority. His technicians were ready to devote themselves to whatever experiments he might dream up, and perhaps the world's most generous philanthropic organization was ready to finance those experiments. Few, if any, figures in the history of science have had a better arrangement.[12]

For summer vacations, Warburg and Heiss would travel (sometimes by horse—a full day's ride) to Nonnevitz, a quaint village on the island of Rügen in the Baltic Sea, where they maintained a small farm of chickens, geese, ducks, and goats. Warburg referred to it as his own "Noah's Ark." He would spend these vacations writing up his research from the previous year. Warburg's papers, as one colleague put it, were "masterpieces of clarity."[13] The Nobel Prize winner Albert Szent-Györgyi once asked Warburg the secret to writing such well-crafted scientific papers. Warburg explained that he would first write a draft that included everything that came to his mind and then set it aside. He would then return a month later and write the paper again without looking at the original draft. If the second draft differed from the first, he would write a third draft, and so on, sometimes rewriting a paper as many as 16 times before arriving at a point where the text was no longer changing.

It wasn't all work at Nonnevitz. Warburg would step away from his writing long enough to take long walks by the sea with Heiss. Warburg was said to be a different person during these vacations, more friendly and relaxed. He got along with the locals better than he did with almost anyone in Berlin and would sometimes stop to chat with them about their crops and the conditions of the soil.[14]

The only thing still missing from Warburg's idyllic life was a Nobel Prize. It would arrive in 1931, in recognition of his respiration research, along with $31,100 in prize money. Emil Warburg had died only months before, but if Warburg was disappointed that his father didn't live to see him achieve the great honor, there is no record of it. When the newspaper reporters arrived at his door, Warburg begrudgingly agreed to a photograph with his dogs, and then had Heiss chase everyone away.

At the awards ceremony in Stockholm, the Swedish chemistry professor Einar Hammarsten introduced Warburg and declared his discovery of the respiratory ferment the culmination of a quest begun by Lavoisier in the eighteenth century. The Nobel was important validation for a man who needed it, and yet it would lose a bit of its luster after Warburg met the scientists who had chosen him for the award. As Hammarsten would later recall with amusement, Warburg could hardly believe the decision had been made by "such an obscure" group.[15]

Shortly after Warburg returned to Berlin, he crossed paths with Einstein at the home of a mutual acquaintance. Warburg told him that he planned to use his Nobel Prize money to buy a white horse with a black stripe that he'd long had his eyes on. Einstein responded that as far as he was concerned, personal possessions shouldn't matter to a scientist. Warburg was stung. Einstein remained one of his idols. Though it's not clear that Warburg said anything in response, he was still bothered by the remark days later. He told an acquaintance that Einstein's dismissal of money and possessions was "nonsense" and that he was happier now that he was more financially secure.

Kaiser Wilhelm Institute for Cell Physiology, around 1940.

Otto Warburg with Prince Wilhelm of Sweden, Nobel Prize reception dinner, 1931.

Arnold Berliner, the physicist who witnessed the moment between Einstein and Warburg, noted in a letter that he planned to tell Warburg that Goethe was on his side of the debate, as Goethe had written that even "if life loses all charm, possessions are still worth something." Berliner appears to have been unaware that the line comes from *Faust* and is spoken by Mephistopheles.[16]

As far as Lotte was concerned, Warburg seemed in the grip of darker forces. He no longer needed the people who "so often annoyed and insulted and lied to him, partly out of their hatred of the Jews, partly out of their envy of his scientific successes," Lotte wrote in her diary. For another person, independence might have been a positive development, but for Warburg it had a dire consequence: he was free to "indulge in his contempt." He became "more and more difficult," according to Lotte, even "vicious" sometimes. In February 1932, less than a year after his institute had been built with Rockefeller Foundation money, Warburg refused to have breakfast with the foundation officials who had come to Dahlem. He did not like late breakfasts, he explained.

Lotte did her best to rationalize Warburg's coldness. "Thinking with scholars is always at the expense of feeling," she wrote. "It is impossible to live an idea and a feeling at the same time. One must be sacrificed." But she couldn't help but see the perverse direction of Otto's life: "He achieves everything he wants and becomes harder and harder instead of happier."

Warburg had just won the Nobel Prize for his discovery of how our cells respire, but Lotte found herself thinking of respiration in a very different way. "It's hard for the little people to breathe the air of great spirits," she wrote. "And so it is often difficult to breathe Otto's air." Then she added, "It sometimes suffocates you."[17]

Warburg's financial backers, including the Rockefeller Foundation, were primarily interested in his cancer research, and with good reason. Cancer rates were continuing to rise across the West. By the

1920s, the growing number of cancer deaths in Germany had become a national fixation. Cancer, in the words of one prominent German doctor of the era, had become "the number one enemy of the state."

To confront this enemy, Dresden's Hygiene Museum created a traveling exhibit called "The Fight Against Cancer." More than half a million people visited it, so many that police would sometimes have to cordon off the huge crowds that would form outside its temporary exhibit halls. Among the highlights of the exhibit was a silent film seemingly designed to terrify all who saw it. At one point in it, a grim reaper appears over a map of Germany, then slowly expands until the sharp blade of his scythe has stretched across the screen.

The acclaimed German doctor Ferdinand Sauerbruch reported that due to the exhibit, his clinic was being overrun by panic-stricken patients showing up with nothing but warts. He called for the exhibit to be shut down. But it was a symptom of the cancer panic rather than the source. The Germans were terrified for a reason: their friends and relatives were dying of cancer in unprecedented numbers. One cancer film of the era was titled *One in Eight*, a reference to how many Germans could be expected to die of the disease.[18]

Cancer was now often referred to as a "disease of civilization," and "civilization" was typically understood in a positive context. Cancer, in this view, was an unfortunate side effect of modern life. As one British cancer expert of the period put it, cancer was "the price we pay for civilization." But to some, "civilization" had a very different connotation. The German Romantics had long preached that modern civilization was corrupt and that superior German traits were being diluted amid the departure from nature and the mixing of races in modern cities. That cancer, along with a number of other diseases, had arisen during the period of industrialization, when Germany left its "natural" state, fit perfectly with this racist fantasy. For the Romantics, cancer wasn't an unfortunate consequence of the modern world. It was more evidence that the modern world was itself a terrible mistake. And the more Germans who died of cancer, the more compelling this perverted logic would appear.[19]

Cancer epidemiology and racist ideology would come together in the person of Erwin Liek, a Danzig surgeon, known as "the father of Nazi medicine," who sought to restore health through a return to natural living and healing. Liek, who wrote two popular books on cancer, was convinced that malignant growths were a problem of "civilization" and "not found among savages." "[T]he simpler and more natural one's way of life," wrote Liek, "the rarer is cancer."[20]

Liek, a eugenicist, maintained that civilization undermined the health of the German race by keeping weak and defective people alive—and thus able to pass on their defects to the next generation. He visited the United States in 1912, working at the Mayo Clinic, among other hospitals, and was delighted to discover that the nation had no social insurance programs. (That the doughy Liek was not in robust health himself—he died in 1935 at age 56—does not appear to have ever given him pause.)

But the problem with modern civilization wasn't merely that it was too forgiving of the feeble. Liek, like Germans who were not Romantics—and many people today—was convinced that modern life had brought with it an almost endless list of carcinogens. His books on cancer discuss dozens of cancer-causing agents: air and water pollution, artificial fertilizers, tobacco, insufficient activity, stress, and the dangers of aluminum pans and utensils, among others. That cancer was more and more common was terrifying in its own right. That almost every aspect of modern life could be said to cause cancer was enough to make anyone panic.

The greatest environmental threat of all, Liek and other German cancer specialists were coming to believe, was the German diet. By the late 1920s, Germans were arguably more concerned with the connection between diet and cancer than any other nation in the world. Some Germans believed that the danger stemmed from what Germans were *not* eating. New evidence showed the links between nutritional deficits and public health scourges. Goiters were traced to a lack of iodine, rickets to insufficient vitamin D. Perhaps cancer, too, would eventually be linked to a specific nutrient that was lost when food was cooked or grains refined.

What was added to food was thought to be even more problematic. The same coal-derived chemicals that had given rise to colorful clothing and to Ehrlich's chemotherapies were making German foods brighter. Bread was being bleached with benzoyl peroxide. Still other chemicals were used to preserve foods that would have gone bad under natural conditions. "The foods produced by our earth are good," Liek wrote, "those denatured by humans are bad."[21]

Liek's books revealed the influence of Frederick Hoffman, the German American insurance analyst whose 1915 tome, *The Mortality from Cancer throughout the World*, concluded both that cancer was a "disease of civilization" and that it was becoming increasingly common in the modern world. Researching and writing the book had required a monumental effort, but Hoffman did not slow down after it was published. Once convinced that cancer was on the rise, he was determined to figure out why.

While he had previously relied on the work of others in his analysis, in 1924 Hoffman started conducting his own population studies. His primary effort was a study of cancer in San Francisco, which then had the highest cancer death rate in the country: 160 cancer deaths for every 100,000 citizens, nearly double the average American rate. The study included thousands of surveys on the lifestyles and eating habits of San Franciscans, both cancer patients and control groups, and a review of some 50,000 death certificates. During the same period Hoffman also conducted surveys from countries across the world, including Mexico, India, Malta, Sudan, Iceland, the Philippine Islands, and New Zealand.

Hoffman was an early adopter and proponent of air travel—he became known as "the flying actuary"—and often collected his data in person. But the nonstop travel and writing—he published over 1,200 papers during the course of his career—began to take a toll. In the spring of 1927, Hoffman collapsed in Newark, New Jersey, and was rushed to the St. Barnabas Hospital, where he was diagnosed with pneumonia. As the concerned doctors looked on, Hoffman fell into delirious fits, reciting a lecture on cancer he had been preparing for his upcoming address in Brussels.

Hoffman's doctors gave him only a 50 percent chance to make it. He left the hospital for Europe a week later. In addition to his lecture in Brussels, he stopped in Berlin to address the German Society for Cancer Research on the use of statistics in cancer investigations. A brief Rockefeller report on the event mentions that Otto Warburg, still in the early stages of unraveling how cancer cells eat, was among the German scientists in attendance.[22]

If Hoffman was gathering cancer statistics in the 1920s with a sense of urgency, it was, in part, because the predictions he made in his 1915 book about the coming cancer epidemic were looking accurate. Hoffman had been aghast in 1915 that approximately 80,000 Americans were dying of cancer each year. By the late 1920s, the number had grown to about 145,000, and Americans, like Germans, were growing increasingly worried. In 1928, West Virginia senator Matthew Neely stood in front of Congress waving a copy of Hoffman's book and declared cancer "more terrifying than any scourge that has ever threatened the existence of the human race."[23]

By the end of his San Francisco survey, Hoffman was convinced that he had found an explanation for the increase in cancer. The cancer problem, first and foremost, was a problem of food—a problem, Hoffman wrote, of the "essential errors" that "underlie the modern diet of civilised races." In 1937, Hoffman published *Cancer and Diet*, a 767-

page book presenting a mountain of new data as well as a review of every source he could find on the subject, dating back to antiquity. In the book, Hoffman noted that the rest of the world had fallen behind the Germans in their research into nutrition's role in cancer. He mentioned Warburg's "epoch making treatise" on the metabolism of cancer cells more than a dozen times.[24]

If Hoffman had moved beyond the long list of possible carcinogens he'd discussed in 1915 to a focus on nutrition, it clarified only so much. There was still a dizzying number of foods and preservatives and cooking methods to worry about, and every expert offered different advice. Germany's traveling cancer exhibition called for a diet high in fruit and fiber and warned Germans against too much fat, sugar, and protein. Liek recommended avoiding meat and sweets (with the exception of honey) and eating dark bread along with fruits and vegetables. Hoffman encouraged his readers to be aware of vitamin and mineral deficiencies and singled out tomatoes and salt as particularly harmful.

Hoffman had arrived at his conclusion that the Western diet was responsible for cancer largely based on the scarcity of cancer in populations that followed traditional diets. Yet when it came to identifying the specific culprits in the Western diet, the comparisons between Western and non-Western populations were less helpful. It might be true that neither the Inuit of the Artic nor the Gabonese of West Africa developed cancer when following their traditional diets, but their traditional diets were remarkably different from one another. The Inuit ate almost exclusively animal products, while the Hunza of the Himalayas ate a diet rich in plant foods.

That an anxiety about carcinogens in food had become entangled in a profoundly racist Romanticism didn't necessarily mean the anxiety was misplaced. It is possible to be simultaneously correct on the facts and wrong on their meaning. But Romantics like Erwin Liek faced an empirical problem as well. For all the concerns about carcinogens, the evidence linking any particular food or nutrient to cancer was not particularly strong. "[I]t certainly seems exceedingly rash to attempt to make us believe that there is any connection between cancer and diet," the Dutch cancer authority H. J. Deelman argued. As Deelman noted, for all the new emphasis on diet, the larger problem of distinguishing the influence of diet from other factors in the environment had hardly disappeared. The cancer-free populations didn't only follow different diets. They had, Deelman wrote, a "wholly different mode of living."[25]

Liek pushed back against his critics as best he could. Though he detested what he saw as the mechanistic science of researchers like Warburg, he didn't hesitate to bring up Warburg's findings as evidence that nutrition was linked to "metabolic disruptions in cancer tissue." And yet there was only so much Liek or Hoffman could say. Deelman's position was scientifically sound. He was not denying that cancer was becoming more common in industrialized countries, only that the diet hypothesis was far from proven.[26]

If Hoffman and the German cancer establishment remained confused about the connection between eating and cancer, they were circling around something fundamental. They had noticed that cancer rates rose in industrialized societies and that the cause appeared to be related to food. With the publication of *Cancer and Diet* in 1937, Hoffman had reason to believe that the rest of the world would follow Germany's lead in focusing on the relationship between diet and cancer.

Perhaps the medical establishment would have taken the cancer-diet connection more seriously had the thinking not fallen into the wrong hands and been forever tainted.

CANCER WASN'T Hitler's only health concern. August Kubizek, the friend from Hitler's early years—the two shared a room in Vienna in 1908—remembered that Hitler was "almost pathologically sensitive about anything concerning the body." And though Hitler's health had been generally good in the 1920s, he had an acute fear of death. "When I'm no longer here" had become Hitler's "mantra," Goebbels noted in his diary at the time.[27]

In 1924, Hitler cut back on meat and alcohol. He explained the decision with a vague reference to his health and weight, but he later credited his turn to vegetarianism to Richard Wagner, whom he idolized. In an 1881 essay, Wagner had written that vegetarianism was the original human diet and that eating animal flesh, along with the mixing of racial groups, had contaminated the human race. The essay inspired one of Wagner's zealous admirers to move to Paraguay, where he could establish a colony free of meat and Jews. Hitler, more ambitiously, wanted Germany to be his vegetarian, Jew-free utopia.[28]

But if Hitler's hypochondria would manifest itself in many different forms, it was cancer that terrified him most. He had long worried that his stomach cramps and severe bloating were evidence of the disease, and with each passing year, his anxiety would deepen. That his fears were shaped by his experience of watching his mother suffer from an experimental cancer treatment is obvious from *Mein Kampf*. In one passage Hitler argues that Germany's military should be no more hesitant to take bold risks than a dying cancer patient who embraces treatment even "if the operation promises only half a percent likelihood of cure." In a critique of the Austro-Hungarian Empire, Hitler wrote, metaphorically, that "its weakness in combating even malignant tumors was glaring."[29]

Cancer was an easy metaphor in the 1920s, as it is today. But the frequency with which Hitler relied on it in his writing and speeches is telling. In some instances, he likens Germany itself to a dying cancer patient. Hitler, in turn, presents himself as the surgeon, "lancing the cancerous ulcer" and "combating malignant tumors." In 1928, his cramps drove Hitler to see a doctor, who diagnosed him with chronic irritation of the stomach and recommended an even more restrictive diet. When the cramps persisted, so did Hitler's fixations on cancer and food.[30]

At some point—it's not clear when—Hitler became an admirer of Erwin Liek. Even before being named chancellor, Hitler would personally offer Liek the position of Reich Physicians' Führer. And in his uniquely confused way, Hitler, too, would also come to see cancer as a "disease of civilization." Perhaps directly influenced by Liek, he suspected that the act of cooking, by taking food out of its natural state, was the root of the problem. He repeatedly returned to this theme during his mealtime rants:

> *Nine-tenths of our diet are made up of foods deprived of their biological qualities. When I'm told that 50 per cent of dogs die of cancer, there must be an explanation for that. Nature has predisposed the dog to feed on raw meat, by tearing up other animals. Today the dog feeds almost exclusively on mixed bread and cooked meat. If I offer a child the choice between a pear and a piece of meat, he'll quickly choose the pear. That's his atavistic instinct speaking. Country folk spend fourteen hours a day in the fresh air. Yet by the age of forty-five they're old, and the mortality amongst them is enormous. That's the result of an error in their diet. They eat only cooked foods. It's a mistake to think that man should be guided by his greed. Nature spontaneously eliminates all that has no gift for life. Man, alone amongst the living creatures, tries to deny the laws of nature.[31]*

The most striking account of Hitler's thinking on diet and cancer in this period comes from Albert Krebs (no relation to Hans Krebs), a regional Nazi leader who was with Hitler in the Atlantik Hotel in Hamburg in February 1932. Hitler, who was campaigning for the German presidency, had delivered a speech the night before. Knowing that Hitler liked to review the published transcripts of his speeches, Krebs planned to show him the morning newspaper. But as Krebs made his way down to Hitler's suite, he heard "rhythmic cries" of "*M'soup!*" In the first room of the suite, Krebs found Hitler's entourage frantically scrambling to prepare another serving for their aggravated führer. It "was like a scene from a French comedy," Krebs wrote.[32]

Krebs continued on to the farthest room in the suite, where he found Hitler "hunched over a round table, looking melancholic and weary, slowly slurping his vegetable soup." When Krebs showed him the newspaper, Hitler shoved it aside. He had something else he wanted to discuss. He began to question Krebs "urgently and with obvious anxiety" about his views on vegetarian diets.

Krebs wasn't sure how to respond, but Hitler wasn't interested in his response. He "launched into a lengthy and detailed lecture" on health food. Krebs knew Hitler well enough not to be startled by his ranting, but that morning Hitler did, in fact, surprise Krebs. For the first time, Hitler took a conversation with Krebs in a personal direction, speaking openly of his fears of illness, even listing the symptoms— sweating, extreme nervousness, trembling, and more—that had led him to adopt his meat-free diet. "This revelation of weakness in an unguarded moment," Krebs wrote, "threatened to expose the hidden truth: that his 'strength' was only an overcompensation."

Hitler also told Krebs of the stomach cramps that he feared were symptoms of cancer. It's possible that the very name "Krebs" had unsettled Hitler. "Krebs" is the German word for cancer. Both "cancer" and "Krebs" are derived from *karkinos*, the ancient Greek for "crab." (Hippocrates is thought to have named the disease, though why cancers made him think of crabs is unclear.) In German, "Krebs" can also refer to actual crabs, as well as crayfish and lobsters, and Hitler was fascinated by these sea creatures. Some scholars have suggested that it was more than a coincidence that Hitler's mind was on cancer even as he gazed at shellfish.

As Albert Krebs sat and listened in amazement, Hitler said that it was the knowledge that he was dying that made his election so critical. If he didn't come to power soon, Hitler said, he wouldn't have enough time left to "finish the gigantic tasks" he had in mind.

With those words, Hitler suddenly regained control and turned his attention to the newspaper and his speech. "The depression was overcome," Krebs wrote. "Hitler the human being had changed back into the Führer."

Adolf Hitler, broken glass-plate negative, date unknown.

CHAPTER EIGHT

"The Eternal Jew"

ON THE MORNING OF May 16, 1933, Max Planck, the president of the Kaiser Wilhelm Society, met with Adolf Hitler in Berlin. It had been more than three months since German president Paul von Hindenburg had made Hitler chancellor, but Planck had been slow to absorb the meaning of Hitler's ascent.

Planck was not alone in this regard. The German recovery of the late 1920s was spectacular but short-lived. With the collapse of the American stock market in October 1929, bankers in the United States called in the loans that had been propping up German industry. The German economy nosedived again, and no one would benefit more from the return to instability than Adolf Hitler. In the 1930 parliamentary elections, the Nazis received 6.5 million votes, giving the party 107 of the 577 seats in the Reichstag. Two years later, some 6 million Germans were out of work, and the Nazis held 230 Reichstag seats.

Hindenburg had appointed Hitler chancellor on the assumption that he and other German conservatives could contain him. Those who had been paying attention knew better. The very sight of the words

"Hitler *Reichschancellor*" left the German journalist Sebastian Haffner in a state of shock. It "was so bizarre, so incredible, to read it now in black on white," he wrote. "[F]or a moment I physically sensed the man's odor of blood and filth, the nauseating approach of a man-eating animal—its foul, sharp claws in my face."[1]

By the end of March 1933, the Nazis' newly built concentration camps were already filling with political prisoners. On April 7, the Reichstag passed the Law for the Restoration of the Professional Civil Service, banning "non-Aryans" (defined as anyone with at least one Jewish grandparent) from government positions, including university posts. Though it would take another five years before virtually all Jews would be purged from public life, German science would never again be the same.

Planck had been on vacation in Sicily when the Civil Service Law went into effect. He chose not to return to Berlin to help manage the crisis even after the Nobel Prize–winning German physicist Max von Laue pleaded with him to do so. Though he had championed individual Jewish scientists in the past, including Emil Warburg and Einstein, Planck, an obedient Prussian, was the wrong man to stand up to a fascist dictator. He would only awaken to the severity of the situation upon learning that Fritz Haber, the director of the Kaiser Wilhelm Institute for Physical Chemistry, had announced he would resign before dismissing the many Jewish researchers in his lab. (Haber was born Jewish but had been baptized, seemingly to advance his academic career.) It was one thing for less prominent scientists to flee Germany; it was another for Haber to leave. Haber had been the architect of Germany's gas warfare program during World War I. He was a celebrated patriot and a pillar of German science, the institution Planck treasured above all else.

Belatedly, Planck decided to act, and he soon found himself sitting across from Adolf Hitler. According to Planck's account of the meeting—written with the help of his wife 14 years later and now looked upon with some suspicion by many historians—Hitler told Planck that he had "nothing against the Jews themselves" but that he had to wage war against them because they were all Communists and thus enemies of the Nazis and the German state. Planck suggested that there were "different sorts of Jews, some valuable for mankind and others worthless" and that "distinctions must be made."

"A Jew is a Jew," Hitler shot back. "All Jews stick together like leeches."

Planck claimed to have tried again, mentioning the importance of certain Jewish scientists to Germany. But Hitler had lost interest and turned the conversation to himself. "People say that I suffer from a weakness of nerves," Hitler said. "That is slander. I have nerves of steel." As Hitler continued in this vein, he began to speak faster and faster, violently slapping his own knee. "I could only remain silent and withdraw," wrote Planck. It was a conclusion, arrived at by millions of other Germans at approximately the same time, that would ultimately doom Europe's Jews.[2]

While Planck did not mention it, in the days after the meeting, word spread that Hitler had assured Planck that no further restrictions would be imposed on the Kaiser Wilhelm Society. Weeks after his meeting with Hitler, Planck appeared at the society's annual meeting and read aloud from a telegram the society had sent to Hitler. German science, Planck declared, was ready "to cooperate joyously in the reconstruction of the new national state."[3]

For Planck, as for many others, the joy would soon be gone. Three months after his speech, Lotte Warburg ran into an "unkempt" Planck "shuffling through the park" looking "miserable." When Lotte told him about the latest firings of Jews from German universities, Planck claimed ignorance but didn't attempt to mask his despair. German science, he said, had become "worthless."

Lotte Warburg would come to despise Planck for his failure to stand up to the Nazis. "Why does he stand by, without even bothering to offer an explanation, as Kaiser Wilhelm Society members are thrown out?" Lotte wrote in 1934. "Why does he go about stooped over, whining and complaining," rather than "damning them all?" She was particularly appalled that Planck had omitted Emil Warburg's name from a public speech about German physics, presumably because he feared even mentioning a Jew.[4]

With two Jewish grandparents, Otto Warburg was unequivocally "non-Aryan" by the terms of the Civil Service Law. But because the Civil Service Law didn't apply to privately funded institutes and because Hitler (under pressure from Hindenburg) had made an exception, initially, for veterans of the war, Warburg had less cause for panic than many of his colleagues, at least in the beginning. Hidden away in his institute in Dahlem, he probably didn't see the stone-faced SS men standing in front of Jewish-owned shops to steer customers away or the menacing mobs of Nazis shouting "Jews out!" in the streets. Warburg might not have heard the reports of brownshirts beating political opponents and sending them off to the newly erected camps. In early 1934, to the amazement of his Warburg cousins, he was not even aware that Germany had withdrawn from the League of Nations the previous October.[5]

But if Warburg remained ignorant in some respects, he did know that Jewish scientists were being chased out of their jobs. In March 1933, even before the institution of the Civil Service Law, the Prussian Academy of Science began proceedings to oust Einstein for the crime of disparaging the new Nazi government. (Einstein chose to resign before the academy had the chance to dismiss him, a move that further infuriated his Nazi antagonists.) Warburg was appalled by the treatment of Einstein. Every member of the academy should have resigned, Warburg told an official from the Rockefeller Foundation later that year. A strong statement projecting unity, Warburg argued, might have prevented all the actions taken against Jewish scientists that followed.

Though Warburg was not a member of the academy, he did resign from the German Physical Society in solidarity with Einstein. According to his own account, Warburg also stormed out of a meeting of the directors of the Kaiser Wilhelm Society institutes, disgusted by the Nazi sympathizers in the room who were prepared to give Hitler control of the society.[6]

Warburg's displays of indignation, of course, made little difference. In the first year after the Civil Service Law came into effect, an estimated 2,600 scientists and other scholars—almost all of them Jewish—fled Germany. Hans Krebs, who was Jewish, had just started in a new faculty position in Freiburg when the new regulations were announced. He continued to go to work for the next few days, unable to fathom that he would really be dismissed. Warburg wrote to Krebs, offering to let him return to his institute, while adding that a move to England did seem the safe choice, given that "we have no idea what is still to come" in Germany.[7]

Krebs chose England. Unable even to return to his lab to pack his belongings, he had them shipped. In the cargo were the Warburg manometers that he would soon use to decipher a series of reactions—now known as the Krebs cycle—that revolutionized our understanding of metabolism.

IF FAR FROM A public champion for the persecuted, Warburg may have been more reckless and confrontational than any other German scientist of the era. The most astonishing aspect of Warburg's story isn't that a gay man from a famous Jewish family survived in Germany; it's that he managed to do so even as he provoked the Nazis at every opportunity.

It wasn't that Warburg was oblivious to the danger he faced. By the end of 1933, he would have already sensed that no one with Jewish ancestors could feel entirely safe in Germany. Emboldened by the rise of the Nazis, anti-Semites from every sphere of German society were emerging from the shadows to flaunt their hatred. In June 1933, a Berlin engineer published a pamphlet calling the Kaiser Wilhelm Society a "breeding ground for Jewish exploiters, oppressors, and Marxists." German science had become "the servant and slave of these vampires," who profited from their ideas and shared their riches with Jewish conspirators. The pamphlet specifically attacked the Jewish leaders of the various institutes. "The Jew Professor Warburg," it stated, "enriched himself with the inventions of his associates while they were starving with their children."[8]

Warburg's first show of resistance was to ban his employees from making the Nazi salute or hanging the Nazi flag at his institute. Through 12 years of Nazi rule, he would defy Nazi orders again and again, just as he did in 1934, when he chased the customs official demanding a "declaration of Aryan descent" out of his institute. On one occasion, a young woman at another institute asked Warburg if she was still allowed to use animals in her experiments, citing reports she had read in the newspapers that it had been forbidden by the Nazis. "You should stop reading the newspaper," Warburg told her.

In another instance, a group of Nazi officials arrived at Warburg's institute to check on whether everyone present had registered with one of the many new Nazi bureaucratic offices. Warburg, according to Lotte's telling, politely dismissed them, telling them that his employees were working hard and had "no time to deal with politics."[9]

On still another occasion, a company of storm troopers arrived at Warburg's institute and said that five of Warburg's assistants needed to leave for the day to take part in a mandatory Nazi march. Warburg asked the leader of the group to identify himself and then told the man that he "would burn his institute" before allowing it to be interfered with in such a way. The storm troopers left and never came back.[10]

Warburg's finest moment during the Nazi years may have been his protection of Erwin Haas, a Jewish researcher who remained at his institute until 1938. When a Nazi official phoned to instruct Warburg to fire Haas, Warburg said it was impossible and that he would resign before doing so. In February of 1935, Warburg mentioned Haas's situation to a Rockefeller official and reiterated that he would quit if he faced any political interference.[11]

David Nachmansohn thought it was ultimately Warburg's aristocratic nature that led him to hate the Third Reich: "It is easy to guess how much he must have despised the Nazi mobs, the vulgarity of their mass meetings, the low stature of the men with whom Hitler surrounded himself." As Warburg put it, he was not going to let "a handful of arbitrary criminals" and Bavarian "noisemakers," as he called the Nazis, tell him what to do.[12]

Warburg's record in the first half of the 1930s was far from perfect. He did let go of one young Jewish researcher, Walter Kempner. In a letter to Kempner's mother, Warburg claimed Kempner had not been working the required hours. Warburg had already fired Kempner once before, in 1929, for making an arrogant remark, of all things. Kempner had only been working at Warburg's institute for a brief period when Warburg fired him again, and Kempner was known to be an eccentric—many years later he would become a diet guru in the United States who sometimes physically abused patients.[13]

Warburg did try to find a position for Kempner in America. And given that the two maintained a warm correspondence for the next three decades, Kempner presumably didn't blame Warburg for his dismissal. But that Warburg was ready to make a stand for Haas and apparently unwilling to do the same for Kempner was likely a reflection of Warburg's broader outlook in the early 1930s. He could tolerate an injustice to a Jewish scientist so long as it didn't impact him in a meaningful way. Kempner had only been back at the institute a short time and wouldn't have been missed. Haas was a great scientist, an important member of Warburg's team.[14]

Had he left Germany like so many others, Warburg could have avoided these moral quandaries. On a number of occasions, he declared that he was on the verge of leaving. In May 1933, Warburg told a Rockefeller official that he saw "no future" in Germany and that he planned to leave as soon as he found a new position elsewhere. According to a Rockefeller memo, Warburg already had an offer from the University of Leeds. That summer Warburg broached the subject of leaving Germany with Max Warburg. "The only country I might consider is England," Warburg told Max, "but can you imagine my having to talk to other professors?"[15]

Worse than mingling with academics would have been departing Germany as a refugee. Warburg once told a Rockefeller official how disturbed he was to have seen his name mistakenly added to a list of displaced German scientists. In retrospect, fleeing Germany appears to have been a moral imperative, but in the early 1930s, German Jewish scientists typically saw it as a source of great shame, an acknowledgment of one's inferiority. And Warburg, by his nature, would have been highly sensitive to this humiliation. His sense of his own greatness coexisted with profound insecurities. He lashed out whenever his work was criticized precisely because the criticism stung so deeply. Near the end of his life, Warburg remained full of antagonism for Richard Willstätter, a famous organic chemist. Willstätter's great offense? He had once been dismissive of a question Warburg had asked at one of his lectures. "Five hundred people laughed," Warburg recalled decades later.[16]

Besides, Warburg saw no good reason to flee. He had served as an officer in World War I—as an uhlan, no less. He was a German through and through. When Lotte expressed doubts about her own future in Germany, Warburg reminded her in a letter that whatever her financial concerns might be, the question of "what happens to your sense of self when you live somewhere you weren't born" would remain.[17]

Warburg, for all his personal faults, knew who he was and how exile would affect him. The famed Austrian writer Stefan Zweig had wrestled with the same dilemma: to live under the Nazi jackboot and hold on to whatever might be left of his prestige and status, or to flee and start anew. Zweig took the latter path and never recovered. "Everything which I had attempted, achieved, learned, enjoyed," Zweig wrote, "seemed wafted away." Not long before committing suicide in 1942, Zweig described his life in exile as a "posthumous" existence.[18]

For a German patriot like Warburg, fleeing Hitler wouldn't only lead to shame and a loss of identity. It would be an implicit acknowledgment of having been mistaken about Germany itself. As Einstein wrote to Fritz Haber in 1933, recognizing Germany for what it had become was "somewhat like having to abandon a theory on which you have worked for your whole life." And if there was one thing Warburg did not do, it was admit to a mistake or abandon a theory.[19]

The reason for leaving Germany—that he was a Jew—made the entire notion all the more absurd to Warburg. Like many Germans of Jewish descent, he had never identified as a Jew and felt no special connection to the larger Jewish community. As Zweig observed, the persecuted Jews of medieval times "had at least known what they suffered for." And who was Hitler, of all people, to send Warburg away? Warburg told Lotte that he had looked closely at a photo of Hitler and concluded that he wasn't even a true German. Warburg suspected Slavic descent.[20]

Warburg also had more concrete reasons for staying. Leaving Germany would have meant giving up his stunning new institute. Though it was part of the Kaiser Wilhelm Society, it was funded by a private US foundation, and Warburg had secured the funding himself. A young researcher claimed to have heard a Rockefeller official offer to take pictures of the institute so that a perfect replica could be built for Warburg elsewhere, but there is no evidence that the Rockefeller Foundation considered building an entirely new institute for Warburg. "[T]here are, to my knowledge, very few departments in the world that are more amply equipped and endowed than those of Meyerhof and Warburg," one foundation official noted. "I doubt if one could find anyone who would be in a position to create an institute for either one of them. The best hope would be for the occurrence of a vacancy" at an existing institution.[21]

The worst part of relocating, Warburg told Krebs, would be losing technicians he had spent years training to carry out his experiments. (Warburg claimed he could communicate with his employees just by glancing at them.) Though speaking about Meyerhof, Warburg was also clearly referring to his own dilemma when he argued that it was especially hard for a distinguished researcher to leave. It "is easy to find a place for an ordinary person," Warburg said, "but it is hard for a king to find a kingdom."[22]

Among the better explanations for why Warburg stayed in Germany after Hitler became chancellor is the simplest of all: during the first year of Nazi rule, he failed to grasp how bad the situation would become. Warburg was convinced that the Nazi experiment would implode as Hitler crossed lines that a civilized country could not cross. Warburg's opinion was common in the early 1930s. Millions of other Germans, including many German Jews, suffered from the same myopia. Hitler, Eric Warburg said in 1933, should be given "enough rope to hang himself."[23]

On the morning of June 16, 1934, Warburg met the Rockefeller official W. E. Tisdale for tea and said that "less ignorant" and "more moderate forces" were "gaining ground." The two continued the conversation over lunch at Berlin's Hotel Continental. After the meal, they spoke outside so as not to be overheard. The German army would come to the country's rescue and restore the monarchy, Warburg said, as the two men paced back and forth in front of the hotel. He gave the Nazis only another six months.[24]

Warburg, anxious about losing his Rockefeller funding, was probably going out of his way to make Germany's future appear less bleak. If he did believe what he told Tisdale, his opinion soon changed. The following May, Warburg told Lotte that the Nazis were "the greatest criminals in history" and that the political situation had "swallowed science whole." He said he'd begun telling "the non-Aryans" he encountered that it was "almost their duty" to do everything possible "to bring these people to their knees and undermine the state." "If you have even the slightest thing to do with these criminals," Warburg continued, "you are despicable."

Warburg also told Lotte that he wanted nothing to do with Germany as long as the concentration camps (still, at the time, used mainly for political prisoners) and secret police remained in existence. But for Warburg, wanting nothing to do with Germany didn't necessarily mean wanting to flee Germany. If anything, it meant the opposite. He was "boiling with rage against Germany," Lotte wrote in her diary. And the "more he hates, the more firm is his decision to stay put." Warburg had closed himself off in his home and institute "like a foreign prince" among the German people. "He tells himself it is nothing more than a test of strength and nerves to see who can outlast the other: them or him." Added Lotte, "And I hope it's him."[25]

Lotte, too, was struggling with the question of whether to stay or flee, and Warburg's advice to his sister is revealing: "He advises us first to stay and then to leave," Lotte noted. "He always contradicts himself."

Later that year, Warburg again mentioned the possibility of leaving for England if he could find a position suitable to his reputation. But he refused to leave, he said, if it would appear that he was being kicked out. Before being humiliated by Nazi lowlifes, Warburg promised to wait until "everything collapses."

Lotte recorded Warburg's words in her diary and then appended a comment of her own: "I fear if he only leaves when everything here collapses, it will be too late."[26]

FOR WARBURG AND others deemed "non-Aryans" by the Nazis, a turning point came late in 1935 with the introduction of the Nuremberg Laws. Through campaigns of violence and economic sanctions, the Nazis had already made life unbearable for most of Germany's remaining Jews. But Nazi hardliners, frustrated by what they saw as a slow pace of change, believed that the existing measures weren't severe enough. Some had begun to take matters into their own hands, forming lynch mobs to attack Jews suspected of having sex with Aryans. The Nuremberg Laws were designed to satisfy this hunger for additional cruelty. The cruelty, in a sense, had always been an end in itself. In the words of one historian, it was "I kick, therefore I am."[27]

One of the two Nuremberg Laws, the Reich Citizenship Law, stripped Jews of their German citizenship and made them subjects of the state. The Law for the Protection of German Blood and German Honor made it a crime for Jews to marry or engage in extramarital relations with the "German-blooded." But these new laws, which the Nazis said would bring clarity to their Jewish policies, had the opposite effect.

In the original draft of the law banning relations between Jews and the "German-blooded," the text had included a significant line: "This law only applies to full Jews," defined as someone with three or four Jewish grandparents. Believing it would limit the international outcry, Hitler left this reference to "full Jews" in place in the draft that was sent to the government press agency. But in the draft sent to the Reichstag, Hitler crossed the line out in pencil. With that single pencil mark, he changed the fate of the hundreds of thousands of Germans who had only one or two Jewish grandparents.

The two different drafts of the law left German officials unsure whom it applied to. Most Nazi Party officials wanted to continue to define Jews, or non-Aryans, by the terms of the 1933 Civil Service Law, which applied to anyone who had even one Jewish grandparent. But those serving in government posts, conscious of Germany's international reputation, called for a more nuanced stance. The Berlin Olympics were less than a year away, and, though it was assumed that the international community would tolerate cruelty against Jews, it was less clear if it would tolerate attacks against the so-called half-Jews and quarter-Jews, who typically thought of themselves as Christians.

The Nazis referred to Germans with only one or two Jewish grandparents as *Mischlinge*, a derogatory term that translates to something like "mongrel." Hitler detested the *Mischlinge*, calling them "monstrosities halfway between man and ape." He once said that half-Jewish blood would not "Mendel out" even after six generations of descendants reproduced with so-called pure Germans. The *Mischlinge* were the living embodiment of Hitler's nightmare, in which tainted Jewish blood mixed with pure Aryan blood with impunity. In the *Mischlinge*, Hitler wrote in *Mein Kampf,* "the vices of the parents are revealed in the sicknesses of the children."[28]

Nevertheless, as the debate over the new law intensified, Hitler wavered. That the *Mischlinge* would eventually meet whatever fate awaited "full Jews" was a given. But Hitler recognized that the persecution of the *Mischlinge* was more than a threat to Germany's international reputation. *Mischlinge* had millions of "German-blooded" relatives who might turn against the Nazis if members of their own families were humiliated and persecuted. Many *Mischlinge* were still serving in the German army at the time, including at the highest ranks.

In the end, Hitler sided with the "moderates." In November, two days after Lotte Warburg wrote in her diary that her brother was waiting for the moment when "everything collapses," a supplementary decree was added to the Nuremberg Laws to distinguish full and partial Jews. Those with two Jewish grandparents, like Warburg, were deemed "first-degree *Mischlinge*," while those with only one Jewish grandparent

were labeled "second-degree *Mischlinge*." Though not subject to all of
the same decrees as "full Jews," *Mischlinge* would still be denied positions of authority and robbed of most of the rights of German citizens. For the time being, they could still be drafted into the army, but
would now be restricted to the lowest ranks. While *Mischlinge* were
not legally barred from studying in universities, various areas of study
were off-limits. In some cases, *Mischlinge* were barred even from their
own churches.[29]

THE FIRST STEP in both respiration and fermentation is a splitting
known as glycolysis: one glucose molecule is broken into two. The biochemical reaction that had captured Otto Warburg's scientific imagination was now a metaphor for his life. He had always been divided in
two, in various ways. He was both a nineteenth-century aristocrat and
a modern scientific visionary, a charming conversationalist and a self-
important dictator. Upon being legally declared a first-degree *Mischling*, or half-Jewish and half-Aryan, he had been split down yet another
fault line.

Though there are no records of Warburg speaking directly about
his new legal status, many other *Mischlinge* did. They were typically
devastated and humiliated to have been classified as second-class citizens. By the terms of the Nuremberg Laws, Warburg could have been
arrested for having an intimate relationship with an Aryan—a crime
known as "race defilement." Warburg, a half-Jew, could have been
arrested even for having relations with a quarter-Jew.

As neither "true" Jews nor "true" Germans, the *Mischlinge* were
caught in the middle of the Nazis' disturbed race fantasies. Bernhard
Lösener, an official at the Reich Ministry of the Interior, once said that
the *Mischlinge* had it worse than the "full" Jews, who at least had a
community to belong to. The *Mischlinge* were stranded between two
worlds: half-Jew, half-Aryan; half-oppressed, half-oppressor.

Even before the introduction of the Nuremberg Laws, the Nazis had made it difficult for Warburg to work without evidence of Aryan heritage. His situation now became all the more desperate. Warburg continued to invite young scholars to his institute, but fewer and fewer would accept the offer. Some thought Warburg's friendly invitations to study at his institute had to be a joke. One English scientist laughed in Warburg's face when he proposed a visit to Berlin. His status as a "half-Jew," Warburg said, made his institute an object of such "disdain" that even being associated with it came with the risk of ruining one's academic career.[30]

On January 13, 1936, the *New York Times* ran a critical editorial on the declining state of the Kaiser Wilhelm Society: "The fate of even such eminences as the physiologists Otto Warburg (supported by the Rockefeller Foundation) and Otto Meyerhof is avowedly precarious." Carl Neuberg, who pioneered the modern study of fermentation and who was among the most prestigious chemists in the world, had remained in Germany even after being dismissed from his position as head of the Kaiser Wilhelm Institute for Biochemistry in 1934. In 1936, in a desperate effort to hang on to his career and life in Dahlem, Neuberg suggested that his equipment be moved to a gatehouse on the grounds of his old institute. Even this feeble request was denied.[31]

For all his defiance when the Nazis attempted to directly interfere with his institute, Warburg told his Jewish colleagues at other institutions to lie low and hope that the Nazis would forget about them. Public statements against the government, Warburg warned, would only make things worse for German Jews. His advice to Meyerhof was to "sit tight and do nothing." But with each passing year, Warburg's claims that the Nazi nightmare would soon be over became less plausible. In the middle of the decade, Hitler was growing more rather than less popular.

Warburg's own fame and popularity spread even as he had become a pariah in Berlin. In 1936, the American radio program *Heroes of Civilization* produced an episode about one of Warburg's minor findings, complete with an actor playing Warburg. That same year, Warburg planned to attend a conference of international scholars in Cambridge, Massachusetts, in honor of Harvard's 300th anniversary. Warburg was delighted by the invitation. "Everything that I have become, everything that I have, I have received from foreigners," he told Lotte. "That's why I completely feel like a foreigner in this country and live as one here, too."[32]

The appearance at Harvard would never take place—Warburg canceled shortly before it started. A ship manifest for the SS *Europa*, which sailed for the United States in August 1936, shows Warburg's name crossed out. The physicist Werner Heisenberg also failed to attend despite planning to. A *Boston Globe* report on the Harvard conference noted that the two scientists missing from the event seemed to be inspiring "more talk" than "all who have appeared."[33]

While the precise circumstances of the cancellation are unknown—Harvard was told only that Warburg had to attend to "unexpected duties"—Warburg was almost certainly forbidden from attending by German authorities. In the buildup to the Berlin Olympics of 1936, the Nazis were busily spreading propaganda to distract from their anti-Semitic policies. Warburg had become politically useful, evidence that a Jew could still live and work in Germany.

Warburg was aware that he served this function for the Nazis. He told Lotte that he suspected there was a note next to his name in a Nazi register somewhere that read, "retain for another ten years for foreign propaganda." By remaining in Germany for so long, he told Lotte, he had made a "devil's pact" with the Nazis.[34]

As Lotte once put it, Warburg's decision to remain in Nazi Germany was made possible only through theater. Her brother, Lotte wrote, had "decided to play the role of the man with no soul."[35]

In becoming a tool of the German authorities, Warburg, perhaps without knowing it, was carrying on a centuries-old family tradition. He was a direct descendant of Simon von Cassel, the sixteenth-century

Warburg family patriarch who was allowed entry into the German town of Warburg to work as money changer and pawnbroker. Church prohibitions against money lending led Jews into such roles, even as they were barred from many other professions. Simon and his descendants were *Schutzjuden*, or "protected Jews." Though still treated as second-class citizens, they were granted special privileges by the local rulers and functioned as leaders of the Jewish community, with Simon's home doubling as Warburg's synagogue. The Warburg family "hung from a golden thread, suspended between gentiles above and Jews below," the historian and biographer Ron Chernow wrote in his book on the Warburg family. "This ambiguous, hybrid status—neither totally Jewish nor gentile—contributed to the deeply schizoid Warburg character."[36]

As late as January 1937, Warburg was still capable of feigning optimism. A Rockefeller official who dined with Warburg that month described him as "less disturbed and uncertain than in the summer of 1934." During the meal, a defiant Warburg described the Nazis as "labile" and mocked their militarism: "Force is not strength, nor is pomp dignity," he said, quoting an English acquaintance.

Later that year, Warburg received permission to travel to Paris for a scientific conference. When he arrived with Heiss, he checked in at the German embassy. Lotte, who was also in Paris at the time, noted that Warburg "dressed very solemnly" for his visit to the embassy, wearing all of his war medals. But when he arrived at the embassy, he found it empty and grew angry.

Lotte had left Germany permanently by that point and was living in The Hague. Warburg filled her in on his life in Dahlem. One day, he said, he was in Max Planck's office together with Friedrich Glum, the general secretary of the Kaiser Wilhelm Society. Standing at the door, Warburg turned to the two other dignitaries of German science and offered a parting remark: "Hitler," Warburg announced, was said to have been "surprised at the cowardice" of the gentile professors who had failed to stand up to him.

Warburg also told Lotte that he now ignored Planck whenever the latter tried to speak with him and that Planck had stopped trying as a result. Lotte herself remained enraged at Planck. "Until the end of his life, he will wear the mask of the noble, selfless, persuasive and true researcher," she wrote in her diary. "And no one will ever discover the monumental cowardice and lack of character that filled his last years."

Planck, however, was the least of Warburg's concerns in 1937. Living anything resembling a normal life in Germany now required concrete proof that one had no Jewish heritage, and various Nazi officials were regularly calling Warburg in for questioning. In response to such queries, Warburg simply lied, telling the police and the officials hounding him that he was a pure Aryan. He had no choice, he explained to Lotte. Life would be "unbearable" if he didn't pretend.[37] Warburg claimed that some of the Nazis believed him and that the ones who knew he was lying didn't dare confront him. It's more likely that they all knew he was lying. "With a name like Warburg" his denials were "a feat that required remarkable nerves," Lotte noted.[38]

It's unlikely that Warburg would have fared much better with a less obviously Jewish name. The Nazis had turned centuries of baptism records from Protestant churches into an immense genealogy registry. The files gathered from dingy church basements sometimes left room for ambiguity, but there were other options in such situations. Warburg's colleagues at the Kaiser Wilhelm Institute for Anthropology could be called on to determine someone's racial purity according to the shape of the individual's head or nose.[39]

Three years earlier, Warburg had vowed that the Nazis wouldn't scare him away. Now he was paralyzed by indecision. He had debated leaving "a hundred times," he told his sister, but he still could not make up his mind.[40]

THE YEAR 1938 began on a particularly unsettling note for Otto Warburg. He could have been arrested by the Nazis for almost any reason, but on January 12 he was confronted with even more alarming news: he was dead. That, at least, is what was reported in *The Times* of London, which ran an obituary for Warburg on that day.[41]

Warburg sent off an angry message to *The Times*, and the paper apologized for the mistake the next day: "We much regret that by an unfortunate misunderstanding we announced yesterday the death of Professor Otto Warburg, the biologist, who is, we are glad to learn, still living."[42]

The man who had died was a different Otto Warburg, a cousin who, in addition to being a distinguished botanist, was a former head of the World Zionist Organization. Warburg was able to find dark humor in the obituary. He pinned the article near his workspace and brought it up throughout the day, saying over and over that he could have written a much better version himself. "It looks as if the obituary was written jointly by poor old Heinrich Wieland and poor old Torsten Thunberg," Warburg joked, a reference to two of his scientific adversaries. But he was said to be genuinely upset that the obituary had failed to mention a number of his most important discoveries. For a man convinced of his own greatness and intent on leaving a lasting mark on humanity, the mere three-paragraph obituary must have come as a terrible blow.

The problems ran deeper still. Referring to his cousin of the same name, Warburg claimed, in an uncharacteristically personal letter to a Rockefeller official, that "the elderly gentleman, whom I do not know and who must be around 100 years old, has caused a lot of confusion in my life. . . . His very existence is practically an impediment to my individuality." Indeed, even the Nazi bureaucrats monitoring Warburg once mixed him up with the other Otto Warburg.[43]

That there were two Otto Warburgs, one an ardent Zionist and one detached from his Jewish background and working in Nazi Germany, was almost too perfect. It was the "schizoid Warburg character" come to life in the form of two distinct individuals. The other Otto

Warburg was not only an impediment to Warburg's individuality but an indictment of his entire person. If the Warburgs were "suspended between gentiles above and Jews below," Warburg's Zionist namesake had reached down. Warburg himself would only reach up.[44]

Whether the obituary was still on his mind several weeks later, on the day Warren Weaver from the Rockefeller Foundation arrived at his institute, is impossible to know. Warburg was sitting in his library, talking to a young researcher, when Weaver appeared at the door. After the researcher left the room, Warburg "with a somewhat cynical smile" told Weaver he had just received a call from Richard Kuhn, the director of the Kaiser Wilhelm Institute for Medical Research, who would be chosen for the Nobel Prize later that year. Warburg said that he had picked up the phone, told Kuhn "I will not speak with you," and then hung up. "I treat them like dogs," Warburg announced, "for that is indeed what they are."

Warburg had good cause to despise Kuhn, who had earlier denounced Meyerhof for his attempts to protect his Jewish employees. Even so, what Warburg said next, that Kuhn and several other Kaiser Wilhelm directors were "rotten to the bone" and scheming only for their own power, shocked Weaver. "One got the impression of a man who, however keen his mind may be when directed toward his scientific research, is nevertheless very near the edge of mental instability," Weaver wrote. That Warburg "is suffering from an intense and perhaps even unsuspected kind of loneliness is clear; that he has a fairly well-developed persecution complex seems almost as clear."[45]

Weaver's assessment wasn't particularly sympathetic, given that Warburg had already faced years of persecution. And Warburg's situation was only becoming more dire. Though he had been allowed to travel with permission in 1937, in August 1938 the Nazis stopped him from attending the International Congress of Physiology in Zurich, where he had planned to present a paper. Warburg sent a telegram to the organizers: "Must cancel attendance," he wrote. "There is no reason."[46]

"No reason" would have been understood as a coy comment on the irrationality of the Nazis. The conference attendees were thrilled by Warburg's subversive message. And Warburg continued to defy the Nazis throughout 1938. He trashed an invitation to a reception in honor of Planck's 80th birthday the moment it arrived. When Warburg's last Jewish employee had left his institute that year, the Gestapo asked Warburg how he had dared to keep a Jewish employee for so long. Warburg responded that the man was a Hungarian Jew and that he had thought the rules only applied to German Jews.[47]

If still capable of lashing out, Warburg no longer did so with the bravado of previous years. Norman Davidson, a young Scottish researcher who had made the surprising decision to study at Warburg's institute during this period, recalled that Warburg "made no secret of the fact that he detested the Nazis." But while Warburg allowed Davidson to keep working on an official Nazi holiday in the spring of 1938, he told him to stay away from the windows. Warburg even switched off the fuses so that Davidson wouldn't accidentally turn on a light.

Warburg's institute, Davidson wrote, was the only one left in Dahlem that had no Nazis on its staff. At the other Kaiser Wilhelm institutes, some staff members regularly appeared in their black SS uniforms and were more interested in discussing politics than science. Davidson remembered once visiting Adolf Butenandt, the director of the Kaiser Wilhelm Institute for Biochemistry. The institute, he recalled, was full of "rabid Nazis." "He received me kindly," Davidson wrote of Butenandt, "but when he asked where I was working in Berlin, and I told him in Warburg's Institute, his attitude to me became very cold indeed." Davidson was "quickly shown the door."[48]

In September 1938, Otto Meyerhof escaped Germany for France. (When the Nazis invaded France, he would make a dangerous journey across the Pyrenees to avoid capture.) Meyerhof had departed just in time. Two months later came *Kristallnacht*. Amid the burning and plundering of synagogues and Jewish businesses, 91 Jews were killed and another 30,000 arrested. Many of the Jews still living in Dahlem were brought to the local police precinct. When no instructions arrived from Berlin on what to do with them, the precinct commander let them go.

There is no evidence that Warburg was harassed during *Kristallnacht*, but only two days later, a traveling exhibit known as "The Eternal Jew" opened in Berlin. The exhibit, which documented the supposed worldwide Jewish conspiracy against Germany, featured caricatures of Jews with grotesquely large noses. It also featured a photograph of Warburg.

The Kaiser Wilhelm Society managed to have Warburg's photo removed from the exhibit the next month after Warburg provided a Nazi official with information about his ancestry and insisted, as he had before, that he was not related to the famous Warburg bankers.[49]

Warburg could feel the world closing in on him now. In February 1939, in a letter to a colleague in England, he denied that he had ever sent the "there is no reason" telegram to the physiology conference and asked for help in squelching the gossip. If he didn't stop such stories from spreading, Warburg added, his actual obituary would soon appear in *The Times*.[50]

The following month, in response to additional Nazi inquiries, Warburg once again stated that he was not a Jew according to the Nuremberg Laws. At approximately the same time, Warburg met with a physiologist who had been offered the position of chair of physiology at Berlin University. After first checking to make sure no one could hear him, he instructed his colleague to stay away.

"Germany," Warburg said, "is heading for a great catastrophe."[51]

"THE ETERNAL JEW," the German variation on "the Wandering Jew," is a mythical figure from medieval folklore, who, like Faust, captivated the German imagination for centuries. According to the legend, Jesus was carrying his cross to Calvary when he paused to rest on a doorstep of a home. The homeowner, a Jewish man, scolds Jesus for loitering on his property and is punished with the curse of eternal wandering. "I go," Jesus says, "but you will walk until I come again."

The Nazis saw the Eternal Jew as a metaphor for the Jew's never-ending malice—a character who, in his refusal to vanish, justified the famous German slogan, "The Jew is our misfortune." But German Romantics of the eighteenth and nineteenth centuries had, at times, humanized the Eternal Jew and marveled at his suffering. In one popular late eighteenth-century poem, Christian Friedrich Daniel Schubart depicts the Eternal Jew standing alone atop a mountain, casting down the skulls of the relatives he has outlived and crying out over his inexplicable survival. "I cohabited with poisonous snakes, and pinched the dark-red crest of the dragon," the Eternal Jew wails, "the serpent stung, but could not kill me; the dragon tormented, but could not destroy me."[52]

Otto Warburg, 1931.

CHAPTER NINE

"The Herb Garden" of Dachau

IN THE 1930S, even as the Nazis were carrying out the first stages of their war against the Jews, they were engaged in another war that seems antithetical to the inhumanity of the rest of the Nazi agenda. This second war was a progressive war, a war to defeat cancer. Shortly after Hitler came to power, the general secretary of the Reich Anticancer Committee promised that the Nazis would engage in "planned cancer combat" unlike anything that had ever come before—and this, at least, was not propaganda.[1]

The Nazis did not start Germany's campaign against cancer—the country had been at the forefront of cancer research for decades—but they accelerated it greatly. Thirty years before the American medical community, the Nazis were encouraging women to examine their own breasts for lumps. At "cancer counseling centers" built by the Nazis, the public learned about the danger of asbestos, pesticides, and other chemicals long before other countries had begun to take such threats seriously. One Nazi regulation even required bakeries to produce whole-grain bread, in part out of the fear that the benzoyl peroxide used to bleach white bread could cause cancer.

That the Nazis were introducing legislation to distinguish half-Jews from quarter-Jews at the same moment they were introducing laws to keep the population safe from dangerous chemicals was not a coincidence. For all its progressive innovations, the battle against cancer wasn't an aberration from the other tenets of Nazism. The entire Nazi project, Stanford historian Robert Proctor argues, can be thought of "as an experiment of sorts—a vast hygienic experiment designed to bring about an exclusionist sanitary utopia." The Jews, though the most feared and hated, were only one of the many supposed pollutants the Nazis planned to eradicate. In *Mein Kampf*, it can seem as though Hitler's unstable mind is incapable even of distinguishing Jews from bodily growths, as though the metaphor collapses under the weight of his rage. "Was there any form of filth or profligacy, particularly in cultural life, without at least one Jew involved in it?" Hitler wrote. "If you cut even cautiously into such an abscess, you found, like a maggot in a rotting body, often dazzled by the sudden light—a kike!"[2]

HITLER'S FOCUS ON diet and cancer only intensified after he became chancellor. In 1933, eager to learn more about health and vegetarianism, he met with an 80-year-old healer known for her herbal and cold-water cures. Later that day, when a Gestapo officer tried to turn his attention to political matters, Hitler grew agitated: "What this old woman told me this morning is far more important than anything I can do in my life."[3]

While Hitler's persistent stomach cramps helped spur his panic—at one point he became so convinced he had cancer that he sat down and wrote out his will—he found plenty of other causes for concern. Even his home and office were objects of suspicion: In September 1934, a physician made his way through Hitler's Berlin headquarters with a dowsing rod, an ancient tool believed to move in response to hidden elements or forces. The rod, according to quacks of the era, could detect cancer-causing currents beneath the ground they called "Earth Rays."[4]

It's not known if the doctor detected any such "Earth Rays" that day, but whatever he found, it failed to put Hitler's mind at ease. The next year, Hitler panicked that his persistent cough and hoarse voice were signs of throat cancer. Upon examining Hitler, a doctor discovered a polyp on his larynx. The doctor told him that it was "a very slight matter," yet Hitler remained convinced he was heading for an agonizing death. Though he calmed down somewhat when the polyp was removed and confirmed to be benign, he summoned the same doctor again later that year, concerned about a new bout of discomfort in his throat. When the doctor failed to find anything wrong with him, Hitler remembered that there might be a more prosaic explanation: he had recently removed a thorn from his finger with his teeth and then swallowed it by mistake.[5]

In the spring of 1936, Hitler met the Berlin physician Theodor Morell at a social gathering and told him his stomach problems were so bad he could barely function. Morell, a bald man with a large belly who was said to sweat profusely and give off a terrible smell, was known for his odd eating habits—he bit directly into oranges rather than peeling them—and was said to sleep with his lower eyelids closed upward. He told Hitler that to be a vegetarian was no longer enough. The führer needed fresh fruits and vegetables, grown in soil treated with the proper manure. Martin Bormann, who was essentially chief-of-staff for Hitler and later was named the head of the Nazi Party Chancellery, became the de facto Reich gardener at Hitler's mountain home in the Bavarian Alps. If Hitler was not in the vicinity, Bormann's fresh produce was loaded onto a plane and flown to him.[6]

Morell also suggested an unusual treatment for Hitler's stomach problems: pills containing a strain of bacteria that had been taken from the intestinal flora of a German military officer. The pills, known as Mutaflor, supposedly cured stomach issues and prevented cancer through interspecies warfare. The good bacteria from an Aryan officer colonized the "living space" of the colon and destroyed the bad bacteria in the process. Hitler's own gut was now a microcosm of the entire Nazi project.

It wasn't only Hitler. Many Nazi leaders were strangely focused on diet and cancer. Goebbels, who often turned to cancer as a metaphor in his propaganda campaigns, was also a hypochondriac and was convinced that his ulcers were signs of cancer. Rudolf Hess, Hitler's deputy führer until he flew to Scotland in 1941, was petrified of cancer and had some of his teeth pulled in the name of cancer prevention. Hess was also passionate about organic farming and started carrying his own organic, vegetarian foods with him when forced to eat out. Hess even brought along pre-prepared dishes to meetings with Hitler at the Reich Chancellery. Hitler, annoyed, told him to cut it out.[7]

Julius Streicher, the Nazi official behind *Der Stürmer*—a tabloid so crudely anti-Semitic that even some senior Nazis were embarrassed by it—helped to establish a tumor research institute in Nuremberg. In 1936, Streicher began championing a cancer drug that could wipe out tumors with the help of extracts from Chinese rhubarb. Though a number of leading German medical figures had already dismissed Streicher's cure as quackery, Hitler was intrigued and asked Karl Brandt, the doctor overseeing his care, to investigate the treatment. Brandt brought the researcher who had developed the new drug to Berlin and forced him to reproduce his experiments. Even when Brandt reported back to Hitler that the drug was a failure, Hitler held out hope. When Streicher continued to insist that the treatment worked, Hitler made Brandt travel to Nuremberg to investigate the matter a second time.[8]

Heinrich Himmler, the head of the SS and among the most powerful men in Nazi Germany, had stomach problems of his own and, like Hitler, feared that his discomfort might be an early sign of cancer. Himmler, too, suspected that diet was at the root not only of cancer but of all of Germany's health problems. "The wrong diet always plays a decisive part in all the troubles of civilization," Himmler once said, "from the loss of teeth to chronic constipation and digestive ailments, not to mention bad nerves and defective circulation."

Himmler, who had studied agriculture and once worked in a fertilizer factory, focused a great deal of his energy on the role of diet in cancer. "The artificial is everywhere; everywhere food is adulterated," Himmler wrote, "filled with ingredients that supposedly make it last longer, or look better, or pass as 'enriched.'" The blame, in his view, fell on "the food companies, whose economic clout and advertising make it possible for them to prescribe what we can and cannot eat."[9]

In 1933, Himmler, then the police president of Munich, established Dachau as one of the first Nazi concentration camps. The camp, situated 10 miles northwest of Munich, initially held political opponents but gradually expanded to include Jews, homosexuals, Roma, and Jehovah's Witnesses. In 1937, prisoners at Dachau were forced to drain marshlands and then to plant and run an industrial-sized organic farm on 200 acres. The Dachau plantation eventually comprised a research station, greenhouses, a spice mill, drying and storage rooms, and an apiary from which the prisoners would harvest organic honey. The herbs and spices grown at Dachau were sold as natural remedies and are thought to have supplied nearly all of the German army's seasonings during World War II.

The Dachau plantation quickly became the largest research center of its kind in the world. As many as 1,000 prisoners could be working in "the herb garden," as the SS men called it, at a given time. It was among the most feared assignments; the emaciated prisoners working in "the herb garden" would regularly drop dead from exhaustion. Others were held by their feet and drowned in the carp pond.[10]

One of the few firsthand accounts of the Dachau plantation comes from Himmler's young daughter, Gurdun, who toured the grounds with her family in the middle of the war: "Today we drove to the SS concentration camp in Dachau. There we then toured everything . . . the large nursery, the mill, the bees, saw how all the herbs were processed . . . all the pictures that the prisoners have made. Magnificent!" Continued Gurdun, "Then we ate, then everybody got a present. It was lovely. A very big operation."[11]

Dachau was only a first step toward Himmler's planned diet revolution. The SS had additional gardens planted at other concentration camps. In January 1939, Himmler oversaw the purchase of 16 additional farms and opened a food and nutrition institute to study "natural methods of agriculture." That same year, the SS, already running its own fruit juice factories, began bottling mineral water.

Himmler was especially focused on what went into the mouths of his SS men. Potatoes, he ordered, should never be salted or peeled. He planned to one day bring "nutrition supervisors" into the SS who would be tasked with weaning his men off meat, among other foods. ("Too Much Meat Can Make You Sick," reads one heading in a Hitler Youth nutrition manual from the 1930s.) Himmler appreciated that it would be a challenge to remake the German diet by replacing "meat and sausages" with "equally tasty foods that satisfy the palate as well as the body." Soybeans, thought to be one solution to this problem, were referred to as "Nazi beans."[12]

While Himmler acknowledged that some of his dietary reforms would have to wait until after the war, certain measures had to be implemented without delay: "The attention of all units must be drawn most vigorously to the toasting of bread," he wrote to a Nazi nutrition inspector in the middle of the conflict.

Himmler's obsession with nutrition wasn't entirely a product of his fear of disease. Like Hitler, he dreamed of a Germany that controlled its own supply of food and could never again be starved. But cancer was never far from Himmler's thoughts. In 1936, his father was diagnosed with stomach cancer. He died after Himmler, on a doctor's advice, chose not to have him operated on.[13]

At approximately the same time, Himmler took an interest in the work of Sigmund Rascher, a Nazi medical researcher examining whether eating plants grown with artificial fertilizers might cause cancer. Rascher hoped to develop a blood test that could detect cancer at an early stage, but to figure out which blood markers might signal cancer, he would need to follow his test subjects for years. Himmler's pris-

Adolf Hitler, broken glass-plate negative, date unknown.

oners, it occurred to Rascher, would make ideal test subjects. On May 26, 1939, Himmler granted Rascher access to Dachau. It was in that dystopian landscape of barbed wire and organic honey that he carried out the very first medical experiments on Nazi prisoners.[14]

IN THE 1930S, Otto Warburg embarked on his own war against cancer, even as his entire world unraveled around him. He had spent much of the previous decade examining cancer cells that overeat and ferment glucose. The next step seemed straightforward: Warburg would cure cancer either by preventing the switch from respiration to fermentation or by interfering with the fermentation process and starving cancer cells.

It was a perfectly logical plan but one far ahead of its time. Though Warburg had already identified the respiratory ferment, both respiration and fermentation remained poorly understood. Before Warburg could cure cancer, he would first need to return to basic science. Specifically, he would need a better understanding of enzymes, the microscopic machines that carry out the work of metabolism.

Warburg became interested in modern enzyme science only after a rapid evolution of his thinking on cells and energy. Throughout most of the 1920s, he had continued to believe that the reaction between iron and oxygen in his ferment was the entire story of respiration. Warburg had been particularly incensed by Heinrich Wieland, his first great scientific nemesis, who had argued that the key to respiration was the reactivity of hydrogen, rather than oxygen. As Warburg saw it, Wieland's experiments, carried out on one of Ehrlich's dyes, as opposed to on living cells, were too speculative. It was Romanticism, Warburg argued, in the process of belittling Wieland's achievements, "to rate the unknown as more important than the known."[15]

Warburg's distrust of Wieland's methods was sincere. A physicist's son, Warburg always favored simple, elegant explanations for natural phenomena. The possibility that hydrogen and oxygen each had to be independently sparked into action to power life made respiration seem far more intricate than Warburg wanted to believe. Warburg had once invoked the story of the magic bullet to make a point about science, just as Paul Ehrlich had in describing his vision of targeted drugs. Only in Warburg's telling, the seventh magic bullet goes astray and kills Max's love because it had been hastily assembled from many different parts. The moral of the story, in Warburg's version, was not that one should never sell one's soul to the devil; it was that overly complicated thinking is the devil's work.[16]

But for Warburg, the biggest problem with Wieland's research was not that it was speculative or too complex. It was that the respiratory ferment Warburg had identified had not been properly acknowledged. Once that acknowledgment came in the late 1920s and Warburg felt properly validated, he would finally shift his gaze beyond his one favored molecule.

Without admitting that Wieland had been partially correct, Warburg now granted that his respiratory ferment—he still avoided the word "enzyme"—was only the last of a series of reactions that make cellular breathing possible. The process begins with the phenomenon Wieland had studied: enzymes ripping hydrogen atoms off of the fat and carbohydrate molecules that we eat. The hydrogen atoms are then stripped of their electrons, which are passed from one molecule to the next on a journey to meet oxygen at Warburg's respiratory ferment. "Wieland and Warburg had been examining opposite ends of a great elephant," the Cambridge biochemist Guy Brown wrote. "Wieland had the trunk where the electrons went in and stated firmly that this was all there was to the elephant; while Warburg had the tail where the electrons came out and thought this was the essence of the elephant."[17]

In retrospect, it's clear why evolution arrived at a multistep system for cellular breathing, rather than allowing hydrogen and oxygen to react. A direct reaction between hydrogen and oxygen would release too much energy at once, as the Nazis were reminded when their prized *Hindenburg* zeppelin turned into a ball of fire and left 36 people dead. But if Wieland and others had long appreciated that specific enzymes function like movers—taking the hydrogen from one molecule and transferring it to another—the movers themselves were not well understood.[18]

In his early research on respiration, Warburg had relied on indirect evidence, learning what he could from the power of cyanide and carbon monoxide to suffocate a cell. He needed a more direct approach to studying the activation of hydrogen, particularly after spending years belittling Wieland's work. He needed, specifically, to isolate the enzymes that transfer hydrogen and to figure out what they are made of.

The task before Warburg was a bit like attempting to single out one ingredient in a smoothie made from countless unknown foods. There were no instructions to follow. Biochemists of the era would first put organic tissue through a meat grinder again and again and then pulverize the cells further until they released their "juice." That juice, in turn, would be spun in a centrifuge to separate molecules by weight and then sent through membranes with microscopic pores to separate molecules by size. The juice might be heated or cooled or shaken or subjected to dozens of different chemical treatments, depending on the enzyme involved. Obtaining enough of any given enzyme within the "juice" might mean going through a mountain of organic matter. One visitor to Warburg's institute in the 1930s recalled that potatoes were arriving by the ton.[19]

Warburg's first success was the isolation of a hydrogen-transferring molecule that, due to its color, would later be called "Warburg's yellow enzyme." When Warburg filtered the enzyme through a material that allowed small molecules to seep through, something peculiar took place: the yellow component slipped away through the filter, and what remained of the enzyme lost its ability to react with hydrogen. The enzyme, Warburg saw, contained two parts: a protein and a smaller nonprotein component—the part that had been lost through the filter—known as a "coenzyme."

Like the iron in Warburg's respiratory ferment, the yellow coenzyme was the reactive component, the secret to the molecule's power. Coenzymes had already been discovered, but no one could say precisely what they were made of. Anxious to determine what the yellow substance was, Warburg began a chemical analysis. What that analysis revealed brought his science to an entirely new place.

Earlier in the century, researchers had made considerable progress in identifying the specific nutrients that could prevent diseases such as rickets or beriberi. These nutrients came to be known as vitamins, but what vitamins did inside our cells to cure diseases remained unknown. Now Warburg, if only inadvertently, had arrived at an answer. The reactive yellow part of his yellow enzyme turned out to be riboflavin, or vitamin B_2.

Through his quests to understand how cells breathe—and how the process goes awry in cancer—Warburg had, unintentionally, discovered why vitamins are so critical for our health. Most vitamins function as coenzymes that make respiration and fermentation possible. Because our bodies can't produce these critical engine parts, we need to obtain them from food.

Warburg left the task of purifying and working out the structure of his yellow enzyme to Hugo Theorell, a Swedish researcher who had come to work at his institute and who would go on to win the Nobel Prize for his findings. Warburg had already moved on to another enzyme. He called it the "between ferment" because it appeared to function as a link between the hydrogen reaction that takes place at the trunk of the elephant and the oxygen reaction that takes place at the tail.

In 1933, Warburg successfully isolated the "between ferment" and saw that it, too, relied on a coenzyme. But here he ran into a significant obstacle. Warburg had relied on horse blood to study the "between ferment." And yet 200 liters of horse blood had yielded only a few milligrams of a mysterious molecule that appeared to be the crucial reactive component. According to Warburg's calculations, to secure enough of the substance to figure out what it was, he would have to slaughter all of the horses in Germany. Warburg, who once complained to Berlin officials that air pollution was going to give his horses pneumonia, needed a new path forward. And he needed it quickly. A group of researchers in Sweden were also making progress on the "between ferment." When Theorell told Warburg he might travel home to Stockholm for Christmas, Warburg responded, in jest, that he would kill him if he revealed any secrets from his lab to his colleagues in Sweden. Theorell had to promise Warburg that he wouldn't even mention the molecule.

Warburg, a scientist who let nothing stop him, was stumped. Then he got lucky. Warburg told his friend Walter Schoeller about his inability to solve the puzzle of the "between ferment." A laboratory director at one of Germany's major chemical companies, Schoeller regularly consulted with Warburg on his cancer research. Since Warburg already knew the melting point and molecular weight of the mystery molecule, Schoeller offered to check an industrial reference book to see if it might yield any clues. He immediately found something much better than a clue: a perfect match. The molecule Warburg sought wasn't mysterious at all. It was nicotinamide. It had been synthesized for the first time in 1873 and was widely used in photography. "Yesterday we could not buy it for any money in the world," Warburg said. "Today we can buy it for 2 marks a pound."

Nicotinamide, it became clear, is the critical component of two hydrogen-transferring coenzymes—now known as NAD and NADP—that are central to both respiration and fermentation. Today, the molecules are an increasingly popular area of interest in both cancer and aging studies. Though most contemporary NAD researchers are unaware of Warburg's contributions to the field, Theorell, who witnessed the discoveries at Warburg's institute, was awed by him. "It is not often that such things happen," he later said, "but let us agree that the few of us who have ever witnessed such an explosion of progress will never forget it."[20]

Warburg's research in the first half of the 1930s, which also relied on his remarkably innovative use of light absorption patterns to determine how coenzymes work, is considered by some to be his greatest accomplishment of all. It wasn't only the breakthroughs themselves that amazed other scientists of the era, but also, as Theorell suggested, the rapid pace at which Warburg made them. As one American researcher who briefly worked at the institute in the 1930s recalled, the operation was minuscule in comparison to that of a large university research department—it consisted only of Warburg and a handful of technicians—and yet it had somehow become "the most famous biochemical laboratory of the time."[21]

Warburg's accomplishments would have been remarkable under any circumstances. But he was not working under any circumstances. He was working under a Nazi regime that was harassing him at every opportunity, questioning even his authority to order alcohol. German science was rapidly falling apart as Jewish scientists fled, and yet the Emperor of Dahlem remained at the height of his powers.

THAT A SIMPLE MOLECULE synthesized in 1873 turned out to be so central to biology was a great surprise. The bigger surprise was still to come.

In the 1920s, the American researcher Joseph Goldberger had discovered that certain foods, such as milk, eggs, and brewer's yeast, could cure pellagra, a potentially deadly disease that leaves its victims with dry, scaly skin, fiery, inflamed tongues, and dementia, among other symptoms. "There is no more pitiful spectacle than the pellagrin," the *New York Times* wrote. "Too feeble to work, he barely shuffles along."[22]

Goldberger's triumph was in convincing the medical world that pellagra was caused by something missing from the victim's diet rather than by a microbe inside the body. (He had to inject himself with the blood of a pellagra patient before anyone would believe it was harmless.) But neither Goldberger nor any other scientist had succeeded in determining which specific molecule cured the disease.

After Warburg's discovery that nicotinamide played a critical role in respiration, the American biochemist Conrad Elvehjem decided to investigate whether it might also be the molecule that cured pellagra. Elvehjem obtained a supply of nicotinic acid, the precursor molecule that turns into nicotinamide in our bodies, from the Eastman Kodak Company and soon confirmed that it was, indeed, the answer to the pellagra mystery.

Nicotinic acid, or niacin, is also known as vitamin B_3. Warburg's study of the molecule carries a sad irony. Though he certainly recognized the scientific importance of his work, Warburg seems never to have appreciated that his research on coenzymes was the world-changing discovery he had always hoped to make. In 1938, 400,000 people were thought to be suffering from pellagra in the American

South. (Some 100,000 Americans died from the disease between 1906 and 1940.) That the little white tablets of nicotinic acid could wipe out the disease was deemed a "miracle" by the *New York Times* in 1939.[23]

Perhaps Warburg couldn't fully recognize his accomplishment because the disease was less common in Germany. More likely, pellagra simply wasn't a sufficiently famous disease to bestow on him the greatness and adulation he sought. Only a cure for cancer would bring the glory Warburg longed for.

Warburg continued his enzyme research after his nicotinamide triumph. During a remarkable stretch in the second half of the 1930s, he isolated and purified one metabolic enzyme after another. That research laid the groundwork for decades of future breakthroughs in biochemistry. But for Warburg, understanding how cells use energy was never an end in itself. It was also a means to understanding and curing cancer. Warburg already believed that cancer arose from damaged respiration. Having found that coenzymes were necessary for a cell to breathe, he wondered if he had also arrived at a major cancer discovery. He wondered, specifically, if feeding cells the key components of coenzymes could keep respiration humming along so that the transition to fermentation would never take place.

As early as 1934, Warburg was spreading the word that he was close to a breakthrough on cancer. At a Kaiser Wilhelm Society board meeting that year, he discussed positive results from rats injected with an unspecified "ferment," noting that it still needed to be tested in humans. It's possible that Warburg, already aware that his safety in Nazi Germany depended on his value to cancer science, was purposefully overstating his findings. In July 1933, Anny Schrödinger, the wife of the famous physicist, told Lotte Warburg that she had heard (seemingly by way of Max Planck) that the Nazis wanted "to keep" Warburg because of his cancer research. But an entry in Lotte's diary at the time suggests that Warburg genuinely believed he was on the cusp of a true breakthrough. He told Lotte, in a letter, what he had told the society:

that he had successfully cured cancer in rats with his "ferment" and planned to move on to human trials. "If Otto goes so far as to say he's very hopeful then there is something to it," Lotte wrote. "Before he would just say: 'The cure will be found someday.'"

The "ferment" Warburg invoked appears to have been a coenzyme, one of the B vitamins necessary for respiration. Warburg continued to study its impact on cancer for years. A 1938 letter sent to Warburg by one of the collaborators on the experiments includes a mention of using the "ferment" to successfully cure 4 of 20 cancer-stricken animals. By that point, Warburg was trying everything.[24] In one series of experiments, he subjected cancer-stricken mice to varying levels of oxygen. In another, he tested different chemicals to see if he could prevent cancer cells from getting the glucose they need in order to grow.

In retrospect, the quickening pace of Warburg's work looks like desperation, as though he was determined to find something, anything, that would show evidence of progress. By the end of the 1930s, Warburg was isolated in Nazi Germany. The small number of Jewish scientists who hadn't yet fled the country were now running out of time. Some would be murdered. Others committed suicide. Those who had managed to hang on to their careers were, like Warburg, *Mischlinge*—half-Jews and quarter-Jews. And though the fate of the *Mischlinge* in Nazi Germany remained uncertain, it was not hard to imagine a bad ending.

CHAPTER TEN

The Age of Koch

As Warburg searched for a cure for cancer in the late 1930s, Louis Pasteur and Paul Ehrlich were not the only two scientists gazing down at him from the walls of his library. The third and final scientist whose portrait hung in Warburg's library was Robert Koch. Though Pasteur is often credited with persuading the nineteenth-century scientific establishment that tiny, invisible animals were responsible for most diseases, the idea continued to seem preposterous to much of the medical world during Pasteur's lifetime. "I am afraid that the experiments you quote, M. Pasteur, will turn against you," an 1860 editorial in the French scientific journal *La Presse* warned. "The world into which you wish to take us is really too fantastic."[1]

Infectious diseases were widely thought to be spread not by living germs but by bad, or contaminated, air. And while Pasteur showed that fermenting microorganisms could be found in diseased plants and animals, he couldn't convincingly demonstrate that the microbes were the true cause of the disease. It was entirely possible, as skeptics of germ theory pointed out, that the symptoms arose first and that the bacteria, like nomads settling in a conquered city, merely took advantage of the newly defenseless tissue. "My heart says 'yes' to bacteria," one surgeon of the era wrote, "but my reason says 'wait, wait.'"[2]

It was Robert Koch, a country doctor carrying out experiments in his home laboratory, who finally won over most skeptics of germ theory. In the 1870s, Koch, like other researchers before him, observed that the blood of a mouse that had died of anthrax always contained the same specific bacteria. They looked like little rods under Koch's microscope and could turn into small bead-like spores. Koch found that if he injected the diseased blood into a healthy mouse, the animal would soon develop anthrax and die. But such experiments were far from conclusive. There are countless different molecules in blood, and in theory, any one of them might have been responsible for transferring the disease. Koch's challenge was to find a way to determine if the bacteria were the true cause of anthrax. To do so, he realized, he would first need to grow the organisms outside of a living animal. He could then inject the newly grown bacteria into healthy mice and observe if the animals developed anthrax.

Koch grew the germs in fluid taken from the eye of an ox. To gather enough mice to test, he and his wife set traps around their barn. In the spring of 1876, after he had seen, in experiment after experiment, that the bacteria alone caused healthy mice to develop anthrax and die, he demonstrated his experiments in front of a group of prominent scientists in Breslau. It was a monumental moment in the history of science, yet it was only Koch's opening act. Six years later, a nervous Koch stood before the Berlin Physiological Society, and speaking slowly and deliberately, made a world-changing announcement: he had found the specific germ responsible for tuberculosis, the disease that killed more people in the Western world than any other.[3]

Koch had not found a cure for tuberculosis. What he had found, as with anthrax, was causality. The "bacilli which are present in the tuberculosis substances not only accompany the tuberculosis process," Koch told his colleagues, "but are the cause of it." A young Paul Ehrlich in the audience that night later described it as "the most important experience" of his "scientific life." Ehrlich would spend the remainder of his career looking for magic bullets, but it was Koch who had found the targets for Ehrlich to shoot.[4]

As news of Koch's tuberculosis discovery spread, he became world famous, perhaps the most respected German alive when Otto Warburg was a child. German merchants sold some 100,000 red handkerchiefs that depicted Koch together with Kaiser Wilhelm and Otto von Bismarck. In the summer of 1884, the *New York Times* described Koch as "the man whose name is at present in everybody's mouth," and his fame would grow greater still in the following years, after he claimed to have found a cure for tuberculosis.[5]

An 1890 article in a British journal wrote of the "Koch boom" and depicted Koch riding on a white horse, his microscope held above his head like a sword ready to strike an attacking serpent labeled "tuberculosis bacilli." The author of the article had recently traveled through a tuberculosis-stricken region of Italy where rumors of Koch's cure led to a frenzy of excitement: "[T]he news that the German scientist had discovered a cure for consumption must have sounded as the news of the advent of Jesus of Nazareth in a Judean village. The whole country was moved to meet him."[6]

Never mind that Koch's tuberculosis treatment proved a great disappointment. By then, Koch was the living embodiment of science's triumph over disease and misery. The Germans had placed their faith in science in the late nineteenth century, and Koch, more than anyone, promised the salvation they sought.

Koch won the Nobel Prize in Physiology or Medicine in 1905, when Warburg was in his early 20s and Hitler was 16. By then he had found the germ responsible for cholera as well. Outside of Germany, Koch is best remembered for a set of criteria, Koch's postulates, long used to determine whether a given microbe could be considered the true cause of a disease. His first postulate states that if a microbe is the cause of a disease, it must be present in every last case of the disease. In 1966, in perhaps Warburg's best-known statement on cancer, Warburg would use nearly identical language:

> *There are prime and secondary causes of diseases. For example, the prime cause of the plague is the plague bacillus, but secondary causes of the plague are filth, rats, and the fleas that transfer the plague bacillus from rats to man. By a prime cause of a disease I mean one that is found in every case of the disease. Cancer, above all other diseases, has countless secondary causes. But, even for cancer, there is only one prime cause. Summarized in a few words, the prime cause of cancer is the replacement of the respiration of oxygen in normal body cells by a fermentation of sugar.*[7]

Later, in the same speech, Warburg made the connection to Koch explicit: "Only today can one submit, with respect to cancer, all the experiments demanded by Pasteur and Koch as proof of the prime causes of a disease."

ADOLF HITLER WAS born in 1889, six years after Otto Warburg, and he, too, came to see himself in the reflection of the most celebrated German man of his youth. Koch, arguably, had a far greater influence on Hitler than on Warburg. In one sense, Koch shaped Hitler's life simply by proving that germs cause disease. Germ theory was not an abstraction for Hitler. He lost at least two siblings to infections and claimed in *Mein Kampf* to have suffered from a serious lung illness as a child. Like many others at the time, he suspected that germs could also cause cancer, the disease that ravaged his mother. As the years passed, Hitler became increasingly fearful of germs and washed his hands more and more often. He was said to hate being touched. He took to quarantining members of his entourage when he feared they were sick and eventually to making everyone he came in contact with prove they weren't sick or contagious.[8]

A fear of germs was common then, as it is today, among people of all political outlooks. As Pasteur once observed, "It is terrifying to think that life may be at the mercy of the multiplication of those infinitesimally small creatures." For Hitler, Koch meant something more than germaphobia. Hitler grew up in the age of Koch, when nearly every last disease could suddenly be tied to a specific hidden cause and, more important, when the work of identifying and eliminating the underlying cause of a disease was grounds for deification.[9]

Koch's language of cause and effect appears in Hitler's writing and speeches again and again. "The cure of a sickness can only be achieved if its cause is known, and the same is true of curing political evils," Hitler wrote in *Mein Kampf*. "To be sure, the outward form of a sickness, its symptom which strikes the eye, is easier to see and discover than the inner cause. And this is the reason why so many people never go beyond the recognition of external effects and even confuse them with the cause, attempting, indeed, to deny the existence of the latter."[10]

Hitler borrowed from Koch's bacteriological jargon from the start of his political career, with obvious targets in mind. In August 1920, at a meeting of the Nazi Party in Salzburg, Hitler described Jewish influence on German life as "racial tuberculosis." "The impact of Jewry will never pass away, and the poisoning of the people will not end," Hitler said, "as long as the causal agent, the Jew, is not removed from our midst."[11]

Whether Hitler was aware that he was invoking Koch as he wrote those words is impossible to know. But Koch was in Hitler's thoughts in September 1939, at the start of the Second World War. That same month, the Nazi filmmaker Hans Steinhoff released a biographical film portraying Koch as a visionary genius with the strength and determination to ignore all those who doubted that bacilli were the true cause of disease. Hitler was delighted with the film and sent Steinhoff a congratulatory telegram. "To give you one-and-a-half hours of pleasure," Steinhoff wrote back, "is a stimulus for me and a deeply felt wish."[12]

In 1941, amid a middle-of-the-night rant, Hitler, hardly a self-aware man, had a flash of insight: "I feel like the Robert Koch of politics," he said. "He found the bacillus of tuberculosis and through that showed medical scholarship new ways. I discovered the Jews as the bacillus and ferment of all social decomposition."[13]

BY 1941, Otto Warburg had been stripped of his title of "university professor."[14] Though he had managed to hang on to his institute, he was now isolated from the outside world and in considerable danger. There were no more visits from Rockefeller Foundation officials, no new foreign students visiting his institute. Warburg had made his deal with the devil, and his payment was coming due. With the start of the war, whatever propaganda value a living Warburg had once held for the Nazis was gone. "Anyone who had official obligations to support the ministry washed their hands of me," Warburg later said.[15]

Worst of all for Warburg, tolerance of half-Jews, the so-called first-degree *Mischlinge*, was running low. According to one census, there were still some 60,000 half-Jews in Germany. Adolf Eichmann, the Nazi official in charge of deporting Jews, wanted the Nuremberg Laws revised so that these remaining first-degree *Mischlinge* could be considered "full Jews" and deported along with the rest.[16]

While some government officials continued to push for the protection of quarter-Jews (second-degree *Mischlinge*), half-Jews such as Warburg, it was widely agreed, could never be assimilated into German society. The only remaining question was whether first-degree *Mischlinge* should be included in the deportations, as Eichmann hoped, or merely forcibly sterilized.

Given the debate about half-Jews taking place in 1941, Warburg's continued presence at the Kaiser Wilhelm Society came to seem all the more outrageous to some Nazi officials. A first-degree *Mischling* was not only still working freely in Germany but running an institution where he commanded a team of Aryans with military-like discipline. The situation was untenable. In the spring of 1941, Rudolf Mentzel, who headed the science division at the Reich Education Ministry, together with another influential Nazi medical official, decided to remove Warburg from his institute so that it could be used as a home for a newly established central cancer agency. Warburg, of all people, was to be sacrificed in the name of cancer science.

On April 5, 1941, the two Nazi leaders met with the secretary general of the Kaiser Wilhelm Society, Ernst Telschow, to explain their intentions for Warburg's institute. Two weeks later, the plan was formally approved. Warburg was dismissed on the grounds that he was "of Jewish offspring." He was to resign and leave the building by June 30, 1941. Warburg had survived in his position longer than anyone in Nazi Germany with two Jewish grandparents could have expected. His time had run out.[17]

The only account of Warburg during this period comes from Antonietta Dohrn, the granddaughter of the founder of the Naples Zoological Station, who recalled that Warburg, while deeply depressed, never relinquished his righteous fury. Dohrn overheard a phone conversation in which Warburg referred to Mentzel as "swine." Asked if he was concerned that his phone was tapped, Warburg said that he hoped it was and that his tormentors wouldn't know what to do in the face of civil courage.[18]

Though Warburg likely spent most of the spring of 1941 in a state of uninterrupted rage, he was still capable of acting strategically. He had connections at the highest levels of German society, and he did not hesitate to call on them. One of those connections, the famed surgeon Ferdinand Sauerbruch, knew Hitler personally. Walter Schoeller—the chemist who determined that the molecule Warburg had found in horse blood was nicotinamide—proved to be Warburg's most important friend. Schoeller was a prominent chemist at a leading chemical company and served on the board of Warburg's institute, but he was not, himself, especially influential in Nazi Germany. Schoeller's wife, Paula, by contrast, was the sister-in-law of Philipp Bouhler, the head of the Chancellery of the Führer.

Bouhler, once deemed the most "shadowy" of all the Nazi leaders, was soft-spoken and had a scholarly disposition. Before joining the Nazis early on, he had studied philosophy and literature at Munich University, where he had also written plays and poems. In late 1939, together with his trusted deputy, Viktor Brack, Bouhler had instituted the Nazis' first systematic killing program, which became known as Aktion T4 after the war. The goal was to eliminate the severely disabled, the so-called useless eaters who were consuming nutrients that might otherwise go to the healthy. It was Bouhler who had the idea to disguise the rooms in which the victims would be gassed as showers.[19]

When Walter and Paula Schoeller reached out to Bouhler on Warburg's behalf, they did not ask him to protect someone of Jewish heritage. They asked, instead, for something stranger, a notion that could only exist in the phantasmagoria that was Nazi Germany: they asked to have Otto Warburg legally cleansed of his Jewish blood by way of a German Blood Certificate, which would have declared him legally equal to a German-blooded citizen. The process was informally known as Aryanization. The term is more commonly used in reference to Jewish businesses that were stolen by the Nazis, but when it suited their purposes, the Nazis could steal a soul as well.[20]

"Warburg's willingness to let his Jewish blood be diluted in this way, and thus to make a pact with the Nazis, incensed colleagues outside Germany," Hans Krebs wrote years later. But by the time Warburg applied for a German Blood Certificate, his life was in jeopardy. Warburg had already lied to his Nazi interrogators, however unconvincingly, about his Jewish heritage. While Warburg's colleagues outside of Germany saw it as a refutation of his Jewishness, for Warburg, who didn't think of himself as Jewish, there was little to refute. Krebs may have also failed to appreciate that Warburg's effort to change his legal standing in Nazi Germany was not at all unusual. Based on 430 interviews with surviving *Mischlinge* and their relatives, the author Bryan Mark Rigg concluded that most *Mischlinge* were anxious to prove their "Aryanhood" at a moment when Germans considered them cowards and monstrosities. "*Mischlinge* internalized Nazi standards even as they tried to fight them," Rigg writes.[21]

The exact number of half- and quarter-Jews who applied to have their legal status changed in Nazi Germany is unknown. Nazi records are incomplete and inconsistent. Petitions for authorization to continue to serve in the armed forces were more likely to succeed than petitions for German Blood Certificates, which were given out only infrequently. According to records of the Reich Ministry of the Interior, by May 1941, nearly 10,000 *Mischlinge* had applied for an upgraded legal status of one type or another, and 263 had been successful. (Scholars now believe the true numbers are considerably higher.) Hitler, growing frustrated with the number of special cases he was asked to consider, once complained that Nazi Party members "seem to know more respectable Jews than the total number of Jews in Germany."[22]

In making the case for Warburg to be given the same legal status as the German-blooded, his supporters noted the pure Aryan heritage of his mother and his military service in World War I. But they also had a more compelling argument: Warburg was Germany's best hope of curing cancer. To a nation desperate to defeat the disease, removing Warburg from his institute was tantamount to removing a great general in the middle of a war.[23]

Bouhler agreed that Warburg was worthy of Aryanization. But even with the support of the Chancellery of the Führer, Warburg's fate remained uncertain. The Aryanization application at the time required photos (front and profile), military records, a written family history, and a statement of the applicant's political views. The review process was known to take months or longer, and Warburg was slated to be removed from his institute in a matter of weeks.[24]

On June 14, Warburg's supporters sent out a desperate plea. The letter appears to have been intended for Hermann Göring, whose power and influence was then second only to Hitler's. (Göring once claimed that it was up to him to decide "who is a Jew and who is Aryan.") The letter indicated that Warburg already had the support of the Chancellery of the Führer and implored Göring to lend his support to Warburg's cause: "To remove Warburg and put someone else in charge of his institute" would "amount to the greatest spiritual robbery in the history of science." Hans Krebs wrote that Göring intervened on Warburg's behalf, but there are no records of Göring's involvement.[25]

Meanwhile, Bouhler put his deputy, Brack, in charge of the case. Like Bouhler, Brack was a quietly ambitious SS bureaucrat. He was known as polite, even meek, at once scheming for power and dreading his rivals within the party. Underneath his drab exterior—he resembled the pitchfork-holding farmer in Grant Wood's *American Gothic*— was a vivid imagination for the macabre. Brack once sent a memo to Himmler suggesting how millions of Jews might be sterilized unknowingly. According to the plan, Jews would be forced to fill out forms at counters in a government office. As the Jews toiled over the meaningless documents, X-ray machines hidden beneath the counters would irradiate their genitals.[26]

One week after the letter to Göring went out and nine days before Warburg was to be removed from his institute, Viktor Brack summoned him to the New Reich Chancellery, the newly built neoclassical home of Hitler's government. The marble floors of the New Reich Chancellery's long galleries were so slippery the guards would sometimes stand by with a stretcher when elderly diplomats visited. As Warburg made his way through the building, his carefully polished Scottish wingtips would have clacked ominously with each step.

Brack would have been wearing his black SS uniform that day, his senior rank reflected by the oak-leaf patch on his collar. Warburg was joined by Ernst Telschow, the chairman of the Kaiser Wilhelm Society. (Warburg, believing Telschow had played a role in his dismissal, refused to share a car with him.) As Warburg sat in the very belly of the Nazi beast, Brack delivered the verdict. He did not grant Warburg a Certificate of German Blood—the application remained under review—but Warburg would be allowed to keep his institute on one condition: he would have to continue to work on cancer. "I did this," Brack told Warburg, "not for you, not even for Germany, but for the world."[27]

Warburg later said that Brack had "probably saved" his life that day, but Bouhler and Brack did not have the power to make such a decision unilaterally. Himmler's appointment book shows that he met with Brack specifically to discuss Warburg that same day. And though Göring had boasted that he decided who was a Jew, the Nuremberg Laws explicitly state that this privilege was, in fact, officially reserved for Hitler.

Though he had credited Brack for saving him, Warburg never doubted that it was Hitler's fear of cancer that made it possible for him to survive, and there is reason to believe that Warburg was right. Hitler took the Aryanization process extremely seriously, busying himself with the minutiae of applications for German Blood Certificates even during the most critical moments of the war. Hitler was particularly concerned with the photographs and would reject anyone who appeared stereotypically Jewish.[28]

Given that Hitler is known to have reviewed applications from *Mischlinge* hoping to be Aryanized in June 1941 and also that Warburg was a Nobel Prize winner working on a cancer cure, it is entirely possible he intervened on Warburg's behalf prior to Warburg's meeting with

Brack. Göring might well have reached out to Hitler directly. There are even grounds for speculation that Hitler, who was in the Chancellery on the day of Warburg's visit, was aware of Warburg's meeting with Brack.

That the highest-ranking Nazis were focused, even for moment, on Warburg on June 21, 1941 is almost incomprehensible. That day was arguably the most critical moment of the entire Nazi project. Operation Barbarossa, the largest military operation in history, was scheduled to begin at dawn the next morning. The eastward push into Soviet territory would be the fulfillment of Hitler's vision. In a single stroke he would secure "living space" and grain fields for generations of Germans, wipe out "Jewish Bolshevism," and demonstrate to the British that the German war machine was invincible.

In the evening of June 21, hours after Warburg left the building, a nervous Hitler paced back and forth with Joseph Goebbels, his propaganda minister, preparing the radio announcement of Operation Barbarossa to be broadcast in the morning. "This cancerous growth has to be burned out. Stalin will fall," Hitler remarked to Goebbels at one point in their conversation.[29]

That Hitler had turned to cancer as a metaphor wasn't unusual. But shortly before going to bed at 2:30 a.m.—the start of Operation Barbarossa then only 1 hour away—the conversation between Hitler and Goebbels took an unlikely turn. With the future of Germany at stake, Hitler and his propaganda minister paused their planning to discuss recent developments in cancer research.

Though Goebbels included few details of the conversation in his diary, he mentioned the name Hans Auler, a researcher who believed he had found a cancer microbe in the 1920s and who had published a book comparing cancer cells to "revolutionaries." The "fact that cancer was even broached seems puzzling," Robert Proctor notes in *The Nazi War on Cancer*. "How did Germany's political leaders find the time—hours before a major invasion—to discuss cancer and cancer research?

Was this idle chitchat, designed to ease the tension, or was there something more at stake?"[30] If Hitler had decided the fate of Otto Warburg earlier that day, it is perhaps somewhat less surprising that he was still thinking about cancer research that night.

There is another possible explanation for why Warburg might have lingered in Hitler's mind that consequential night. Though the name had changed to "Barbarossa," the secret plan to invade the Soviet Union had initially been known by another name: Operation Otto.

NOT LONG AFTER Hitler went to bed, 3 million German troops advanced into Soviet territory. "The most destructive and barbaric war in the history of mankind was beginning," writes the historian Ian Kershaw. As the Germans raced eastward, they destroyed Stalin's underprepared units wherever they encountered them. After the first days of the invasion, much of the Soviet air force had been wiped out—as many as 1,000 planes were destroyed. It looked as though the Germans might reach Moscow in a matter of weeks.[31]

Barbarossa marked not just the start of a new stage of the war, but the start of the Holocaust as well. As German forces advanced through Eastern Europe, SS death squads, together with German police and soldiers, went from town to town, shooting political enemies and Jews wherever they found them. At first they shot only men. Soon enough, Jewish women and children were being shot as well. One death squad commander complained that shooting civilians was placing an "emotional strain" on his men. To help manage the stress, Germans who shot children were provided with extra alcohol.[32]

On December 14, 1941, six months after the two had discussed the fate of Otto Warburg, Viktor Brack met again with Himmler. It was likely during this meeting, some historians have concluded, that the two men began planning the killing of all European Jews.

CHAPTER ELEVEN

"I Refused to Intervene"

ON JANUARY 20, 1942, Nazi leaders gathered at a lakeside villa in Berlin to plan the "Final Solution of the Jewish question." Half-Jews, or first-degree *Mischlinge*, they concluded, were to be treated as full Jews, while quarter-Jews would be assimilated into Aryan society. Although Warburg had managed to keep his position at his institute, he remained a first-degree *Mischling* and so was on the wrong side of the new divide.

The decisions made that January afternoon might have meant the end for Warburg, but some officials continued to push for sterilizing rather than deporting half-Jews. On March 6, senior Nazi officials gathered again, this time for a conference dedicated to the "Final Solution" of the *Mischling* question. At this second conference, overseen by Adolf Eichmann, attendees debated the finer points of mass sterilization. Would it be realistic, the pragmatists wondered, to sterilize tens of thousands of first-degree *Mischlinge*? How would they find enough hospital space? Another conference attendee noted that sterilization would not make the *Mischlinge* headache go away, at least not until all of the sterilized *Mischlinge* had died out. One Nazi in attendance later referred to the plan of waiting for first-degree *Mischlinge* to die as a "setback" he was prepared to endure.[1]

By the end of this second "Final Solution" conference, the *Mischling* question still had no answer, but Warburg's situation had nevertheless changed dramatically. On the same day that the conference convened, Germany's High Command of the Armed Forces designated Warburg's institute "of military importance." Whether the timing of the order was more than a coincidence is unknown.[2]

The order made Warburg's life at least somewhat more secure, but he was far from safe. He remained a first-degree *Mischling*, and his tormentors were not yet done with him. In the fall of 1942, Rudolf Mentzel, the Nazi science official who had tried and failed to oust Warburg from his institute the year before, began sending out queries to various German scientists about Warburg's still outstanding application for a German Blood Certificate.

Mentzel was almost certainly hoping to find information he could use against Warburg, but if the responses to Mentzel that have survived are any indication, the Nazi bureaucrat ended up profoundly disappointed. No respectable researcher could deny the importance of Warburg's contributions to science. One response, from the Nobel Prize–winning German chemist Adolf Windaus (who did not know Warburg personally), noted that Warburg was considered "the most important physiological chemist alive." The letter, sent in October 1942, might have forced Mentzel to give up his campaign against Warburg's Aryanization. That same month, Warburg was made a member of the Reich Committee for the Fight against Cancer, further solidifying his standing in Nazi Germany.[3]

That Warburg's application for equal status with someone of German blood was even being considered in the fall of 1942 is "amazing," according to Beate Meyer, of the Institute for the History of German Jews. In the summer of 1942, a new order had put an immediate end to such applications, and very few exceptions were made.[4]

Whether Warburg's application was ever accepted is unclear. Hans Krebs wrote that Warburg was formally reclassified as a second-degree *Mischling* (someone who had only one Jewish grandparent), but there are no known documents that confirm this. It is possible that Warburg achieved something even more unlikely than Aryanization: equal status with the upper crust of Nazi Germany not as a suddenly transformed Aryan but as a despised Jewish "mongrel."

BY THE SUMMER OF 1943, Warburg and Heiss would have grown accustomed to the piercing air-raid sirens and rumbling engines of Allied bombers overhead. They may have gone into the cellar of their home or to public air-raid shelters. If they ever dared to look into the night sky, they would have seen the blinding flashes of light followed by streaks of tracer fire from German antiaircraft guns.

Warburg, still searching for a cancer cure, made another important discovery during this period: some of the enzymes required for fermenting glucose could also be detected in the blood of rats. These enzymes had no function in the blood, so Warburg concluded that they must have leaked out of other tissues. Warburg was particularly interested in one enzyme, zymohexase (now known as aldolase). He found that he could detect it at much higher levels in the blood of a rat with cancer than in the blood of a healthy rat.[5]

Warburg's discovery would give rise to important diagnostic tests after the war. A 1963 textbook credited Warburg and his employee, Walter Christian, with opening up an entire field with the finding. But Warburg wasn't interested in new lab tests. He wanted to defeat cancer, and he suspected that the zymohexase in the blood of rats with cancer wasn't leaking from the tumor itself but from muscle tissue. The cancer, he reasoned, was somehow recruiting enzymes out of healthy muscle tissue to participate in its campaign of destruction.

Following a model of drug development pioneered by Paul Ehrlich, Warburg set a new plan of attack: He would take zymohexase from

humans and inject it into healthy rabbits. The rabbits would then produce an anti-zymohexase serum that would block the enzymes and
"check the cancer in its fermentation processes."[6]

It was an outlandish idea, and Warburg had little time to pursue it.
One morning in the summer of 1943, he arrived at his institute to find
that it no longer had windows. The staff spent weeks picking up shards
of glass and scattered bits of cement. When another bomb fell and did
still more damage to the building, Warburg chose to leave Berlin. That,
at least, was the explanation he provided for his move. It's also possible that Warburg, even after outmaneuvering his most dedicated Nazi
foes, could no longer withstand the poisonous atmosphere in Dahlem.
After the war, when a Russian scientist arrived at his institute to pillage
the remaining chemicals, the door was answered by an unidentified
man. Warburg had fled Dahlem, the man told the Russians, because
he was a Jew.[7]

Whatever his reasoning, Warburg was in need of a new location for
his institute, and thanks once again to his influential connections, he
found one in a scenic region of lakes and woods some 30 miles north
of Berlin. As Jews across Europe were being packed onto trains and
transported to death camps, Warburg, a homosexual with two Jewish grandparents, relocated to an elegant mansion on the grounds of a
sprawling country estate.

Known simply as Liebenberg, the estate had once been owned by
the closest friend of Kaiser Wilhelm, and the kaiser himself had spent
many of his happiest days there. (In 1898, Theodor Herzl, the founder
of modern Zionism, traveled to Liebenberg in an effort to persuade the
kaiser to support a Jewish homeland in Palestine.) Hermann Göring
regularly visited Liebenberg to shoot deer and wild boar. The primary
residence, an enormous castle, was occupied. Warburg's institute was
set up in the estate's lake house mansion.

Though nearly all German resources were reserved for the war
effort, the Nazis nevertheless had the mansion renovated for Warburg,

complete with a new roof and a transformer station. In addition to his scientific equipment and lab animals, Warburg brought along his books and antique furniture. (When additional trucks were needed, Warburg instructed one of his employees to bribe the rationing office with a bottle of distilled alcohol.)[8]

Liebenberg was an absurdly fortunate landing spot for Warburg under the circumstances. Though it's unlikely that Warburg was aware of it, a cousin, Dr. Betty Warburg, had already been murdered at the Sobibor extermination camp along with her mother, Gerta. But the Emperor of Dahlem, now in exile, did not feel lucky. Warburg, anticipating Germany's imminent collapse and surrender, retreated to his vacation home on the island of Rügen with Heiss until work at the lake house was completed. With Warburg away, his technicians were left to carry out their experiments with little guidance. Warburg planned to oversee the work via daily telephone calls, but by the last years of the war, it was difficult to secure an open line. The problem was resolved only after Heiss delivered chocolates to a local telephone operator.[9]

Warburg was accustomed to summer breaks at Rügen, but with the exception of the First World War, he had never been away from a laboratory bench for so long as an adult. He used the time to work on a book on heavy metals and respiration. When Hans Krebs later read the first draft, he was horrified. Warburg had filled a book ostensibly about the role of iron and other metals in cellular function with nasty asides about celebrated scientists.

Warburg hardly needed a reason for petty attacks, but his situation surely embittered him even more. In December 1943, he was denounced in an anonymous letter to the Gestapo for sabotaging the war effort—a crime punishable by death. Among other offenses, the letter accused Warburg of being anti-Germany and pro-England, of refusing to work, and of using the institute's gasoline—a scarce resource then restricted for war-related efforts—for trips to his vacation home with Heiss. Warburg's behavior, the letter said, was not merely detest-

able but "asocial"—a loosely-defined charge the Nazis used to perse-
cute anyone who didn't conform to public norms.

The accusation of "asocial" conduct put Warburg's life in grave danger
yet again. By 1944, some half-Jews were being assigned to labor camps.
Others, accused of behaving like full Jews, had been condemned to death
for the crime of "undermining the war effort." Warburg's enemies at the
Education Ministry were still in power, and by this point in the war, the
Nazis needed little excuse to murder anyone, let alone a first-degree *Mischling*. Upon learning that he had been denounced and was under investigation, Warburg's dark mood turned darker still. Already predisposed to
great paranoia, Warburg now had entirely legitimate grounds for suspect-
ing everyone around him of being his enemy. Though every member of
Warburg's staff likely had more than enough evidence to have Warburg
arrested, Warburg was convinced that Fritz Kubowitz, the man who had
worked with him on many of his most important experiments, including
some that led to his Nobel Prize, had written the letter.[10]

Warburg was almost certainly correct in assuming the denunciation came from one of his employees. The postmark reveals that it had
been mailed near Liebenberg, and the specific examples of his crimes,
such as using gasoline, would only have been known to close associates.

The letter denouncing Warburg was soon followed by a second letter, sent to the Kaiser Wilhelm Society, that included the allegation
that Warburg was a homosexual. Warburg, isolated in his palace in the
woods, was now overcome with Lear-like rage and indignation. Most
of his employees had been with him since the start of his career. He had
taught them chemistry. Their names appeared side by side with his own
on his scientific papers. When Warburg said he couldn't leave Germany
in the 1930s, he had cited not his magnificent building but the impos-
sibility of replacing these men.

Theodor Bücher, a 30-year-old German researcher training in biochemistry at Warburg's institute, was on good terms with Warburg
and attempted, with little success, to calm him down. He "felt him-

self persecuted," Bücher wrote, "even persecuted to death." In Bücher's assessment, Warburg had become a deeply unhappy man, torn apart by the clash between "feelings of inferiority" and a "raw egocentrism." Bücher saw Warburg try to fight back against his demons and admired the effort it took. But by 1944, the demons could not always be tamed.[11]

The investigation that followed Warburg's denunciation was not the only open investigation of Warburg at the time. Seven hundred miles away in Sweden, Warburg was being considered for another Nobel Prize, for his work on coenzymes. But in 1944, international fame and prestige was of no use to the Nazis, and in any case, Hitler had forbidden Germans from accepting the award.

With his life again in jeopardy, Warburg again reached out to Walter Schoeller, his colleague who was married to Philipp Bouhler's sister-in-law. By then, the limits of what Schoeller could do for a persecuted Jewish scientist were all too clear. In September 1942, the Nazis had arrested Wilhelm Traube, one of Germany's most distinguished organic chemists, and planned to deport him. Schoeller was among the scientists who made pleas on Traube's behalf, but he was too late. The Gestapo officers who arrested the 76-year-old Traube had beaten him so badly that he died in prison.

Warburg, as usual, would have far more luck than most. Philipp Bouhler, the man at the helm of the Chancellery of the Führer, visited Liebenberg himself to confirm that Warburg was working on his cure for cancer. According to one account, Warburg managed to fool Bouhler by having new experiments hastily set up.

Though the details of the Liebenberg inspection remain murky, Bücher wrote a detailed account of the aftermath. As soon as Bouhler left the building, a still-rattled Warburg invited Bücher for a walk. They strolled across the estate's endless lawns in silence for several minutes, and then the 60-year-old Warburg turned to the young man by his side and asked him a question he was not prepared for: "Do you consider me to be asocial?"

Bücher would have understood "asocial" in the Nazi context. Warburg was asking Bücher if he thought of him as a misfit. A startled
Bücher could only muster that Warburg might show more in the way
of noblesse oblige. With that, Warburg's momentary show of vulnerability was over, and he changed the subject.[12]

WHATEVER EXPERIMENTS he might have staged during the inspection of the institute in Liebenberg, Warburg appears to have gone back
to working in earnest afterward. According to Bücher, every member
of the lab, including Warburg, continued to carry out experiments in
search of cancer therapies even as the Allied bombing squadrons roared
overhead. For stretches of 1944, Warburg not only worked but maintained a semblance of ordinary life at Liebenberg, sometimes visiting
the palatial primary residence, where the regularly scheduled "music
evenings" continued even as Germany crumbled.

Once a week, Warburg reportedly had a roast goose prepared for a
group of Dutch prisoners of war who had been stationed at Liebenberg.
(Heiss likely did the roasting.) Toward his own staff, by contrast, Warburg felt only animosity. He was no longer speaking to Kubowitz at
all. The tension, Bücher wrote, would soon arrive at "a horrifying end."

By late 1944, millions of Soviet soldiers had gathered on the banks
of Poland's Vistula River. Hitler, meanwhile, had organized a national
suicide program in the form of the *Volkssturm,* a poorly equipped people's army comprised mostly of teenage boys and old men who were
expected to stand as a last defense against Germany's enemies. All German men ages 16 to 60 who weren't currently in the military were
called upon to enlist.

Because Warburg's lab had been designated a war institute, he had
the power to save his employees by informing authorities that their
work was necessary for his research. And Warburg did manage to protect Bücher and Heiss. But other members of the lab were less fortu-

nate. Their fate was in the hands of their boss, a man who had solid evidence that they had betrayed him.

Fritz Kubowitz, the employee Warburg believed had denounced him, ended up in the *Volkssturm I*, an assignment that was, for many, a death sentence. After being sent to the front, Kubowitz reached out to Warburg, then at his vacation home on Rügen, for help. Warburg appears to have been genuinely distraught. He began to scribble short diary entries in the back of one of his laboratory notebooks, a highly uncharacteristic act. In an entry dated February 24, 1945, Warburg justified his harsh decision on Kubowitz to himself: "You cannot denounce your superior today and ask him to save your life tomorrow."

Erwin Negelein, the man listed as a coauthor on Warburg's groundbreaking cancer papers in the 1920s, also ended up in *Volkssturm I*. On March 17, 1945, his emaciated wife arrived at Liebenberg and pleaded with Warburg to send a replacement for her husband at the front. Warburg, convinced that Negelein had played a role in denouncing him, turned her away. "I refused to intervene," Warburg wrote. "Why should I try and send somebody else instead, in order to protect her husband?"[13]

Three days later, Warburg received a government order to relocate his equipment and sensitive documents before the enemy arrived. With air-raid sirens blazing at all hours and the Red Army only about 100 miles away, he retreated with Heiss to Rügen.

On April 16, 1945, the Russians crossed the Oder River and pushed west to Berlin. Within two weeks, the Red Army controlled Liebenberg and had emptied the lake house of Warburg's equipment and treasured antique furniture. When Stalin's soldiers made it to Rügen, they found Warburg at home. According to a secondhand report, Warburg spoke first, announcing himself as the "famous Nobel Prize winner, Professor Otto Warburg from Berlin." The soldiers put down their guns.

In 1933, when the Nazis came to power, there were more than 100 scientists at the prestigious Kaiser Wilhelm institutes who qualified as Jewish, according to Nazi definitions at that time. Otto Warburg was the only one to maintain his position until the very end.[14]

WHILE SOME NAZI initiatives against cancer, such as mass screenings for cervical and uterine cancers, had to be abandoned in the last years of the war, others, including experiments involving food dyes, hormones, tobacco, and asbestos, continued without interruption.

In a memo to senior Nazi medical officials sent early in March 1945—when the Allies had Berlin surrounded and were preparing to end the Nazi regime—Himmler proposed a new cancer project: he wanted them to look into the mystery of why concentration camps had "no people with cancer." Himmler even asked the medical officials to calculate how many cancer cases could have been expected among the 700,000 prisoners, given the rates in the general population. Within three months of sending off this memo, Himmler had been captured. He poisoned himself and died while being examined by a British doctor.[15]

Cancer would remain on Hitler's mind to the end as well. On November 5, 1941, with the prospect of an easy victory against the Red Army looking increasingly unlikely, Hitler sat down for lunch at his headquarters in eastern Prussia with several guests, including his dentist, Hugo Blaschke, a man who disliked Hitler not because he was a mass murderer but because his teeth were so awful.

Any sane military commander would have been focused exclusively on the war at that moment. Hitler had something else to discuss. Over the course of the meal, Hitler declared that Caesar's soldiers had followed vegetarian diets and that the Vikings would never have managed their legendary expeditions if they had been able to preserve meat and thus eat it on their journeys. He also claimed that humans had likely lived longer in the past, perhaps from 140 to 180 years of age, and that the decline began when sterilized food "replaced the raw elements in [their] diet."

Blaschke was a vegetarian himself and believed that the shape of human teeth revealed that our ancestors were herbivores. But even he was left "speechless." And Hitler was only warming up. People live longer in countries like Bulgaria, where polenta and yogurt are popular dishes, Hitler told his perplexed guests. (Yogurt, according to a popular theory, could cleanse the colon of cancer-causing bacteria.)[16]

Hitler then launched into a diatribe in which he managed to mangle Erwin Liek's thinking on diet and cancer with his own unique mélange of nonsense and fantasy:

> *The fact that man subjects his foodstuffs to a physicochemical process explains the so-called "maladies of civilization." If the average term of life is at present increasing, that's because people are again finding room for a naturistic diet. It's a revolution. . . . It's not impossible that one of the causes of cancer lies in the harmfulness of cooked foods. We give our body a form of nourishment that in one way or another is debased. At present the origin of cancer is unknown, but it's possible that the causes that provoke it find a terrain that suits them in incorrectly nourished organisms. . . . Nature, in creating a being, gives it all it needs to live. If it cannot live, that's either because it's attacked from without or because its inner resistance has weakened. In the case of man, it's usually the second eventuality that has made him vulnerable. A toad is a degenerate frog. Who knows what he feeds on? Certainly on things that don't agree with him.*[17]

At the time, Hitler was following the raw-foods diet developed by the Swiss physician Max Bircher-Benner, the muesli inventor who believed that raw foods offered more direct access to the sun's energy. (Himmler was also an admirer of Bircher-Benner.) Hitler didn't force his guests to follow his diet of muesli, gruel, and linseed mush. But even as the Nazis were torturing and gassing their victims by the hundreds of thousands in concentration camps, he would sometimes lecture the meat eaters he dined with about the horrors of animal slaughterhouses.[18]

The muesli didn't make Hitler any healthier. In addition to his usual stomach problems and terrible headaches, his vision was failing. He feared he was going blind. Despite Blaschke's best efforts, more and more teeth were falling from the führer's rotting gums, forcing him to eat only soft foods. Electrocardiograms indicated progressive heart disease.

At the end of 1944, doctors found another growth in Hitler's throat and removed it. There was no evidence of cancer. It would have mattered little if there had been. With no hope left for victory, Hitler retreated to his two-story concrete bunker beneath the Reich Chancellery. Several years earlier, he had repeated his claim to Robert Koch's mantle:

> The discovery of the Jewish virus is one of the greatest revolutions that have taken place in the world. The battle in which we are engaged today is of the same sort as the battle waged, during the last century, by Pasteur and Koch. How many diseases have their origin in the Jewish virus![19]

Now, trapped in his bunker, Hitler sometimes dug at imagined bacteria on his flesh with a pair of golden tweezers.

Hitler had become a horrible specter, lurking in his German underworld. His ghastly appearance—he was severely hunched with pale yellowish gray skin and lifeless eyes—startled visitors. His left hand trembled so badly he had to restrain it with his right. "Often he would just sit there with painfully distorted features," Hitler's valet Heinz Linge remembered.[20]

Christa Schroeder, one of Hitler's secretaries, remained with him in the bunker until almost the end. Schroeder had enjoyed Hitler's diatribes during better times. She recalled how he would fall "into poetic rapture" when talking about how his food had grown: the farmer's arm sweeping through the air as he drops the seeds, the seeds sprouting "into a green sea of waving stems." That image alone, Hitler once told Schroeder, "should tempt man back to nature and her produce."

Hitler was now often too feeble to engage in his infamous histrionics and declamations. "The things he talked about became gradually more flat and uninteresting," Schroeder remembered. He had stopped ranting to his secretary about "racial problems" or even "political questions." In the last months of his life, Hitler had the strength left for only three topics: dog training, the stupidity of the world, and diet.[21]

On April 29, 1945, with the Red Army down the street from his bunker, Hitler married his longtime girlfriend, Eva Braun. The following day, he ate a vegetarian lunch of spaghetti with a raisin-cabbage salad and then retreated to his study with his new bride. At 3:30 p.m., a loud bang echoed through the underground structure. One of Goebbels's children was playing nearby. When he heard the gunshot, he let out a shout: "That was a bullseye!"[22]

Hitler had left behind an ocean of blood, but inside his death chambers, the corpses left few stains. Zyklon B, the poisonous gas initially developed as an insecticide, kills by suffocation. As the molecules attach to the iron of Warburg's respiratory ferment in place of oxygen, the fire of life flickers and goes out.

PART III

The Seed and the Soil
(Postwar)

Faustus:
And what wonders I have done all
Germany can witness, yea, all the world; for which
Faustus hath lost both Germany and the world . . .

—CHRISTOPHER MARLOWE,
DOCTOR FAUSTUS

CHAPTER TWELVE

Coming to America

OTTO WARBURG WAS no longer legally defined as a half-Jew after the war, but he could not be made whole again. Warburg had always been divided in a psychological sense, a nineteenth-century aristocrat at the frontiers of twentieth-century science, a man of extraordinary self-belief who, at any given moment, might unravel in a fit of doubt-fueled fury. Warburg had now been divided by history as well. The Germany of his youth, the Germany of scientific dominance and Jewish Nobel laureates that had existed before 1933, was gone forever.

Warburg was even split along geographical lines after the Allies carved Germany into four occupation zones. His summer house on the island of Rügen was in the Russian sector. So, too, was Liebenberg, the estate to which his lab had moved near the end of the war. His primary home and former institute, meanwhile, were in Dahlem, and so within the American sector.

Because Warburg was at Rügen at the war's end, he spent the summer of 1945 under Soviet rule. In the immediate aftermath of the war, the plundering of German science, including German scientists, was among Stalin's highest priorities. That made Warburg a valuable asset, and various Soviet military and intelligence officials worked to recruit him. While some high-ranking officers tried to entice Warburg with flattery and kindness, others, according to an American military report, "were of the less agreeable character."[1]

In June 1945, dozens of Soviet soldiers arrived at Warburg's lab at Liebenberg. They worked quickly, packing up the equipment and tossing Warburg's books down the stairs. The disassembled centrifuges were put into straw-lined crates. Warburg's treasured manometers went into boxes. All of it, along with the mattresses and virtually everything else that could be removed, was carried into waiting vans.

If the Russians intended to hold Warburg captive, they did not do a particularly good job of it. In September 1945, Warburg made his way back to Dahlem. Eric Warburg, who was working with an American intelligence unit to prevent the Soviet Union from recruiting or kidnapping German scientists, set out in search of his cousin. He learned that Warburg was hiding from the Soviets in the home of another professor in Dahlem. When Eric arrived at the house, he waited for his cousin by the front door. When he looked up, he saw Warburg approaching, ghost-like, through a long dark hallway. Warburg had virtually disappeared to the outside world at the start of the war in 1939. As he made his way forward through that dark hallway, he was also remerging into a new life.

"He was about the most unsentimental person I have ever known," Eric Warburg wrote, "but there was a fraction of a tear in one of his blue eyes as he said to me in his Berlin accent, 'I always knew the war would really be over only when you stood before me.'"

Warburg's sentimental moment was short-lived. He "immediately got down to brass tacks," telling his cousin he needed 40 liters of gasoline as soon as possible so that he could collect the books and scientific instruments he had left behind in the Russian sector. Eric Warburg responded that the trip was "absolutely forbidden" according to the new fuel regulations the Allies had introduced. "I couldn't care less,"

Warburg said. As usual, Warburg got what he wanted. His cousin authorized the gasoline, and Warburg went to recover his goods.[2]

That the Soviet military allowed Warburg to pass in and out of the territory it controlled might have been part of the effort to recruit War- burg, which continued after his return to Dahlem. In Warburg's telling of the story, Georgy Zhukov, the famed Red Army general who had taken Berlin, personally intervened to have Warburg's horses returned from Rügen, where they had been confiscated. Warburg also claimed, later on, that the Soviets had offered him an institute in Moscow with 100 employees. Warburg declined and thereafter liked to boast that neither Hitler nor Stalin had managed to move him. Though Warburg was not the most reliable narrator of his own life at any point, especially in his last decades, the stories about the Soviet Union's efforts to recruit him appear to be accurate. While there is no direct evidence that sup- ports Warburg's claim that Zhukov took him to dinner and asked him how he could help, in January 1946 the Soviets did give Warburg a BMW automobile as a gift.[3]

Warburg refused all invitations to visit Moscow—he feared, with good reason, that he might never return—but the Red Army did get something in exchange for its generosity. In 1946, Warburg published an article in Russian in the Soviet journal *Biokhimiya*. According to Roger Adams, a chemist serving as an adviser to the American military who prepared a five-page report on Warburg in early 1946, Warburg also kept the Russians at bay by offering them new cancer drugs, along with specific instructions on how to use them. Adams's report does not specify the nature of the drugs. Warburg probably gave the Red Army B vitamins, which he believed could help maintain respiration and pre- vent cells from shifting to fermentation, thus preventing cancer.[4]

The Americans, who had turned Warburg's institute into the offices of their local military government, were interested in Warburg as well, or at least they were interested in keeping Warburg out of Stalin's reach. Adams hoped to fend off Soviet advances by giving Warburg a position as a scientific consultant to the US Military Government's Field Information Agency, Technical (FIAT), which had an office in Berlin. The opportunity required Warburg to sit for an interview with a public safety official, who also produced a report on Warburg:

> *Subject further stated that the German people are too stupid politically to achieve self-government on a democratic basis. It will take a very long time to change this situation, at least twenty years, and in the meantime Germany must have a government imposed by the Allies. Subject stated that he had never voted or belonged to any party because he considered all such activity a stupidity. Subject further stated that militarism is most natural for the Germans and admitted that he considered his own military service as a very pleasant time and a life of the soldier worthy and honourable, "provided the army in question was used in the interest of humanity and civilization."* [5]

The public safety official concluded that Warburg was "completely unsuited for the position" on the grounds that he lacked "known liberal, statesmanlike qualities and views etc."

Adams ended his own report on Warburg on a more sympathetic note: "Throughout last fall, despite his considerable fear and dread, Dr. Warburg retained always his dignity and wry sense of humor." Adams added that Warburg was "a very tired man primarily concerned with seeking some possibility of peace" and also that he was "in grave danger of being kidnapped" by the Soviets.

In a letter to Lotte sent in January 1946, Warburg said that he had received offers from abroad, but was still hesitant to leave Germany: "[A]s you know from 1933, I am no fan of immigration." He also said that he was considering the life of a wandering scientist who moved from one lab to the next. Because he would contribute wherever he went, Warburg explained, he would never wear out his welcome in any one location.[6]

While making this rather transparent attempt to save face, Warburg was, in fact, anxious to find a new home and a new place to work. It is not clear if he understood how the global scientific community felt about him. According to one account, Britain's scientists were "so fed up with Warburg's conduct during the war and especially with the indecent aspersions he has cast on English scientists in his field that" they "would not knowingly permit him to land on English soil."

Warburg reached out to the Rockefeller Foundation, but the officials who had once championed his science now wanted nothing to do with him. In retrospect, the foundation's decision to support German science in the 1930s appeared foolish, if not traitorous. Warburg's institute now stood as a testament to the foundation's shortsightedness. After surveying his colleagues, one Rockefeller official summarized the general sentiment within the organization about Warburg after the war: "All give 'several reasons'—not identical, but similar—for being cool to the suggestion of bringing Warburg to the U. S." The list of reasons included "prima donna-ness," "Europe needs its scientist," "career about finished," and "collaboration."

The nature of the suspected "collaboration" was not specified. It was likely a reference to Warburg's Aryanization, but there was also a rumor circulating that Warburg had volunteered his services to the German military at the start of World War II. If Warburg had done so, it might reasonably be understood as an attempt to secure his own safety in a dangerous situation. The worst of the Nazi atrocities had not yet begun, and many *Mischlinge* were volunteering at the time in an effort to solidify their place in German society.[7]

Even as his own actions during the National Socialist period were being questioned, Warburg was making accusations against others. He told a Rockefeller Foundation official that Ernst Telschow, the Kaiser

Wilhelm Society official he blamed for nearly causing him to lose his institute in 1941, was "the worst Nazi." Warburg also told the American Military Government that Wilhelm Eitel, an influential Kaiser Wilhelm Society member, had been a "staunch National Socialist" who conspired against professors who spoke out against the Nazis. (Though Warburg's assessment of Eitel was entirely accurate, Eitel was nevertheless brought to the United States, where he continued his successful career.)

In August 1946, Warburg and three other Kaiser Wilhelm Society scientists gathered to review the case of Otmar von Verschuer, the biologist who led the Kaiser Wilhelm Institute of Anthropology, Human Heredity, and Eugenics during the war. Verschuer was under investigation by the Americans, and his guilt wasn't difficult to discern. He had mentored Joseph Mengele, the notorious Nazi doctor of Auschwitz. (During the war, Mengele had sent Verschuer the eyeballs from his corpses for further study.) Three of the four committee members, including Warburg, concluded that Verschuer was a "racial fanatic" and linked to the crimes of Auschwitz.[8]

That Warburg could correctly assess the guilt of others did not mean that he had suddenly gained perspective. When the biophysicist Max Delbrück visited Warburg in Berlin in 1947, he found him still indignant that the Soviets had confiscated his equipment and papers. It was a "scandal," Warburg said. He was particularly upset that they had taken his father's scientific papers. "[T]his was at the time when Berlin was in ruin and still practically smoking," Delbruck later said. "But that the Russians had taken his father's scientific papers, that was really a scandal!"[9]

Warburg did show glimpses of true humanity after the war. He made efforts to help find a new position for Carl Neuberg, the Jewish pioneer of fermentation research who had been the director of the Kaiser Wilhelm Institute for Biochemistry before the Nazis took over. In a letter, Warburg told Neuberg that he had called on German officials to return Neuberg's institute to him. Doing so, Warburg wrote, would be the first time Germany addressed "the injustice done to scientists" with "deeds rather than phrases." Warburg added that he "could think of nothing which could do more to restore the prestige of Dahlem" than Neuberg's return.

But Warburg's periodic acts of kindness were overshadowed by his obtuseness. At the same time he was singling out the Nazi criminals he hoped to see punished, he agreed to provide an affidavit on behalf of Viktor Brack, who was then being tried for war crimes at Nuremberg. The affidavit, requested by Brack's attorney and presented during his trial, recounted how Brack had intervened to save Warburg's life. Warburg's brief statement ends on a nauseating note: "Considering that Brack did this at a time when racial hatred and war psychosis had reached their climax in Germany, one has to admire the courage with which Brack advanced the cause of tolerance and the peaceful work of science against the basic principles of National Socialism."

According to the German historian and Otto Warburg authority Petra Gentz-Werner, it is highly unlikely that Warburg knew anything about Brack's horrific crimes when he put his name to the affidavit. But at the very least, he knew that Brack was a high-ranking Nazi official. The affidavit was perfectly in keeping with Warburg's narcissism. In Warburg's moral framework, good and evil could only be measured based on how one had treated him.[10]

WARBURG CLAIMED THAT he was prepared to become a wandering scientist, yet for a time he had nowhere to wander. That changed in November 1947, when Robert Emerson, a prominent American botanist, invited Warburg to spend six months in Urbana at the University of Illinois. Though Emerson had studied photosynthesis under Warburg in the 1920s, he had later committed a cardinal sin: he had concluded that Warburg was wrong about the number of photons needed to power a photosynthetic reaction.

Afterward, Warburg dismissed Emerson's work at every opportunity. But Emerson, a descendant of the transcendental philosopher
Ralph Waldo Emerson, was Warburg's opposite, an unusually kind
man who believed that the greatest contribution science could make
was to inspire scientists themselves to lead ethical lives. After the war,
he put his career on hold to help resettle the Japanese Americans who
had been interned in camps by the US government. He even brought
a Japanese American scientist who had been imprisoned to Illinois to
work as his assistant. (This man, Shimpe Nishimura, would also conclude that Warburg was dead wrong about photosynthesis.)[11]

Emerson appears to have genuinely believed that once Warburg
arrived in Urbana, the two of them could work together to resolve the
photosynthesis controversy once and for all. He did not know Warburg
nearly as well as he thought. Emerson would regret the decision even
before Warburg arrived, in the summer of 1948. Warburg insisted that
he needed to bring both a scientific assistant as well as Heiss—who
had "a clean political record as determined by American authorities,"
Warburg noted. The request itself was reasonable enough, but in the
months following his acceptance of Emerson's offer, Warburg repeatedly changed his mind about the details of his travel arrangements.
Emerson, who had to negotiate Warburg's every request with university
officials, did his best to accommodate the Nobel laureate. But by May
of 1948, with Warburg's arrival still a month away, Emerson was reaching the limits of his extraordinary generosity:

> [J]ust trying to arrange for the visit has kept me busy
> for a large part of the winter. . . . Last report I had
> was that he [Warburg] and Heiss might leave by
> June 1st. I hear they have 400 kilos of baggage and a
> poodle, on all of which they expect the Univ. of Illi-
> nois to pay transportation. It will turn out that the
> reason Warburg wants to leave Germany is because

> *the American administration has been unable to get*
> *any more of that good German dog-food, made of*
> *pure beef-steak, the only thing the poodle will eat.*
> *There will be Hell to pay when he finds that in*
> *America they feed horse-meat to dogs! And imagine*
> *the problem of finding housing for Warburg, Heiss,*
> *and a poodle!* [12]

Warburg and Heiss—it was only to be the two of them in the end—arrived by plane on June 26, 1948, with six enormous crates. The flight was 12 hours late, and Warburg was "visibly annoyed," according to a witness at the scene, that no airport officials were available to help with his cargo. Then it began to rain. "Warburg arrived in the midst of the heaviest thunderstorms I have experienced in my fifteen years in Urbana," the biophysicist Eugene Rabinowitch wrote. "[T]his proved to be an augury."

For the American scientific community, Warburg's arrival was a major happening. In September, the *New York Times* ran an article about Warburg's efforts to cure cancer by blocking the enzymes of fermentation. In November, a photo of Warburg seated at his laboratory bench at the University of Illinois appeared on the cover of the journal *Science*. To the outsider, it looked like a happy story: the famed German biochemist coming to America after the war to help resolve a great scientific debate. But despite the positive press, the visit itself was going even worse than Robert Emerson could have imagined.

Unaccustomed to working in a lab that he could not control like a Prussian military officer, Warburg complained about everything. And though fairly fluent in English, he refused to speak anything but German, leaving Emerson scurrying to find a German-speaking lab assistant. Warburg had agreed to give lectures while in Urbana but was now refusing to do so, likely because he was embarrassed about his English. He said he would only comment on the lectures of other professors. [13]

Warburg was given quarters in a faculty center, yet Heiss was assigned to a room in a University of Illinois fraternity house. Warburg visited the house one day—a moment rife with comic potential—and was predictably repulsed. He insisted that Heiss be moved. Heiss was set up with a cot in Warburg's room, but as soon as Emerson put out one fire, another would start. "I can never foresee what his next impossible demand is going to be," an exasperated Emerson wrote to a colleague. In another letter, he said that he had fallen behind in his work "because of the pressure we are under to provide for Warburg, who has a way of setting the entire laboratory on its head almost every day."[14]

Things were already going poorly, and then they got worse. As winter arrived, the heat was turned on in the laboratories. Warburg said it was far too hot and demanded that it be turned off. The other researchers were forced to carry out their experiments in their overcoats. Warburg, meanwhile, continued to come up with new explanations for why Emerson and his colleagues were arriving at results that contradicted his own measurements. One day their algae had not been grown in the proper light; the next day they had chosen the wrong chemical for their buffer solution. With each new complaint, Emerson inched closer to his limit. Though he never took the local public transportation, after one of Warburg's outbursts Emerson reportedly got on a bus and rode it around campus in an effort to calm himself.

James Franck, a Nobel Prize–winning physicist who had trained under Emil Warburg, was among the researchers involved in the photosynthesis debate. "Everyone knows the man is a genius, but he also has the misfortune of being crazy," Franck wrote to a colleague, adding that "the craziness is only clear to those who really have to deal with him."

Given that he knew Warburg's father, Franck likely grasped the deeper roots of his refusal to budge on the photon debate. Warburg had always believed that biology could achieve the mathematical elegance of physics. "In a perfect world," Warburg once said, "photosynthesis must be perfect." Based on the understanding of photosynthesis at the time, a total of 4 electrons needed to be passed to 2 atoms of oxygen to complete the cycle of reactions. And because Einstein, with the help of Warburg's father, had discovered that each photon set only 1 electron free, any number of photons higher than 4 was more than refutation of Warburg's measurements. It was a violation of the known laws of the universe.

Warburg once even turned to Einstein himself for guidance, asking him why he assumed that the energy of only 1 photon was needed to set an electron free. Einstein, it seems, did not take the question very seriously. "Well, that's a lot," Einstein said.[15]

In spite of everything he had endured, Emerson made a concerted effort to help Warburg find a permanent position at another American institution. The search did not go smoothly. When the biochemist Martin Kaman came to meet Warburg, Emerson introduced him as a "physicist." Kaman was mystified. Emerson later explained that he had no choice but to lie because Warburg didn't respect other biochemists and would have quickly dismissed him if he had known the truth.

Warburg was entirely incapable of recognizing the efforts Emerson was making on his behalf. Instead, he grew paranoid that Emerson was going to "denounce" him as a "dangerous Communist" to the US State Department and thereby prevent him from having his visa renewed. In a letter to Walter Kempner, the researcher he had fired after the Nazis came to power, Warburg wrote that Emerson was known to be "a psychopath." Warburg asked Kempner, who was then at Duke, if he might be willing to warn the State Department about Emerson so that the denouncement wouldn't be believed.

That Warburg was genuinely fearful of a denouncement was perhaps a sign of the psychological scar left by his life under the Nazis. It had only been five years since Warburg had been denounced to Nazi authorities by one of his employees. In the same letter to Kempner in

Otto Warburg at the University of Illinois, 1949. Dean Burk points at the manometer Warburg holds. Robert Emerson looks over Warburg's left shoulder.

which he shared his paranoia, Warburg himself drew a connection to his past. "As you can see," Warburg wrote, "I'm still fighting for a place to belong, even if only a little bit."

If Emerson could not find a new lab for Warburg, he remained hopeful that he and Warburg could at least resolve their photosynthesis debate. After six months, Warburg's visit ended with a bizarre competition that today would seem like a reality TV event: Warburg and Emerson each carried out their own photosynthesis experiments, with the results presented to two impartial judges. Though most other scientists who had bothered to run their own experiments believed that Emerson was correct about the number of photons, one of the two judges, Dean Burk, of the National Cancer Institute, was Warburg's former student and a tireless champion of his science.

The competition ended in a stalemate. Warburg departed Urbana in January 1949, claiming he had completely triumphed in a drama "watched by all America." He did not bother to say goodbye to a thoroughly dismayed Emerson.[16]

WARBURG'S VISA WAS renewed at the end of his stay in Urbana, and he moved on to the National Cancer Institute in Bethesda, Maryland, to work with Burk. With a loyal supporter by his side, Warburg was far more content. It's easy to see why Warburg took to Burk. Warburg had many admirers throughout his life, but no one had shown him as much love or devotion or had embraced Heiss so openly. Burk, an amateur artist, painted portraits of both Warburg and Heiss, and he maintained an independent correspondence with Heiss for decades.

Perhaps best of all, from Warburg's perspective, Burk went to war against Warburg's opponents with as much ferocity as Warburg himself. If anything, Burk was the more dedicated warrior. In his last decades, Warburg would sometimes consider backing away from scientific battles. He once told Burk there was no use in publishing yet another polemic that would "lead nowhere." Whenever Warburg expressed these uncharacteristic sentiments, Burk would urge his mentor to fight on for the "cause," often inciting Warburg with updates on how his enemies were maneuvering against him.

After six months with Burk, Warburg moved again, this time to the Marine Biological Station in Woods Hole, Massachusetts. He brought along the German-speaking assistant Emerson had found for him and put his services to good use. Still uncomfortable lecturing in English, Warburg had his assistant deliver his photosynthesis lecture. Warburg sat in the audience and listened. When the assistant said that Warburg "believed" that only 4 photons were needed, Warburg leapt up from his seat: "*Vot* do you mean, 'I believe'? I know."[17]

Warburg spoke English on at least one other occasion at Woods Hole. During a discussion with a group of American scientists, Warburg and fellow German James Franck got into a sparring match over photosynthesis. "Warburg was shouting at Franck, 'You are *wronk,*'" a witness recalled, "and Franck was responding 'You are *wronk.*'" Another scientist had to step in to settle them down. But there were at least some less acrimonious moments during Warburg's stay at Woods Hole. One day, Warburg attended a picnic on a nearby island with a group of physiology students. In a photo taken at the picnic, Warburg sits at the beach in long sleeves and pants, surrounded by shirtless young scientists in their swimming trunks. He appears remarkably relaxed. One member of the group recalled Warburg sitting by a tree eating lobster as Heiss tended to his every need.[18]

Warburg did not want to spend the rest of his life moving from lab to lab, but he was still having no luck finding a more permanent position in America. It didn't help that he was 65 and considered by some to be too old to hire. But given Warburg's fame and accomplishments, his age could have been overlooked. The more serious obstacle to finding a position in America might have finally dawned on Warburg during a dinner party held at Woods Hole, shortly before he returned to Germany. An unidentified woman at the party, a wife of a Caltech professor, turned to Warburg and asked him why he had remained in Germany "when the Nazis were doing such bad things."

"I wanted to protect my co-workers," Warburg lied. "What could I have done?"

The woman had an idea: "You could have committed suicide!"[19]

Warburg and the other dinner guests sat stunned. Someone had finally informed the Emperor of Dahlem of his missing clothes.

CHAPTER THIRTEEN

Two Engines

WARBURG RETURNED TO Germany in good spirits. "America was a great adventure," he wrote to an acquaintance in December 1949. "I really made a lot of friends there, and no one believed that I would ever leave."[1]

Later that month, the same *New York Times* journalist who had written about Warburg's cancer research the year before published a page-one story on his photosynthesis studies. The article stated as fact that only 3 or 4 photons are necessary to power photosynthesis and suggested that Warburg had overturned previous thinking on the subject by way of "highly ingenious new techniques." The outrageous claim was presented in support of another: that Warburg, through a series of "epoch-making" photosynthesis experiments, had discovered a new way to grow algae that would vastly expand the world's food supply.

Making it possible to grow food more efficiently had always been one of Warburg's goals. The *New York Times* declaring that he had triumphed must have been enormously satisfying for Warburg, and more good news was soon to come. On May 8, 1950, the American military returned Warburg's Dahlem institute to him in a ceremony attended by General Maxwell D. Taylor, the US military commander in Berlin. One of the few members of the Kaiser Wilhelm Society deemed untainted by the Nazis, Warburg received new equipment and funding at a moment when many German scientific labs were barely functional. "The Palace of Cell Physiology has emerged from the ashes, more beautiful than ever," Warburg wrote to an American colleague in December 1951.

In June, when Martin Klingenberg, a physical chemistry student from Heidelberg, arrived at the institute, Warburg took him on a tour. Klingenberg was amazed by the state-of-the-art equipment. Compared with his lab in Heidelberg, Warburg's institute seemed like "paradise."[2]

At approximately the same time Warburg reoccupied his institute, he received a delivery from Sweden that would shape his thinking on cancer for the rest of his life. The shipment, sent by the scientist George Klein, contained a type of cancer cell that grows in abdominal fluid. Warburg had previously worked with razor-thin slices of tumors, but such samples inevitably included noncancerous cells as well. The cells from the abdominal fluid, by contrast, gave Warburg nearly pure cultures.

Warburg placed the new cancer cells in a glass vessel and attached it to his manometer. He had been studying the respiration of cancer cells for nearly 30 years, but what he measured this time was noticeably different from what he had measured before. In the 1920s, Warburg had found that respiration with oxygen could continue at a steady rate even as fermentation increased. And yet, those results never quite fit with Warburg's theory—based on the conclusions of Pasteur—that fermentation and respiration always maintain a seesaw-like relationship, fermentation going up as respiration went down.

The new experiments made this problem go away for Warburg. The cells from the abdominal fluid hardly needed oxygen at all. They seemed to rely almost exclusively on fermentation. They fermented, Warburg wrote, like "wildly proliferating Torula yeasts." As Warburg saw it, these new experiments were a huge advance, the long-awaited evidence that he had been right all along. When Klein's boss later asked Warburg for a recommendation on Klein's behalf, Warburg happily agreed. "George Klein has made a very important contribution to cancer research," Warburg wrote. "He has sent me the cells with which I have solved the cancer problem."[3]

In December 1950, Warburg and Heiss traveled to Stockholm for the 50th anniversary of the Nobel Prize awards, where Warburg was to give a speech about cancer research. Before Warburg's lecture, he and Heiss set up four charts with data from experiments on the metabolism of cancer cells. As Warburg spoke, Heiss paced with a wooden staff and periodically pointed to the charts. When he was done speaking, Warburg informed his audience of Nobel laureates that he had just told them all they need to understand about the biology of cancer. The rest, Warburg said, was "garbage."[4]

If it was increasingly rare for Warburg to doubt himself, as late as 1952 it was still technically possible. That year he traveled to Copenhagen to give a series of lectures at the Carlsberg Laboratory. The Danish biochemist Herman Kalckar took Warburg and another scientist on an outing to the castle in Elsinore where *Hamlet* is set. It was a beautiful day—sunny and "crystal-clear," Kalckar recalled—and as they toured the famous castle, Warburg seemed himself. Peering through an iron grate into the dark expanse below, Warburg noted that it was "a perfect place" for the "Midwest Gang"—his nickname for Robert Emerson and the other scientists at the University of Illinois who disagreed with him about photosynthesis.

It was a typically Warburgian comment, but later that day Warburg turned to Kalckar and asked a question that was not at all characteristic: "Why do I encounter so much alienation from British and American biochemists?"

Kalckar, startled, tried to be diplomatic. He suggested that Warburg's "opinionated footnotes" might be part of the problem.

"Give me one example," Warburg said.

Kalckar obliged, pointing out that Warburg had called the legendary twentieth-century biochemist Sir Frederick Gowland Hopkins a "Romantic"—an insult suggesting that Hopkins would speculate beyond what the experimental data supported.

"Did I really?" Warburg asked. He seemed genuinely surprised that he had said such a thing. He told Kalckar that he had always admired Hopkins. "Warburg's open confession was remarkable," Kalckar wrote. "Perhaps Hamlet's castle played its part."[5]

Whatever came over Warburg that afternoon, it was a rare occurrence in the final decades of his life. The next year Warburg learned of new evidence that supported his understanding of cancer, and he grew still more certain that he had been right all along. In the 1920s, Warburg had been able to demonstrate only that an increase in fermentation is a basic feature of most cancers and that he could cause cells in a dish to ferment in the manner of cancer cells by poisoning them with chemicals that interfered with their use of oxygen. Warburg had never shown that depriving cells of oxygen could lead to an actual cancer in an animal. In 1953, the American researchers Harry Goldblatt and Gladys Cameron claimed to have done exactly that. Though Goldblatt acknowledged that the experiments couldn't prove that oxygen shortages caused cancer in people, he pointed out that cancer is often found in places in the body where the blood supply is limited and cells have less access to oxygen.

Warburg was delighted with the finding of Goldblatt and Cameron and mentioned it often in his own articles and letters. He had long believed that chemicals and radiation caused cancer by damaging a cell and leaving it unable to use oxygen properly. The new research suggested that a disruption in the delivery of oxygen to a cell could cause cancer, even if the cellular machinery itself wasn't damaged. The origins of a deadly cancer might be as innocuous as a plugged gland.

However the use of oxygen was impaired, the result would be the same: a need for more energy from fermentation.

Goldblatt and Cameron's discovery, and that of others in the early 1950s, helped Warburg fill in the sketch of cancer he had first proposed 30 years earlier. When cells turned to fermentation to generate enough energy to survive, he now claimed, it would lead to a "struggle for existence." The cells least capable of replacing the power of respiration with fermentation might die quickly, but the more flexible cells would cling to life, even as they gasped for air.

The "struggle" to survive in the suffocating environment might be a prolonged process, which explained why cancers often arose years after someone had been exposed to carcinogens. But over time, as the cells took up glucose and divided, the best fermenters would win the Darwinian competition and multiply until they came to dominate. Alas, the victory of the superior fermenters, as Warburg had argued from the start, would come at a terrible price. Because the energy supplied by fermentation alone cannot maintain the structure of a "differentiated" cell—a cell that is specialized for a particular tissue—the survivors of the oxygen drought end up as primitive "undifferentiated" cells that eat and grow without pause. "[T]he differentiated body cell is like a ball on an inclined plane, which would roll down except for the work of oxygen-respiration always preventing this," as Warburg once explained it. "If oxygen respiration is inhibited, the ball rolls down the plane to the level of dedifferentiation."

Warburg believed that all cells had lived by fermentation alone before the earth's atmosphere had filled with oxygen. The return to fermentation, as he imagined it, amounted to evolution in reverse. Cancer was a cell's journey backward in time to its deepest origins. Even "Einstein descended from a unicellular fermenting organism," Warburg noted.

Beyond Warburg's original and most fundamental observation—

that cancer cells swallow more glucose and ferment more than non-cancerous cells—all of his arguments were speculative, precisely the type of theorizing Warburg would never have tolerated from others. But during the last two decades of his life, he presented this vision of cancer as though he himself were Einstein explaining a basic theory of physics. "If the explanation of a vital process is its reduction to physics and chemistry," Warburg wrote, "there is today no other explanation for the origin of cancer cells, either special or general."[6]

WARBURG WAS NOW so convinced he was right about cancer that the mere thought of considering other approaches seemed ridiculous to him. In 1953, the Nobel Prize–winning German biochemist Adolf Butenandt reached out to Warburg about founding a new cancer research center in Germany. Warburg told Butenandt that there was no point, given that he had already solved the cancer problem.

And so the events of the last week of 1953 could only have come as a terrible shock for Warburg. While most Americans were celebrating the holidays with family, a number of the nation's leading scientists were huddled in Boston at a cancer symposium. One of the scientists in attendance was Sidney Weinhouse, an unassuming researcher working at a cancer institute in Pennsylvania. Weinhouse had recently studied respiration in cancer cells with a new tool, radioactive carbon molecules, that made it possible to make even more precise measurements.

When it was Weinhouse's turn to present his data at the Boston symposium, he did not dispute Warburg's most fundamental claim, that cancer cells ferment more. But he did dispute Warburg's argument that fermentation was a cell's response to a struggle to generate energy with oxygen. The cancer cells Weinhouse studied appeared to be breathing just fine.

Weinhouse's conclusion was essentially a confirmation of what Warburg had found in the early 1920s—that respiration continued even as cancer cells fermented—but it was a stark refutation of Warburg's more recent finding on the cancer cells taken from abdominal fluid. Warburg was claiming that fermentation replaced respiration in cancer cells, Weinhouse that fermentation accompanied respiration. The next week a science newsletter summarized Weinhouse's findings under the headline "Cancer Theory Overthrown." Dean Burk immediately wrote to Warburg to let him know about the insurrection. "Weinhouse," Burk wrote, "is your 'cancer Emerson.'"[7]

For Weinhouse, it might have been nothing more than a minor correction to an important scientific discovery. But Warburg couldn't tolerate corrections. As far as Warburg was concerned, Weinhouse had declared war. A year later, Warburg proclaimed his victory in that war in a speech delivered in Stuttgart:

> What was formerly only qualitative has now become quantitative. What was formerly only probable has now become certain. The era in which the fermentation of the cancer cells or its importance could be disputed is over, and no one today can doubt that we understand the origin of cancer cells if we know how their large fermentation originates, or, to express it more fully, if we know how the damaged respiration and the excessive fermentation of the cancer cells originate.[8]

The speech was translated by Burk and appeared in the February 1956 issue of *Science*. The next month, the *New York Times* published a story under the headline "German Physiologist Is Sure That He Has Discovered the Cause of Cancer." The article began by describing Warburg as "probably the most distinguished figure in contemporary cancer research." The remaining paragraphs read as though written by

Warburg himself. Under the heading "Case Proved," the article stated that when "fermentation has completely supplanted respiration, the normal cell has changed to a cancer cell."[9]

Weinhouse, if lacking Warburg's righteous passion, did not intend to back down. In August 1956, *Science* published Weinhouse's response to Warburg's speech. Weinhouse tried to soften the blow with a mention of "the great debt" biochemists owed to Warburg, but his insistence that the science did not support Warburg's theory of how cancer begins could only be softened so much.

Science allowed Warburg to respond to Weinhouse in the same issue, and he was, predictably, less diplomatic. Warburg did make an effort to walk back his boldest claims, suggesting that "damaged" respiration could include cancers in which respiration continued but failed to "turn off" fermentation. Much of the disagreement, Warburg wrote, was "a dispute about words."

What Warburg would not walk back was his deeper claim, that an increase in fermentation was the most fundamental characteristic of cancer. "The problem of cancer is not to explain life, but to discover the differences between cancer cells and normal growing cells," he wrote in his usual mix of elegance and condescension. "Fortunately, this can be done without knowing what life really is. Imagine two engines, the one being driven by complete and the other by incomplete combustion of coal. A man who knows nothing at all about engines, their structure, and their purpose, may discover the difference. He may, for example, smell it."[10]

As Warburg restarted his career in Dahlem in 1950, cancer therapy was stuck in the same place it had been decades earlier. Radiation was used to blast malignant cells, and surgeons wielded their knives against tumors, but the medical world still lacked the thing it needed most: cancer drugs that extended patients' lives. In the 1950s, such drugs began to emerge. They are now known as chemotherapies, but they were far from the precision weapons Paul Ehrlich had imagined when he coined the term at the turn of the twentieth century. The drugs attacked rapidly dividing cells, cancerous or not. They ravaged the body, leaving patients frail and nauseated, sometimes barely clinging to life. Chemotherapy, Siddhartha Mukherjee writes, turned out to be "a ghoulish distortion of Ehrlich's dream."[11]

But then, it was not Ehrlich but another famed German scientist, Fritz Haber, who had prepared the way for modern chemotherapy. The director of the Kaiser Wilhelm Institute for Physical Chemistry, Haber initiated and oversaw Germany's gas warfare program during the First World War. (While Warburg was at the Eastern Front, Haber's gas unit took over the lab space that had been set aside for Warburg at the Kaiser Wilhelm Institute for Biology.)

Haber, said to be the father of modern biological warfare, is sometimes portrayed as an evil mastermind. With his shaven head and pince-nez, he looked the part. Though not considered as brilliant as some of his colleagues in Dahlem, Haber was a remarkably creative chemist with a Warburg-like belief in his ability to overcome technical obstacles. Warburg once told his sister that Haber purposefully woke himself in the night to think so he could continue to come up with new ideas. The ideas were frequently bizarre. After World War I, Haber spent years trying to extract microscopic gold particles from the ocean in a scheme to pay off Germany's debts.[12]

If capable of villainy, Haber was a complicated villain. Beneath a gregarious exterior lay an anguished, almost Kafkaesque outlook—he once said that he felt as though he lived under "a great boulder" that was lifted every once in a while so that he could get a bit of air and sun. Though a "full Jew" according to Nazi criteria, Haber, widely considered a German hero after the war, could have held on to his job in 1933—at least in the short term. But unlike so many others in prominent positions, Haber was not willing to look away. Rather than fire his many Jewish employees, he resigned from the Kaiser Wilhelm Society and fled Germany.

Months later, the deeply patriotic Haber remained shaken and mystified by his sharp turn of fate. "I never did anything, never even said a single word, that could warrant making me an enemy of those now ruling Germany," he wrote to a colleague in December 1933. Soon after, Haber died of heart failure. His work would outlive him, most infamously his research on gases. Intended to be used for insect control, it led to the development of Zyklon B, the gas the Nazis would later use to murder a number of Haber's own relatives.[13]

Among the weapons to come out of Haber's gas unit during World War I was an agent that smelled like garlic and that spread in a shimmering pale yellow mist. The weapon, which became known as mustard gas, was a source of terror for soldiers. It seeped into their uniforms and left their skin itching and burning and blistering, as though they were being cooked without heat. Gas masks were largely useless against it. And yet, most exposed to mustard attacks survived. Hitler himself had been hit with mustard gas in the last months of World War I. In *Mein Kampf* he described the experience as "a pain, which grew worse with every quarter hour." "My eyes," Hitler wrote, "had turned into glowing coals."[14]

As early as 1919, a pair of researchers had noticed that soldiers who survived their encounters with the mustard agent often had more than scarred and pocked skin to remind them of their ordeals. The chemical poison obliterated the blood-forming cells of the bone marrow, leaving victims unable to rally the army of immune cells needed to fend off infections. The finding was published in an obscure journal and mostly ignored and forgotten.

Milton Winternitz, dean of the Yale School of Medicine from 1920 to 1935, didn't forget about it. A chemist, Winternitz had worked on mustard compounds himself during World War I in a unit studying how to defend against Haber's gases. During the next world war, Winternitz was appointed chairman of the Committee on the Treatment of War Gas Casualties. In that capacity, he asked two Yale pharmacologists, Louis Goodman and Alfred Gilman, to study mustard agents for their therapeutic potential.

While reviewing the literature on survivors of mustard attacks, Goodman and Gilman hit on something critical. For someone suffering from cancers of the blood, such as leukemia or lymphoma, in which immune cells will not stop multiplying, a limited supply of immune cells was the opposite of catastrophe. It was their only hope of survival.

As a first step, Goodman and Gilman injected a related mustard compound into mice with blood cancers. It quickly wiped out the animals' immune cells and did so without eliciting the blistering caused by skin exposure. On August 27, 1942, not long after Zyklon B was first released into the Nazi gas chambers, the mustard compound was injected into a 47-year-old Polish immigrant known only as "J.D." who had exhausted all other treatment options for his non-Hodgkin's lymphoma. Within four days of his first injection, J.D. was moving more. He told a nurse that his throat felt better. He continued to improve over the following weeks. One month after his first injection, his medical records indicated that "all cervical and axillary nodes [of the cancer] were gone."[15]

J.D.'s cancer returned within two months, and he died shortly thereafter. But his response to the mustard compound was a turning point in the history of cancer medicine. In defending his work on gas weapons, Haber had argued that poisoning soldiers with gases was no worse than shooting at them with bullets. The end result, painful injuries or death, was the same. Cancer scientists had now arrived at the inverse conclusion. The toxic effects of chemotherapy could be justified because the drugs achieved the same end result—saving a life—as a magic bullet.

Precisely how the first chemotherapies for cancer worked was not well understood at the time. Scientists now know that mustard compounds form bonds that prevent rapidly dividing cells from replicating their DNA. The chemotherapies that soon followed stop cells from dividing by interfering with their use of a particular B vitamin needed to synthesize new molecules. Though it is often overlooked even today, these drugs are metabolic therapies. They work on enzymes that Warburg understood to be central to cancer, and they disrupt the metabolic pathways that are turned on as a cell transitions to fermentation.

Weinhouse had shown that Warburg was wrong on the specifics, but chemotherapy turned out to be a validation of Warburg's broader belief in the fundamental importance of metabolism.

CHAPTER FOURTEEN

"Strange New Creatures of Our Own Making"

In some respects, Warburg's postwar life represented a profound shift from his prewar experience. Without Rockefeller funding, he had fewer resources than he had been accustomed to. In an effort to help, Eric Warburg set up a fund in England so that Warburg could purchase scientific books. Warburg used at least some of the money to support his riding habit. "One can't get decent riding breeches and boots in Germany any longer," he explained to his cousin.

There were other changes for Warburg now as well. Upon its reestablishment in 1948, the Kaiser Wilhelm Society had been renamed the Max Planck Society. In 1953, Warburg's institute was formally rebranded the Max Planck Institute for Cell Physiology. Warburg, having witnessed Planck's less than stellar resistance to the Nazis, continued to wear his old badge, depicting the kaiser, at official events. Still furious at his former employees for denouncing him, Warburg also needed an entirely new staff after the war. Later on, when Hans Krebs asked Warburg why he had fired all of his employees, Warburg told him that, in addition to being disloyal, they had grown senile. Krebs pointed out that the employees in question were 15 to 20 years younger than Warburg himself. "Different people age at different rates," Warburg said.[1]

But the most remarkable aspect of Warburg's postwar life was not how much it had changed, but how little. Five years after the Holocaust, Warburg might as well have been in 1933. He was working on cancer and photosynthesis again and engaged in bitter feuds with everyone who disagreed with him again. He and Heiss mounted their horses every morning and rode through Dahlem as they always had. They took winter walks through the lakes and woods on the outskirts of Berlin and vacationed at a spa resort. Warburg even took up a new hobby, becoming an avid sailor.

Warburg's readiness to move forward wasn't unusual in the 1950s. While many of the most notorious Nazis were captured and tried, the vast majority, including those who had played leading roles in the extermination programs, moved seamlessly back into German life. At least 25 cabinet officials in the new West German Republic had once been members of Nazi organizations. As late as 1957, almost 80 percent of the officials in the justice ministry were former Nazi Party members. In the words of Ralph Giordano, a first-degree *Mischling* whose family spent the last months of the war hiding in a friend's basement, Germany's willingness to let so many Nazis reenter society amounted to a "great peace with the perpetrators." The guilt of the Germans under Hitler was only Germany's "first guilt." "The second guilt" was "the repression and denial of the first."[2]

THE GERMAN PUBLIC'S FEAR of cancer was also largely unchanged after the war. Germans were still deeply concerned about the harm of chemicals in the food and air, and these fears would help give rise to a robust postwar environmental movement. Most postwar German environmentalists were certainly not Nazis. But given the Nazi obsession with environmental pollutants, it's hardly surprising that a number of individuals with troubling pasts ended up at the forefront of the movement.

Among the most famous postwar German environmentalists was Alwin Seifert. His 1971 book, *Gardening and Ploughing without Poison*, sold a quarter of a million copies in Germany and inspired the younger generation to choose organically grown foods. Seifert's many fans presumably didn't know that under the Nazis he had been known as the Reich Landscape Advocate or that his arguments in favor of organic agriculture had influenced the thinking of Hitler, Himmler, and Hess, among others.

The Nazi leadership had been drawn to Seifert for a reason. He sometimes spoke of the threat of nonindigenous plants colonizing German soil in language that echoed Nazi rhetoric about the Jews. Until relatively recently, historians were still debating whether Seifert was a committed Nazi or had merely turned to Nazi terminology to market his ideas. Whatever the answer, in 2009 the German historian Daniella Seidl made a troubling discovery: Seifert's hands were dirty with more than naturally fertilized soil. He had been a regular visitor to the Dachau gardens, Himmler's concentration camp–cum–organic paradise. Seifert, who advised Himmler on various aspects of the Dachau gardens, even brought several prisoners from Dachau to work at his home.

The slogan of the German environmental movement in the 1950s was "Leave our food as natural as possible!" It was coined by Werner Kollath, a former Nazi who had written a textbook on racial hygiene in which he lamented that legislation had failed to keep inferior people from reproducing. Like Hitler, Kollath believed that all humans had once been vegetarians and that Germans should give up meat for plants. Kollath's prescription for making the German diet more natural "had deep roots in the Third Reich," the historian Corinna Treitel wrote.[3]

Kollath, who is also known for popularizing the concept of "whole foods," published a book on "diseases of civilization" in 1958 that recounted much of the evidence that Erwin Liek had presented 30 years earlier. He was far from the only person making the argument at the time. The notion that cancer was a "disease of civilization" went through a broad revival in the first decades after World War II, thanks to an emerging body of research on cancer rates in migrant populations. Native Japanese women, it was found, were far less likely to have cancer than women of Japanese descent living in the United States. Black Americans of West African heritage had cancer rates far above the rates of West Africans. Cancer, these studies suggested, couldn't be entirely explained away by aging or genes or bad luck. Living in the wrong country appeared to be the greatest risk factor of all.

These findings would lead postwar cancer authorities in Europe and the United States to conclude that 70 percent or more of cancers were caused by environmental factors, a category that includes diet. At the time, at least, this wasn't considered bad news. If most cancers could be traced to modern lifestyles, then in theory, as Oxford epidemiologist Richard Peto once put it, "there are ways in which human beings can live whereby those cancers would not arise." This was the hesitant conclusion of a 1964 World Health Organization report, which noted that "the majority of human cancer" seemed "potentially preventable."[4]

But if the evidence that cancer was a disease of "civilization" had grown more compelling, there was still little progress on the question of which particular aspects of "civilization" caused the disease. After the war, more attention was devoted to the threat of artificial chemicals, both in the air and in our foods and drinks. The Germans had obsessed over the dangers of artificial chemicals for decades. America now began to catch up, and the man sounding the alarm was a German immigrant.

Wilhelm C. Hueper was a physician and researcher. Before leaving Germany as a young man in 1923, he had studied the dangers of paraffin breast implants. Like so many German doctors of his generation, Hueper was fixated on rising cancer rates, and he was intimately familiar with the power of toxic chemicals. He had fought for Germany during the First World War, delivering Fritz Haber's gas weapons to the front lines.

Hueper was particularly interested in the hazards of chemicals in dye factories, which had been linked to bladder cancer in 1895 by another German researcher. The same aniline dyes that had inspired Paul Ehrlich to dream up magic bullets later inspired in Hueper visions of a global catastrophe: what if the dyes, and, for that matter, all of the other chemicals studied by German scientists, were only the beginning? The cancer rate had "markedly increased in all civilized countries," Hueper maintained, because "new synthetic substances" were entering modern life "in never-ending number."

In 1934, Hueper was hired by a DuPont laboratory in Delaware to study the threat not only of aniline dyes but also of pesticides, Freon, Teflon, and a long list of other chemicals that had made their way into the lives of ordinary Americans. He soon discovered that he could induce bladder cancer in dogs by feeding them a compound found in chemical dyes. But though the discovery was widely recognized, Hueper was forced out of his job after only three years. He claimed that he was repeatedly blocked from carrying out the type of experiments that might draw tighter links between DuPont's chemicals and cancer and that he was prevented from publishing many of his results. It likely didn't help that Hueper, known for his "abrasiveness" and "Teutonic" manner, tended to alienate colleagues as easily as Warburg.[5]

Hueper also resembled Warburg in his inability to stop working. He knew there was already a large body of scientific literature linking specific industrial chemicals to cancer, but no one had created a comprehensive review of all of the possible threats workers faced. In early 1938, Hueper decided to take on the project. Still between jobs, he worked at his dining room table, filling out file card after file card—marking each with information about a distinct carcinogen—and stacking them into towers of doom. The result, published in 1942, was *Occupational Tumors and Allied Diseases*, an 896-page catalog of cancer-causing poisons.

Though Hueper's book was initially ignored, the concern that artificial chemicals were responsible for many cancers was nevertheless growing more pervasive in America—Hueper's own unsigned editorials in the *Journal of the American Medical Association* helped generate the fears. His list of enemies, both in industry and in government, continued to grow, and yet the possibility that he was right about cancer was impossible to dismiss. Due, in large part, to an increase in lung cancer, the overall cancer death rate in Western countries was still on the rise. In 1948, Hueper became the first director of the Environmental Cancer Section at the National Cancer Institute, solidifying his status as the "father of American occupational carcinogenesis."

Though his book was written in a measured, scientific style, Hueper was capable of rhetorical flourishes. He once referred to the carcinogens as "biological death bombs" that might prove "as dangerous to the existence of mankind as the arsenal of atom bombs prepared for future action." He had a talent for "shock and alarm," a friend once recalled.[6]

Hueper achieved a certain amount of fame in 1962, the year the marine biologist and science writer Rachel Carson wrote about him in *Silent Spring*. The book, which documented how industrial chemicals were destroying the earth and the health of humans and other species, sold hundreds of thousands of copies and inspired an entire generation to think more seriously about the environment. President Kennedy was among Carson's admirers and cited her work in opening an investigation into the links between chemical pollutants and human illnesses. One year after the book's publication, Congress passed the Clean Air Act. Next came the Wilderness Act, the Clean Water Act, and the Endangered Species Act. Carson's book changed America in a way that few others ever have.

By her own account, Hueper was her inspiration for *Silent Spring*, and he served as one of her primary sources for the book. "Dr. H. says he thinks the time now is right for the book for people are beginning to want the facts," Carson wrote to a friend as she embarked on the project.

Though *Silent Spring* is in no way anti-Semitic or racist, it can read as though it was written in Germany in the 1930s. Carson described "sinister" and "devious" chemicals that lurk around us at all times. They "lie long in soil, entering into living organisms, passing from one to another in a chain of poisoning and death." Sometimes the chemicals "pass mysteriously by underground streams until they emerge and, through the alchemy of air and sunlight, combine into new forms that kill vegetation, sicken cattle, and work unknown harm on those who drink from once pure wells." Even the title of Carson's chapter on cancer, "One in Every Four," was an unwitting echo of a Nazi-era cancer propaganda film known as "One in Eight."[7]

Silent Spring cites Hueper comparing the fight against cancer to the fight against infectious diseases at the end of the nineteenth century, when "the brilliant work of Pasteur and Koch" had established the "causative relation" between germs and many diseases. Cancer could be traced to hidden chemicals, Hueper noted, just as tuberculosis and cholera had been traced to hidden organisms. Hueper, like Hitler and Warburg, had looked into the mirror and found Robert Koch gazing back at him.

Hueper was familiar with Warburg's thinking on cancer. In his 1942 tome on environmental carcinogens, Hueper reviewed Warburg's cancer research along with other theories. The two men likely crossed paths in 1949, when Warburg was a visiting scientist at the National Cancer Institute. And so Hueper's influence on *Silent Spring* might account for Otto Warburg's prominent place in the book—Warburg is the first scientist introduced in Carson's chapter on cancer, and his theories are discussed at length.

As Carson explained, artificial chemicals could be causing cancer by damaging how cells breathe, thus driving the transition from respiration to fermentation. Carson also reviewed Warburg's terrifying hypothesis that multiple small doses of a chemical poison might be even more dangerous than a single large dose. A large dose would kill a cell, but the small doses from pesticides or food preservatives could slowly poison cellular breathing—and put a cell on the path to cancer—without eliminating it.

By the same logic, Carson argued, Warburg's explanation of cancer's origins could explain why various treatments, such as radiation, could both treat and cause cancer. Such therapies might destroy weakened cells that already struggled to breathe, but at the same time, they would damage the respiration of previously healthy cells. "Measured by the standards established by Warburg," Carson wrote, "most pesticides meet the criterion of the perfect carcinogen too well for comfort."[8]

IF WARBURG AND HUEPER crossed paths in 1949, Hueper might have influenced Warburg more than Warburg influenced Hueper. On the surface, at least, Warburg could hardly have been more different from Germany's environmental gurus. He was the very archetype of the experimental scientist that the Romantic "back to nature" movement had long despised for reducing life to a chain of biological mechanisms. But while he remained as devoted as ever to experimental science, Warburg also had a Romantic streak, and he, too, would come to see carcinogens everywhere he looked.

Warburg's anxieties predated Hueper. Lotte noted Warburg's intense fear of death in her diary in 1926. In the early 1930s, not long after his new house was built, he became concerned about the tar used to set his parquet floor. "He constantly talks about the tar smell because the tar workers get cancer," Lotte wrote in her diary in 1930. Though Heiss was against ripping up their new home—"He fights for his parquet," Lotte noted—Warburg prevailed and had the tar removed.

After the war, though, Warburg's paranoia grew considerably more pronounced. He is thought to have been shaken by the loss of two of his sisters, Lotte and Käthe, both of whom died at the end of the 1940s. Though he had been a regular cigar smoker, Warburg, who turned 67 in 1950, now quit. He refused X-rays from a doctor and became neurotic about exhaust from cars.

Warburg also began sending off petitions to the health minister of West Germany. He called for West Berlin to be surrounded by trees, which would form a protective "green belt" around the city, and urged the government to ban dangerous substances, including food colorings and chemicals used in flavoring and preservatives. In the early 1950s, he helped push through a new law requiring that labels on canned foods list all chemical additives. He said that previous laws designed to keep food safe had failed because exceptions would be made "for every poison that was profitable."[9]

At some point after the war, Warburg bought a field near his house so that his garden could be expanded. Rather than use pesticides, Krebs wrote, Warburg "encouraged the nesting of tits," and he would keep an eye out for caterpillars and other pests. In addition to an assortment of vegetables, he grew raspberries, pears, apples, apricots, strawberries, and red currants. Warburg's animals—rabbits, turkeys, hens, ducks, geese, and others—supplied all the manure.

"Altogether his whims, fancies, and anxieties about food were at times rather exasperating to Heiss, who did all the cooking," Krebs, who maintained his friendship with Warburg after the war, wrote. It appears to have been an understatement. Though Warburg hired a gardener to tend to the grounds, it was Heiss who carried the burden of his paranoia. Warburg only drank water from a well in his new field, and Heiss would have to retrieve water and inspect it for impurities. Heiss reserved one night a week for baking—his bread coupled with his homemade goat butter was said to be especially delicious—and

Jacob Heiss, date unknown.

Warburg forbade him from buying the flour in a store for fear that the bleach was a carcinogen.

According to neighbors, Warburg's institute itself was coming to resemble a "farm" in this period. Warburg was even using one of his centrifuges to make cream and butter from the milk he would procure from a special herd of cows. Eating out, meanwhile, was nearly impossible. Warburg would arrive at restaurants with a tin containing his own Heiss-baked bread and a tea bag. (He was sometimes scolded by the staff for bringing in outside food.) A colleague remembered Warburg ordering a piece of cake one evening. He chewed a forkful for a moment, paused, and then spat it out. He had detected a flavor linked to cancer, he explained.

Even Warburg's allies began to find him insufferable. The biochemist David Nachmansohn, one of Warburg's great admirers, recalled that when he saw Warburg in the late 1950s, he "took pains to avoid discussing cancer or photosynthesis." Nachmansohn added that he was not alone in this strategy.[10]

SILENT SPRING HAS A sad epilogue. Carson was undergoing treatments for breast cancer while writing the book. She died two years after its publication, before she could witness the tremendous influence of her work.

Hueper lived until 1979 and continued to warn of environmental carcinogens until the very end. Shortly before his death, he received a prestigious award from the National Institutes of Health. At the time, it seemed an overdue celebration of a man who had dedicated his life to trying to save people from cancer. But as Stanford historian Robert Proctor documents in *The Nazi War on Cancer*, Hueper, "the man who, more than any other, brought the cancer hazards of pollutants in our food, air, and water to scientific attention," had a "secret."

After World War I, Hueper had briefly joined a proto-Nazi militia, and his Nazi sympathies appear to have remained undiminished after he moved to America. When Hitler became chancellor in 1933, Hueper began searching for positions in Nazi Germany and sent a query letter to Bernhard Rust, the head of the Reich Education Ministry. Rust was the right choice. He had personally overseen the firing of Germany's Jewish scientists. (His deputy, Rudolf Mentzel, later tormented Warburg.) In his letter to Rust, Hueper cited his brother, a Nazi functionary, as a reference and signed off with "Heil Hitler!"[11]

Hueper wasn't simply daydreaming about a new life under Hitler's rule. In the summer of 1934, he returned to Germany with his wife, hoping to fill a position once held by a Jewish scientist. "[O]penings had become available because of the Hitler turmoil," as Hueper put it in the unapologetic memoir that he left unpublished.

Yet Hueper could not find a satisfactory position and returned to the United States. In 1936, he gave a speech on race and disease in which he noted the risks of healthy races mixing with races that carried more genetic defects. It was translated into German and published in a racist journal. Twenty years later, Hueper suggested that dark-skinned people were more resistant to carcinogens and so better suited to working with hazardous chemicals.

Hueper's Nazi sympathies can come as a shock to those who remember him as a crusader for public health. As Proctor puts it, "How could the hero of *Silent Spring* have found hope in the Nazi movement?" But Hueper's past does not, by itself, invalidate Carson's book. Nor is it entirely fair to judge Carson for where the book fell short. Hueper had long denied the role of smoking in cancer, and following his lead, Carson overlooked the most potent carcinogen of her era. Her detractors have also pointed out that the fears she raised about the pesticide DDT were exaggerated and that the ban on DDT that her work helped promote would lead to vastly more deaths from malaria in less developed countries. But in the first decades after the war, at least, that artificial chemicals were responsible for most preventable cancer was as good a hypothesis as any for the rising cancer rates in the industrialized world.[12]

Not everyone thought that artificial chemicals were the best explanation for the rise of cancer in postindustrial societies. In the 1960s, the notion that yet-to-be identified viruses were responsible for most cancers was revived as well. But chemicals in our foods and products, unlike cancer-causing viruses, were easy to find. By the mid-1970s, the fear of chemical carcinogens was driving many Americans into a Warburg-esque state of paranoia. "Strange new creatures of our own making are all around us, in our air, our water, our food and in the things we touch," said Russell Train, the administrator of the Environmental Protection Agency. Americans, he warned, were playing "a grim game of chemical roulette whose result they would not know until many years later."[13]

Other than Hueper, no one person in America likely did more to instill the belief that chemicals cause cancer than Bruce Ames, a highly regarded Berkeley biochemist and geneticist. Ames began investigating the question in 1964 after reading the ingredients on a bag of potato chips that his son had asked him to open. He came up with an ingenious test. Ames first engineered a strain of bacteria that couldn't grow without consuming a particular amino acid. He then placed the bacteria in a medium without that amino acid and subjected the cells to whatever chemical he wanted to test. Since the bacteria lacked the genes to grow in such an environment, whenever a new colony arose, it was evidence that a mutation had formed. The more new colonies that formed, the greater the mutation rate. A greater mutation rate, in turn, meant a greater chance that the given chemical could cause cancer.

When it was shown that the same chemicals that made it possible for the bacteria to form new colonies would typically also cause cancers in lab animals—at least at extremely high doses—the Ames test, as it was soon called, was widely adopted. Americans were "already exposed to an estimated total of 25,000 synthetic chemicals," and "hundreds of new ones" were being "introduced each year," the *New York Times* reported in 1975. In another article, headlined "The Parade of Chemicals That Cause Cancer Seems Endless," the *Times* bemoaned that once a given chemical was regulated or banned, it "disappeared from public debate, to be replaced shortly by the next 'carcinogen of the month.'"[14]

A coherent and genuinely terrifying picture of how "civilization" caused cancer was coming into focus after a century-long search for answers.

It took only a few more years for that picture to fall apart.

CHAPTER FIFTEEN

The Prime Cause of Cancer

In 1965, the American scientists Clint Fuller, Andrew Benson, and Harlan Wood unexpectedly received mail from Germany. Inside each envelope was something even more unexpected: a personal check from Otto Warburg for 4,000 deutsch marks (about $1,000 at the time). As the accompanying letter explained, Warburg wanted the three scientists to attend a symposium in Strasbourg, a scenic city on the French-German border, where they would have the honor of responding to his latest photosynthesis papers.

The Americans' surprise quickly gave way to puzzlement. Warburg, by then, was embroiled in yet another debate about photosynthesis—this time over the question of whether the oxygen given off by photosynthesis is derived from water or carbon dioxide. Warburg was sure the answer was carbon dioxide. The men he had invited to the symposium—along with most others in the field—were sure it was water. Still, 4,000 deutsch marks was a lot of money and Strasbourg was a beautiful city. And so all three Americans flew to Europe, a little confused but happy to tell Warburg that he was dead wrong, if that's what he wanted.

The day before the symposium, Fuller, Benson, and Wood met a German professor who was to chair the event and asked what they should anticipate. Warburg, the professor explained, had planned a "hanging party." The three Americans were there only so that they could "surrender publicly" and acknowledge that Warburg had been right all along.

Needless to say, the visitors were taken aback. They had no intention of surrendering to Warburg, let alone publicly. The next day, the symposium attendees gathered on the steps of the convention hall to greet Warburg, as though he were a foreign dignitary gracing them with his presence. When he finally arrived, he did so in style. As Fuller later recalled, a long black Mercedes pulled up with the license plate that read "MP-B1" (Max Planck Society–Berlin One). The chauffeur emerged from the car first and opened a back seat door. As the crowd looked on with anticipation, Warburg stepped out of the vehicle in a beige suit with what Fuller described as a "10-gallon Cowboy hat" on his head. At age 81, Otto Warburg had finally gone full diva.

Warburg took a moment to look Fuller up and down before shaking his hand: "Herr Fuller, you are much too young!" Warburg said. Fuller thought it was a compliment, though he wasn't entirely sure.

The symposium began and Warburg presented first. Heiss accompanied him onto the stage carrying two rolled-up charts. Warburg and Heiss unrolled the first together and held it up for the audience. There was not much to see. It was merely a picture of one of the glass vessels that Warburg used to make his measurements. Warburg explained that he could accurately measure photosynthetic reactions with this simple vessel in a matter of hours.

With that, it was time to unroll the second chart. This one was different. It featured a long series of complicated equations based on photosynthesis experiments carried out with radioactive carbon—the same technique Weinhouse had used to supposedly "overthrow" Warburg's cancer research. Such studies, Warburg explained, could take days to complete and would achieve much less precise measurements.

When his profoundly biased presentation of the two different approaches was complete, Warburg looked out at the audience: "It's up to you to make your decision."[1]

When their own turns came, Fuller, Benson, and Wood ignored Warburg's talk and presented their own, entirely contradictory data. After the symposium, Warburg showed no signs of displeasure. Together with the mayor of Strasbourg, he hosted an elaborate dinner and said nothing of what had transpired that afternoon. He even invited the three Americans who had just contradicted his every word to visit his institute.

It's possible that Warburg was in an unusually generous mood that evening. More likely, he had achieved a state of self-assuredness so complete and all-consuming that he was no longer even capable of considering that he might be wrong. The strong reaction of his colleagues was itself evidence that he was right. The greater the discovery, Warburg wrote to Burk, "the greater is the resistance."

Shortly before his death in 1959, Emerson discovered that there are two light reactions in photosynthesis, meaning that a measurement of more than four photons did not contradict Einstein's explanation of how photons and atoms interact. It was as clear a sign as possible that Warburg had been mistaken all along, but when a German biochemist later came across one of Warburg's old notebooks, he discovered that Warburg had a different response to the news. "Finally, Emerson has confirmed my results," Warburg had written.[2]

THE YEAR AFTER THE "hanging party," Otto Warburg spoke at a gathering of Nobel laureates in the Bavarian town of Lindau. He spent most of the nearly hour-long lecture making the same exaggerated claims about the role of damaged respiration in cancer that he had been making ever since he had found the exceptionally high rates of fermentation in cancer cells of abdominal fluid. Citing the widely accepted estimates of the era, he pointed out that about 80 percent of cancers could be prevented. From Warburg's vantage, the lack of emphasis on prevention—as opposed to treatment—was especially galling, considering that, as he put it in his talk, there is "no disease whose prime cause is better known."[3]

The "Lindau Lecture," as the talk became known, was Warburg's final performance, and it received mixed reviews. His broader message about prevention and metabolism was as important as ever, yet the world had grown tired of his outrageous antics. In a dismissive article about the lecture, the German magazine *Der Spiegel* suggested that Warburg was now making false claims about cancer on an almost annual basis. Warburg had once told Hans Krebs that the secret to winning academic battles was to outlive one's scientific opponents. He hadn't considered that it may be possible to live *too* long and to see your once dominant position slip away and your theories forgotten.[4]

As the Nobel laureates in the audience in Lindau surely knew, Warburg's notion of a "prime cause" of cancer was misleading. Warburg was using the language of Robert Koch, but he had subtly changed the terminology. Even if Warburg had been entirely correct that cancer arose when cells replaced respiration with fermentation, the "prime cause" wouldn't be the transition to fermentation itself but whatever caused that transition to take place.

The researcher most qualified to identify the prime causes of cancer in the 1960s was not Otto Warburg but Richard Doll, the celebrated British epidemiologist who had linked smoking to cancer in the 1950s. Doll's path to cancer research had been fortuitous. At 17, Doll had decided to become a mathematician, only to ruin his chance of winning a scholarship to Cambridge by drinking 3 pints of 8 percent alcohol ale the night before an exam. Doll elected to become a physician instead. While training at St Thomas' Hospital in London in the 1930s, he and his fellow students took annual trips abroad to observe how medicine was taught in other countries. In 1936, they traveled to Frankfurt, Germany. One evening, while drinking with a group of German medical students, Doll shared his concerns about anti-Semitism. His German companions insisted that only a Jew would say such things and forced Doll to stand atop the table so that they could measure his ankles—thick ankles and flat-footedness were said to be typically Jewish traits.

It wasn't even the most shocking moment of Doll's trip. Before attending a talk on radiology, Doll had been warned that the lecturer was "a keen Nazi" and would expect everyone in the room to stand up and say "Heil Hitler" when he entered. Doll refused to stand and salute. The presentation proved as horrific as he could have expected. The lecturer showed a drawing of X-rays attacking cancer cells, only it wasn't a typical medical illustration. The X-rays were depicted as Nazi troops. Their target: cancer cells adorned with Jewish stars. It "didn't require many experiences of that sort to realize that there was something evil that had to be eliminated from the world," Doll said.

Though known for his patrician bearing, Doll was a radical. In 1941, while serving in the war as an English medical officer, he applied to become a paratrooper so that he could confront the Nazi evil head-on. He was rejected and ended up serving in an infectious disease ward in Cairo. He had every reason to think he had missed his moment, but only three years after the war, history would offer Richard Doll another chance. England, having lost more than 300,000 soldiers and civilians during the war, was now under attack from a different sort of enemy that was killing its citizens by the tens of thousands. Lung cancer, a relatively rare diagnosis prior to the 1920s, was turning into a full-fledged epidemic. Between 1930 and 1944 alone, lung cancer deaths among English men had increased sixfold.[5]

That no one knew why lung cancer deaths were becoming so much more common only made the trend all the more terrifying—or at least terrifying to many, if not most, people. Some medical authorities at the time made the same argument long used to deny that cancer was on the rise in the late nineteenth and early twentieth centuries. The entire scare was said to be an illusion created by new diagnostic techniques. Chest X-rays, in particular, were thought to be revealing lung cancers that had always occurred at contemporary rates but had previously gone undiagnosed.

Smoking had become far more common in the decades before the increase in lung cancer rates, and, among those who believed the epidemic was real, smoking was a leading suspect. But that smoking caused lung cancer was not considered obvious. The medical attitude of the 1930s and 1940s was famously revealed in a sarcastic comment made by the American surgeon Evarts Graham, who pointed out that nylon stocking sales had increased during the same years that smoking had increased. Graham's skepticism turned out to have been wrong, as his own research would later show, but he was making an important point. That two trends coincide—in this case cigarette smoking and lung cancer—does not mean that one causes the other. And as the skeptics pointed out, cigarette smoking was only one of a number of possible carcinogens that had grown more common in tandem with lung cancer. There were more pesticides on plants than ever before. Driving (and thus inhalation of car exhaust) had also increased dramatically during the same period. More cars meant more roads paved with carcinogenic tar. If smoking was a suspect for the crime of lung cancer in the late 1940s, it was standing in a police lineup together with a lot of other dangerous-looking suspects.

In the late 1940s, the British government first turned to Austin Bradford Hill, a highly respected epidemiologist and statistician, for help in solving the lung cancer mystery. Hill wanted a medical doctor on his team and knew that Doll, though not yet 40, was adept with numbers. Had it been feasible to do so, Doll and Hill might have run a true experiment, randomly assigning subjects with similar backgrounds to different groups. One group would inhale more car exhaust; another would smoke more; still another might consume more chemicals. The "control" group would carry on with their lives as normal. Such studies, known as randomized controlled trials, are the only way to definitively show cause and effect, and Hill himself had helped to design the method. But such a study, in addition to being profoundly unethical in this case because it could have caused great harm to the participants, would have to be conducted over many years, perhaps decades, before it could provide meaningful data on the causes of cancer. Doll and Hill, in need of a more realistic approach, turned to epidemiology, the study of health and disease patterns among populations.

For all the successes of Pasteur and Koch in tracing infectious diseases to specific microbes, it was epidemiology that did the most to ease the burden of the diseases in the nineteenth and early twentieth centuries. Doll's scientific idol was John Snow, the English physician who had founded modern epidemiology in 1854 by locating the source of a deadly cholera outbreak in London. Snow solved the puzzle by mapping where the cholera victims lived and where they had obtained their drinking water. The dead, it turned out, had all been drinking from the same water pump. The question for Doll and Hill, and for England, was whether a similar technique could work for a chronic disease like cancer, which arises over many years and can have many different causes.[6]

As a first step, Doll and Hill selected two groups of hospital patients in and around London. One group consisted of 649 people who had been diagnosed with lung cancer. The other, the control group, was made up of hospital patients of similar ages who did not have cancer. Both groups were then asked to respond to surveys about their lives. The questions about smoking accounted for only one of the nine sections of the survey. Doll, a cigarette smoker himself, favored the car exhaust and road tar hypotheses.

The surveys were carried out by social workers. When Doll obtained the results, he recorded them with his fountain pen and did the calculations himself, like an accountant filling out a ledger of profits and losses. By the time he was done, he had given up his own cigarette habit.

The study had found that those who smoked 25 or more cigarettes a day were 50 times more likely than nonsmokers to get lung cancer.

It took many more studies and many more years to convince the world that cigarettes were responsible for the increase in lung cancer deaths. Wilhelm Hueper, who continued to insist that artificial chemicals were the true cause of cancer, would emerge as one of the most outspoken skeptics of the smoking hypothesis.

Doll and Hill, as well as the American researchers Ernst Wynder and Evarts Graham, who carried out a similar study at approximately the same time, are now celebrated as the first to definitively link smoking to cancer by means of a modern scientific study. Yet, as was so often the case when it came to cancer, the Germans were more than a decade ahead. In 1939, the German doctor Franz H. Müller had published his own sophisticated epidemiological study and arrived at an even stronger conclusion. In 1950, Hill and Doll wrote only that smoking was "an important factor" behind the rise in lung cancer deaths; Müller had concluded that it was "the single most important cause."

Müller's paper was only one of a number of groundbreaking studies of smoking and cancer published by Germans during the Nazi era. And though their efforts largely failed to curb smoking, the Nazis launched the world's most aggressive anti-smoking campaigns, complete with workplace bans, counseling centers for addicts, and nicotine-free cigarettes. In typical fashion, the Nazis would find a way to blend this initiative to protect life with beliefs designed to destroy it. Smoking was said to be the habit of depraved Jews and Communists, as well as Blacks and the Romani people. Jewish capitalists were blamed for the spread of tobacco use across Europe. Like the people who smoked it, nicotine itself was considered a threat to Aryan genetic material.

Hitler had smoked as many 40 cigarettes a day as a young man. He claimed he had given up smoking and "tossed his cigarettes into the Danube" upon realizing how much money he was wasting. It's more likely that he was too poor to support his habit. In 1942, he said his decision might be credited for "the salvation of the German people." Had he continued to smoke, Hitler reasoned, he would have died years before.[7]

IN THE DECADES AFTER Hill and Doll published their landmark study of smoking and cancer in 1950, Doll conducted a series of innovative cancer epidemiology studies and helped to identify a number of new carcinogens, including asbestos. And so when, in the early 1980s, the US government wanted a better understanding of why so many Americans got cancer, Doll was the obvious person to turn to. Together with his colleague Richard Peto, he carried out an extensive review of all of the available literature. The result was a 112-page report, "The Causes of Cancer," published in 1981. It would stand as the definitive word on the subject for years.

Like Warburg, Doll and Peto believed that 75 to 80 percent of cancers were avoidable. That was not a surprise. Nor was it surprising that they attributed 30 percent of American cancer deaths to smoking. The truly startling finding was not what caused cancer but what did not. By Doll and Peto's estimate, the artificial chemicals in our air and food accounted for only 2 percent of all cancer deaths. They attributed another 4 percent of cancers to toxic chemicals in the workplace.

Doll has been accused of being a shill for big business, but he wasn't the only cancer authority pushing back against the chemical hypothesis at the time. In the 1970s, Bruce Ames's test for carcinogens made him a hero to environmentalists. But in the 1980s, Ames and others began to test natural chemicals as well. The natural chemicals, it turned out, could also cause the bacteria to mutate and grow. There were even more carcinogenic chemicals in plants than in the pesticides sprayed on the plants. The findings could sometimes make a mockery of the entire field: a single cup of coffee had 19 distinct carcinogens in it. After running one test after another, Ames concluded that

99.9 percent of the "toxic chemicals" we are exposed to are "completely natural." "Pollution," Ames admitted, "seems to me to be mostly a red herring as a cause of cancer."

Ames, too, would be accused of siding with corporate interests over ordinary Americans. But there was always a much better argument to be made in defense of the chemical carcinogen hypothesis. When Ames found that the natural world was full of toxic chemicals, his point wasn't that plants cause cancer, but that his own eponymous test couldn't tell us what did cause the disease. Doll and Peto, likewise, knew that their survey-based epidemiological methods were far from perfect—after all, people don't necessarily know what chemicals they might have been exposed to over many years. Doll and Peto looked at animal studies in their analysis as well, but whether the high doses of a chemical that could cause cancer in a rat could also cause cancer in humans at much lower levels of exposure was far from certain.[8]

Doll and Peto concluded that only a small percentage of American cancers were caused by artificial chemicals because they had no convincing evidence to the contrary. But in "The Causes of Cancer," they attributed less than a third of cancer deaths in America to smoking. That left most of the hundreds of thousands of cancer deaths each year still in need of an explanation.

CHAPTER SIXTEEN

Cancer and Diet

UNSURPRISINGLY, Warburg saw his "Lindau Lecture" as a triumph. "The path proposed in Lindau seems to be succeeding," he wrote to a colleague. "If it turns out to be so, the opposition to the Dahlem work will be silenced; it will be similar to what happened to Pasteur. After he had healed the dogs, everyone agreed with him."[1]

The following year, with Dean Burk's help, Warburg published a revised version of his lecture under the title *The Prime Cause and Prevention of Cancer*. The published edition included a preface in which Warburg stressed the importance of vitamin supplements for cancer prevention. Warburg's thinking on vitamins was related to his concerns about chemical poisons. Artificial chemicals, he believed, caused cancer by interfering with the enzymes that allow cells to use oxygen. Vitamins provide critical components of those same enzymes, meaning that a lack of vitamins could cause the enzymes to malfunction just as a poison might.

Warburg's interest in supplements wasn't new. In the 1920s, he had developed a mineral tablet that was sold by a German pharmaceutical company. And he had been researching vitamin-based treatments for cancer since the 1930s, when he made his major discoveries about coenzymes. But in his last years, Warburg was newly convinced that consuming large quantities of vitamins could prevent cancer. In 1968, a colleague sent Warburg a letter asking for his thoughts on one particular vitamin-based therapy. Warburg suggested that the formula needed 10 times as much vitamin B_2.[2]

Warburg might have been making extreme and unsupported claims, but the idea that a lack of particular vitamins could lead to cancer wasn't unreasonable. Richard Doll and Richard Peto considered the relationship between vitamins and cancer in the "Diet" section of "The Causes of Cancer"—which included the natural ingredients of food, as opposed to artificial chemicals added to food. Doll and Peto found the data on vitamins and cancer inconclusive, but unlike Warburg, they also considered other possible ways that people's diets might cause cancer. They went through the possibilities one by one, from the fat and fiber content of food, to natural chemicals that might act as carcinogens, to how cooking might turn an otherwise safe ingredient into a potential hazard.

It was a well-intentioned effort, but figuring out how many cancers could be linked to our diets turned out to be far harder even than determining the hazards of artificial chemicals. Because it is nearly impossible to run an experiment that controls what individuals eat and drink over many years, Doll and Peto relied primarily on epidemiological studies, which merely track what people choose to eat and whether they develop health conditions. Doll had used the same approach to study smoking and cancer, but such "observational" studies depend on study subjects accurately recording what they have consumed each day, and a separate body of literature suggests that the vast majority of people are unable to do this.

An additional obstacle, Doll once pointed out, is that "we don't eat individual things." Most of our meals include many different foods and varying mixes of fats, proteins, and carbohydrates. The black-and-white world of smokers and nonsmokers is impossible to reproduce in studies of nutrition and cancer. This is one reason why so many different foods have been shown to cause cancer in one study and to prevent it in the next.

Even when the data are accurate and one food or drink can be successfully isolated from the rest of the diet, it remains difficult to make definitive statements about diet and cancer. That people who eat a lot of vegetables are found to be less likely to get cancer in a study might mean that eating vegetables lowers one's cancer risk. It might also mean that people who eat lots of vegetables tend to eat less of another food or exercise more and that it's the absence of the other food or the additional exercise that is responsible for the protection against cancer. Researchers make statistical adjustments to account for these possibilities, but no amount of mathematical maneuvering can eliminate all of the possible confounding factors. As UCLA epidemiologist Sander Greenland once put it, identifying smoking as a cause of cancer was a "turkey shoot." Assessing most other environmental causes is more like shooting without aiming.[3]

The final part of the "Diet" section of "The Causes of Cancer" appeared under the heading "Overnutrition." Amid the incredibly complex scientific discussion in the report, the idea that cancer could be a problem of eating too much sounded almost laughably simplistic, but Doll and Peto took the idea seriously, writing that overnutrition should perhaps have been the first part of their diet discussion given some of the striking findings.

The idea that eating too much or carrying too much weight made someone prone to cancer wasn't new. In his 1810 book *Observations on the Cure of Cancer*, the Scottish surgeon Thomas Denman wrote that it was "thought, if not proved, that those who become corpulent or fat" are more susceptible to cancer and also that cancer was "a more rapid and intractable disease" among the overweight. Over the next century, many other cancer authorities arrived at the same conclusion. "Probably no single factor is more potent in determining the outbreak of cancer in the predisposed, than excessive feeding," the British cancer expert W. Roger Williams wrote in 1908.[4]

The belief that cancer had something to do with eating too much and excess weight gained traction in the early twentieth century via the rodent feeding studies done by Peyton Rous and others, which found that low-calorie diets were protective against cancer. These studies faded to the periphery of mainstream science after the First World War, but overeating emerged as the most popular of the many food-related cancer theories in Germany in the 1920s and 1930s. Much of the German public's interest in the topic can be traced to a slew of unscientific books touting "hunger cures." But more serious work also considered the quantity of food one ate to be an essential part of the modern cancer story.[5] Frederick Hoffman, the insurance actuary who became the world's authority on cancer statistics, considered dozens of different ways that our diets might contribute to cancer in his 1937 book *Cancer and Diet*. And though Hoffman typically acknowledged his doubts about each hypothesis, he ends the book with a surprisingly definitive statement on overeating and cancer:

> *I consider my own duty discharged in presenting the facts as I have found them, which lead to the conclusion that overnutrition is common in the case of cancer patients to a remarkable and exceptional degree, and that overabundant food consumption unquestionably is the underlying cause of the root condition of cancer in modern life.*[6]

In their discussion of overnutrition, Doll and Peto singled out Albert Tannenbaum, a cancer researcher in Chicago who had studied the hazards of uranium as part of the Manhattan Project. While working on cancer experiments that were unrelated to diet, Tannenbaum noticed that mice that weighed less appeared to get cancer less often. He spent much of his career studying the phenomenon. In one experiment, Tannenbaum gave a group of 50 female mice a diet of 2 grams of standard rations each day. A second group of 50 female mice received the same 2 grams of food plus as much cornstarch—a carbohydrate formed from glucose—as they cared for. After 100 weeks, the mice that had binged on cornstarch had 26 mammary tumors. None of the mice on the restricted diet had a tumor.

Early twentieth-century researchers had performed similar experiments and arrived at similar results, but in Tannenbaum's next experiments, he did something that earlier researchers had not been able to do. Before running the studies, he first exposed two groups of mice to a powerful chemical carcinogen. After 60 weeks, mice on the restricted diet had developed 11 skin tumors. The mice who also ate cornstarch had 32 skin tumors. Chemical carcinogens alone could cause a mouse to develop cancer, Tannenbaum saw, but they appeared to do so more readily when animals were well fed.[7]

Tannenbaum found that underfeeding mice prevented or delayed the growth of every type of cancer he studied. The less Tannenbaum fed his mice, the greater the effect. With each new experiment, he grew more convinced he was onto something significant—his colleagues joked about how often he could be found weighing his mice. But though confirmed by other scientists, Tannenbaum's results failed to bring nutrition studies into the mainstream of cancer research. Such studies didn't exactly seem like cutting-edge science in the early twentieth century, let alone in the postwar period. And the skepticism of feeding studies was reasonable. No one could say whether the low-calorie diets that benefited mice would also benefit people.

In an effort to see if his findings were in fact relevant to humans, Tannenbaum searched through the records of insurance companies in the late 1940s. What he found likely didn't surprise him: the more someone weighed when past middle age, the more likely the person was to die of cancer. Avoiding excess weight "might result in the prevention of a considerable number of cancers in humans," Tannenbaum wrote, "or, at least, the cancer process may be delayed in time of appearance."[8]

THAT EATING TOO MUCH appeared connected to cancer was a perplexing plot twist for Germany, a nation that began the twentieth century panicked about both cancer and having enough to eat. In fact, the abundance of nutrients in many Western societies in the second part of the twentieth century was not unrelated to Germany's long-standing anxieties about its food supply.

Warburg had always wanted to grow more food by making photosynthesis more efficient, yet the scientist who found a way for Germany (and the world) to have far more food was not Warburg but Fritz Haber, the Kaiser Wilhelm Society chemist who had pioneered the use of gas weapons during the First World War. Haber's plan was to secure the German food supply by solving the nation's nitrogen problem.

Outside of scientific circles, nitrogen may be less appreciated than the three other primary components of the organic world—carbon, oxygen, and hydrogen—but we cannot live without it. Animals get their nitrogen by eating plants or by eating animals that have eaten plants, and almost all plants take their nitrogen from the earth. Without enough nitrogen, crops will not grow and humans will not eat. As the world's population grew in the nineteenth century, agricultural production increased and the nitrogen in the earth's soil rapidly dwindled. For the farsighted scientists waking up to the problem of nitrogen depletion, it was an existential crisis, not unlike the threat of global warming today. In the early twentieth century, nitrogen "was the weakest link in the chain of life," in the words of Haber's biographer Daniel Charles, "a substance more scarce than water, sunlight, or any other nutrient."

Germany did what it could. Manure is excellent fertilizer because it contains nitrogen, and in the nineteenth century, Europeans were transporting bird excrement from Pacific islands by the boatful. But birds could defecate only so often. A new answer was needed, and it was found in the mountains of Chile, where nitrogen (bonded to oxygen in the form of nitrate) could be mined. It was a lifesaving, but temporary, solution. By the end of the century, Europe was importing more than a million tons of Chilean nitrate each year. It was only a matter of time before the supply ran out.

For European scientists of the era, the lack of nitrogen in the soil was a maddening crisis. Nitrogen, after all, is everywhere. It makes up almost 80 percent of the air we breathe. Yet this nitrogen is a gas formed from pairs of nitrogen atoms locked together in a tight embrace. Scientists had no way to separate the 2 nitrogen atoms from each other and thus no way to make them bond with oxygen and form the molecules plants rely on to grow. The nitrogen in the air, Charles wrote, was "as nourishing as the seawater surrounding a thirsting sailor."[9]

In 1908—the year Warburg first measured the breathing of sea urchin eggs—Haber created a device that placed nitrogen gas in the air under so much heat and pressure that the atoms finally gave in and released their twins. Inside the device, hydrogen molecules stood by, waiting to bond. The resulting nitrogen-hydrogen compound, ammonia, naturally reacts with oxygen to form the nourishing molecules plants need.

The breakthrough didn't immediately revolutionize worldwide agricultural production. By the time German industry had managed to mass-produce ammonia, World War I had started. Rather than going into German fields in the form of artificial fertilizers, the ammonia was used to make explosives for the military. But in time, more and more of the nitrogen in Haber's ammonia did find its way into the earth, and the impact could hardly have been greater. In the opinion of the Czech Canadian scientist and author Václav Smil, Haber's ammonia-producing process was "the most important invention of the twentieth century." Without it, Smil argues, the world population could never have grown from 1.6 billion in 1900 to 6 billion by the end of the twentieth century. In 1997, Smil estimated that at least 2 billion people were then alive thanks to Fritz Haber.[10]

Hitler had planned to solve Germany's food crisis by conquering Eastern Europe. Had he not been longing for war, perhaps he would have noticed that one of the Jewish scientists he had chased out of Germany had already found a way to provide Germany with far more bread than all the grain fields of the Ukraine ever could.

Warburg, too, might have taken a lesson from Haber—a lesson that extended beyond food production. Though Warburg thought that diet was hugely important, his interest was limited to the hazards of artificial chemicals and the protective effects of vitamins. He had no interest in the connection between overnutrition and cancer. Haber's success with nitrogen and the subsequent increase in agricultural production was a reminder of the fundamental relationship between nutrients and growth. And if ignored by Warburg and almost all other cancer scientists during his lifetime, the lesson was not lost on everyone.

In an 1889 article in *The Lancet*, the English surgeon Stephen Paget pondered why a cancer might spread to one tissue of the body rather than another. "When a plant goes to seed," Paget wrote, "its seeds are carried in all directions; but they can only live and grow if they fall on congenial soil." The "seed," in Paget's construction, is the cancer cell. The "soil" is the body—the environment in which the seed grows.

The first researcher to draw a clear connection between glucose in the diet (in the form of carbohydrates) and Warburg's glucose-hungry cancer cells may have been the pioneering cancer researcher Anna Goldfeder, who carried out her own series of feeding studies in the late 1920s. Goldfeder reported that tumors grew more quickly when animals ate high-carbohydrate diets and mentioned that Warburg's research might explain the phenomenon. "The organism is to be regarded as a soil for tumor growth," Goldfeder wrote.[11]

If Warburg missed Goldfeder's paper, he had plenty of other

chances to think about aspects of a cancer's soil other than oxygen. In 1926, on Warburg's recommendation, the Kaiser Wilhelm Society had invited the Danish cancer scientist Albert Fischer (no relation to Emil) to come to the Kaiser Wilhelm Institute for Biology as a visiting scientist for three years. Warburg worked in the same building as Fischer and knew him well. In 1929, Fischer suggested that the cells of a body could be analogized to plants. Cancer cells are like "weeds," Fischer said. They outgrow their neighbors because they are better at taking up the food in their environments.[12]

Ernst Freund, the young Austrian doctor who had found a connection between elevated blood glucose and cancer in the 1880s and then attempted to cure Crown Prince Friedrich, also thought in terms of seeds and soil. In 1938, the year Germany invaded Austria, Freund fled to England, where, at age 75, he obtained funding for a new cancer research program. By then, Freund had moved on from his single-minded focus on the role of elevated glucose in cancer, but he remained convinced that cancer was a systemic problem that could only be understood in the context of the metabolism of the entire body.

As Freund saw it, the microscope itself had led scientists astray by allowing them to peer into the cancer cell and thus to lose sight of the larger picture. "[I]nstead of confining ourselves to an investigation of the action of the malignant cell on the organism," Freund wrote, "we ought to investigate the action of the organism on the malignant cell."

Shortly before his death in 1946, Freund published a book, *Metabolic Therapy of Cancer*, in which he urged cancer scientists to think beyond the cancer cell and to consider the soil of the body in which it grows. "[I]t is only natural to link growth of any kind with nutrition," Freund wrote. "The farmer ascribes a good or bad crop primarily to the nature and composition of the soil, i.e., the nutrient, though he must make allowances for the quality of the seed; and every layman realizes the value of diet to growth."

"Curiously enough," Freund continued, "medical science takes a different attitude."[13]

IN "THE CAUSES OF CANCER," Doll and Peto paused to consider medical science's attitude toward overnutrition. Why, they wondered, had Albert Tannenbaum's feeding studies made "little impact on cancer research," given the evidence showing that the quantity of food an animal consumes can have "profound effects" on cancer risks? Part of the problem, they suspected, was that low-calorie diets were impractical for most people. But there was a deeper issue as well: "[W]e still have no clear idea of the mechanisms whereby dietary restriction protects laboratory animals," Doll and Peto wrote. Tannenbaum had struggled with the same issue. Understanding how a low-calorie diet prevented cancer "might really give an insight into carcinogenesis itself," he noted in 1959. And yet there remained "no explanation."[14]

The lack of certainty on the role of diet is reflected in the conclusions of "The Causes of Cancer." Doll and Peto estimated that diet accounted for 35 percent of American cancer deaths. That figure made diet the single greatest cause of cancer, and a 2015 review found that Doll and Peto's findings have held up remarkably well. But it was little more than an educated guess. The actual percentage of American cancer deaths caused by diet, they stated, might be as low as 10 or as high 70. "Diet is a chronic source of both excitement and frustration to epidemiologists," they wrote.[15]

Neither Tannenbaum nor Doll and Peto considered that Warburg's discovery of fermentation in cancer might offer a way to explain the fundamental connection between seed and soil, between overeating cancer cells and overfed bodies. The oversight is perhaps understandable. Warburg himself could not see the link, and though the two phenomena might have seemed related, no one could fully explain the relationship. What was missing at the end of the postwar period was a metabolism synthesis, a way to combine Warburg's discovery of fermentation in cancer cells with a long history of observations on food consumption, body weight, and cancer.

That synthesis would eventually arrive, and with it, perhaps, the best answer of all to the question of why cancer became so common in the Western world in the nineteenth and twentieth centuries. But before Warburg's discovery of how cancer cells eat could help solve a long-standing cancer mystery, something odd happened: everyone forgot about it.

PART IV

Pure, White, and Deadly (The Twenty-First Century)

Faustus:
How am I glutted with conceit of this!
Shall I make spirits fetch me what I please,
Resolve me of all ambiguities,
Perform what desperate enterprise I will?
I'll have them fly to India for gold,
Ransack the ocean for orient pearl,
And search all corners of the new-found world
For pleasant fruits and princely delicates.

—CHRISTOPHER MARLOWE,
DOCTOR FAUSTUS

CHAPTER SEVENTEEN

Lost and Found

IN HIS 80S, Otto Warburg occasionally did something that he would have once found unthinkable: he played hooky. When the weather was nice, Warburg would call on his young glassblower, Peter Ostendorf, to drive him to the nearby Havel River. For the rest of the afternoon, the two would sail peacefully in the slow-moving waters in Warburg's Nordic Folkboat, Warburg lecturing his young employee on his favorite books or the history of the region during the Holy Roman Empire.

If still "married" to science in the 1960s, Warburg had made another important discovery: there was more to life than the laboratory. Warburg and Heiss befriended two elderly women in Dahlem and sometimes had flowers from their garden sent to them. In return, the women would share their tickets to the philharmonic. Always a dog lover, Warburg grew especially attached to his nearly 200-pound Great Dane, Norman, who could be seen traveling to and from the institute in the back of Warburg's black Mercedes. In the evenings, Norman rested on a couch in Warburg's bedroom, where he took his "nightcap": a piece of Warburg's best chocolate.[1]

Warburg remained very much himself. He hated it when people

raced their sailboats and would purposefully steer into the middle of the river to disrupt them. Ostendorf, who generally remembers Warburg fondly, could do nothing but sit in Warburg's boat, embarrassed. Warburg also purchased a new country home in the North German island of Sylt. He chose the location, he once explained, because "the district is the most thinly populated in the whole of Germany." With few neighbors to agitate him, Warburg directed his ire at the noise from planes overhead.[2]

Less work for Warburg was still a lot of work. In the late 1960s, he continued to publish some 5 to 10 papers every year on photosynthesis and cancer. But for all his willful obliviousness to modernity, Warburg could sense that his own research was becoming outdated. In 1953, James Watson and Francis Crick had studied Rosalind Franklin's X-ray crystallography images of DNA and deciphered the structure of the molecule. Modern molecular biology was suddenly coming into focus. Genes are made of DNA molecules, which function like a code in the nucleus of a cell. This code is read by other molecules to create, or express, proteins. To study the inner workings of the cell without taking into account the new molecular biology came to seem absurd, like trying to make sense of a complicated computer program without knowing anything about coding.

Warburg was not at all pleased to find himself in this new scientific era. It didn't help that Dahlem's notorious Kaiser Wilhelm Institute for Anthropology, Human Heredity, and Eugenics had quietly morphed into a modern genetics institute. But the problem ran deeper, all the way to the core of Warburg's scientific being. In Warburg's view, molecular biology's new obsessions with genes was adding a layer of complication to every discussion and, in the process, obscuring fundamental observations. It was, Warburg said, "as though before one could say anything about the plague bacillus, one would first have to determine the sequence of the nucleic acids."[3]

The German academic Wolfgang Lefèvre witnessed Warburg's

anachronistic existence in the late 1960s. At the time, Lefèvre was a student at the Free University of Berlin, the institution the Americans had built across the street from Warburg's home in Dahlem after the war. He remembered that he and other students would sometimes stay up all night drinking and smoking while engaging in heated political discussions. When they finally stumbled out for fresh air, it would already be morning, and there, across the street, would be Otto Warburg atop his horse. Heiss would invariably be by the animal's side, holding the reins as Warburg went round and round his property.[4]

In the spring of 1965, Dean Burk wrote a letter to Warburg describing how a colleague had taunted him at a recent conference: "Dean, why don't you throw away your manometers and get down to some real biochemistry." At approximately the same time, Peter Pedersen, a biochemist at Johns Hopkins, spotted a Warburg manometer left out in the hallway with the trash. The era of metabolism research was over. According to Pedersen—one of the few cancer researchers who would continue to pursue Warburg's ideas about cancer and energy—there was "little or no interest" in Warburg's cancer research by the late 1960s.[5]

As molecular biologists studying cancer turned their attention to DNA and cancer-causing genes, Warburg's interest in how cells extract energy from food seemed not only outdated but backward. A 1972 article in the German magazine *Der Spiegel* claimed that Germany had fallen behind in the study of cancer DNA and viruses and that it was entirely Warburg's fault. Warburg had "unilaterally" sent cancer research in the wrong direction, a biochemist at Frankfurt University told the magazine.[6]

If any trace of Warburg's metabolic explanation of cancer still lingered, it would be gone by the middle of the next decade, after University of California, San Francisco (UCSF) researchers J. Michael Bishop and Harold Varmus identified the first cancer-causing gene. Such genes, known as oncogenes, had turned out to be normal human

Otto Warburg, date unknown.

genes, only in mutated form. For the first time, Boveri's initial hunch that chromosomes held the secrets to cancer could be explained in the language of modern science.

Two conceptions of cancer, one metabolic, one genetic, emerged from the study of sea urchins in the first decade of the twentieth century. The genetic conception had now prevailed. Warburg's name was virtually erased from cancer science. In 1988, one biochemist was shocked to discover that young scientists of the day had not even heard of Warburg.[7]

The genetic explanation did not contradict the widely held belief that most cancer deaths were caused by our environments and potentially preventable. Smoking or toxic chemicals or diet could still be said to be "causes" of cancer. The emphasis on mutated genes merely offered a more precise way of explaining how environmental carcinogens inflicted their damage. But the new molecular biology might have undermined the broader push for cancer prevention just the same. Because mutations can arise by chance alone as dividing cells copy their DNA, any given cancer could now be dismissed as bad luck.

Prevention also felt less urgent in the new oncogene era. After Bishop and Varmus, a cure for cancer was thought to be only a matter of time. The gene they had identified, SRC, was one oncogene, but there were sure to be many others. To win the war on cancer, researchers would need only to identify these genes and learn how to turn them off. The magic bullets might still be missing, but cancer scientists had finally found their targets.[8]

EVEN AS MEMORIES OF Otto Warburg were fading, one scientist's unlikely journey back to Warburg was already beginning. In the summer of 1985, nine years after Bishop and Varmus published their groundbreaking discovery, Chi Van Dang completed his medical residency at Johns Hopkins and decided to do his clinical fellowship in oncology at UCSF. Dang and his wife, Mary, crammed as much as possible—including their Persian cat—into their red Toyota Tercel and headed West, tracking their route, as best they could, on AAA maps.

The drive from Baltimore to San Francisco was only one more leg in Dang's long journey. Dang, today the scientific director of the international Ludwig Institute for Cancer Research, grew up in Vietnam in a family of 10 children. There was no TV in his home, but there was something to watch. After dinner, his father, the country's first neurosurgeon, would often take out the 8-millimeter projector and show the children films of surgical techniques. With no better options for entertainment, Dang and his siblings sat and watched. "Some of it was pretty gruesome," Dang recalled.[9]

In 1967, with the Vietnam War raging, Dang's parents sent him and an older brother to live with an orthopedic surgeon in Flint, Michigan, who had met Dang's father in Vietnam. Dang, 12 at the time, felt guilty about leaving his family behind in a war zone, but he adjusted to his new life well. Though he was sometimes harassed for being Asian, he was relieved that few of his new classmates in Flint realized he was Vietnamese.

Dang went on to the University of Michigan, where he studied chemistry. He never planned to become a cancer researcher, but like so many young people interested in medicine and science at the time, he wanted to be a part of the revolution in molecular biology. It felt like a "golden moment," Dang recalled. And what better place to experience that moment than UCSF, home to Bishop and Varmus, the two researchers who had put the molecular biology revolution into motion.

The adjustment to San Francisco took some time. Dang would drive up the steep hills, only to find his Toyota, a stick shift, rolling back down whenever he tried to switch gears. It was a good reminder that a journey into cancer research can feel Sisyphean at times, yet Dang remained committed. Though he officially became stateless when South Vietnam ceased to exist in 1975, he could not have been more firmly rooted in medical science. Six of Dang's nine siblings would also become doctors. Another became a dentist.

While learning to treat cancer patients, Dang got his chance to interview with Bishop and Varmus themselves for a postdoctoral research position. As he sat down for the meeting, it struck him that he still knew alarmingly little about the new world of molecular biology. When Varmus asked him what he wanted to work on, the most Dang could muster was "oncogenes." Varmus nodded and asked him which particular cancer-related gene interested him most. Dang paused. It was a perfectly reasonable question, but one he couldn't answer. Dang told the famous scientist the truth: he had no idea. "I barely knew anything," Dang said much later.

Varmus pointed Dang toward UCSF researchers then working on MYC, one of a number of the newly discovered oncogenes. At the time, little was known about MYC other than that when overexpressed— meaning there were too many copies of its corresponding protein—it could drive cells to grow and multiply. As Dang and others began to study MYC, they saw that it isn't merely one more signal in a chain. It codes for a type of protein, known as a transcription factor, that binds to the DNA, turning scores of other genes on and off and dramatically reshaping the behavior of the cell. Wherever Dang looked in the cancer cell, he seemed to find traces of MYC's influence.

After his postdoc at UCSF, Dang returned to Johns Hopkins, still determined to identify the many genes under MYC's sway. Lacking today's modern sequencing tools, the search proved maddeningly slow. Work that would now take days could stretch on for months, even years. But Dang and the postdocs in his lab pushed ahead. Like cartographers tracing the contours of a newly discovered land, they mapped each chain of linked reactions, or signaling pathways, on diagrams that were reproduced and hung on the walls of cancer labs around the world. The diagrams were breathtakingly complicated. Dang and his generation of cancer biologists were not simply discovering a new land, but a land of interconnected mazes—or mostly interconnected. The metabolic enzymes responsible for supplying food and breaking down nutrients for energy did not even appear on Dang's diagrams. They had their own separate diagrams, and those became increasingly hard to find in molecular biology labs in the 1980s.

Dang was thus surprised when in 1997, after more than a decade of tracing MYC's influence on cells, he found that it spread to the enzyme lactate dehydrogenase (LDH). As one of the key players in fermentation, LDH was not supposed to be linked with an oncogene. It was one of the housekeeping enzymes that belonged on the other diagram, the metabolism diagram. Dang had followed MYC to a place that wasn't even on his original map.

Dang might have put the finding aside, but he had long felt that it was important to pay attention to discoveries that don't seem to make any sense. Such findings, Dang says, can "teach you something." So, instead of dismissing metabolic enzymes, Dang began to read as much as he could about them. That reading quickly led him to Otto Warburg, who had isolated LDH and explained its role in fermentation some 60 years earlier.

The literature from Warburg's day confirmed what Dang had found: LDH activity was elevated in cancer. MYC, when overexpressed, wasn't only driving cells to divide more often than they should. It was also driving them to eat and ferment more than they should.

"The perception was that metabolism was just there to support everything else," Dang said. And yet the more Dang read and reflected, the less sensible it seemed to think of metabolism as somehow separate from the rest of the cancer cell's activities. Cancer was like a building project, and the different teams on the project had to work together. "It has to be a coordinated process so that you can build things in an orderly way," as Dang put it. "The bricks and the cement don't somehow magically end up where they're supposed to be."[10]

DANG'S PAPER ON MYC and LDH, published in 1997, was initially met with great skepticism from most of his colleagues. The lack of enthusiasm for his work came as no surprise. Dang knew that the study of cellular metabolism was out of fashion. But then something happened that did truly surprise Dang. A colleague took him aside and told him that the negative reaction to his findings might also have something to do with the lingering resentment older Jewish researchers felt toward Warburg for his decision to stay in Nazi Germany—and perhaps the fact that he survived with relatively little trouble.

It was not an entirely outlandish idea. Warburg himself came to believe that Jewish researchers had turned against him for precisely this reason. And it's true that various Jewish scientists had challenged

Warburg's claims about cancer and photosynthesis after the war. But then, countless scientists, Jews and Gentiles alike, resented Warburg for all sorts of good reasons. And it was not only the Jewish scientists who thought that Warburg was wrong about both photosynthesis and cancer.[11]

It's possible that Warburg's reputation made it easier for some researchers to leave his metabolic understanding of cancer behind as the new era of molecular biology set in. In 2000, Robert Weinberg, a pioneering oncogene researcher, coauthored a seminal paper, "The Hallmarks of Cancer," which listed six fundamental changes in a cell that make cancer possible. The increase in glucose consumption and fermentation characterized by the Warburg effect was not one of them. Weinberg, who also failed to so much as mention Warburg in the first edition of his highly regarded 2006 cancer textbook, makes no secret of his distaste for Warburg. "I confess to harboring very negative feelings about Warburg because of his Nazi affiliations," he said. Weinberg explained that Warburg's story was personal for him, as his own parents had fled Germany in 1938.

Even so, Weinberg insists that his resentment of Warburg did not influence his work. Weinberg ignored Warburg, he explained, because he does not believe that "altered carbohydrate metabolism of cancer cells" is "the root cause of their aberrant behavior." Besides, virtually everyone was ignoring Warburg and metabolism. "By the time I became deeply immersed in cancer research, Warburg was more of a historical relic," Weinberg said, "almost a forgotten footnote in history."

In Weinberg's view, the anger at Warburg was not widespread and was not, ultimately, about his past in Nazi Germany. What most scientists disliked about Warburg, rather, was his "simplistic depiction of cancer" and "his imperious, Prussian style of handing down judgments" from his "self-imagined throne."[12]

Weinberg's assessment is probably correct. In fact, many of Warburg's most loyal supporters after the war, including David Nachmansohn, Hans Krebs, and Harry Goldblatt, were Jewish. The best explanation for the "anti-Warburgian disease entity," as Warburg once referred to the sentiment against him, was not Warburg's political judgment before and during the war. It was his scientific judgment after the war.[13]

IF CHI VAN DANG HAD been the only scientist rediscovering Warburg in the late 1990s, it's possible that the supposed "anti-Warburgian disease entity" would have continued to shape scientific opinion. But as Dang was reading through scientific papers from the 1930s, Craig Thompson, now the president and CEO of the Memorial Sloan Kettering Cancer Center, was beginning his own circuitous journey back to Warburg. Like Dang, Thompson came of age during a period when Warburg's science was thought to be obsolete. "Nobody wanted to do biochemistry because all the great discoveries had been made," Thompson said. "The feeling at the time was, 'Let's do cool stuff.'"

Thompson has a reputation for brashness. He played soccer at Dartmouth, but quit the team when his coach asked him to focus more on training and less on science. Shortly thereafter, Thompson decided he was done not only with soccer but with college as well. In 1972, he applied to Dartmouth's medical school as a 19-year-old sophomore and was accepted—even after getting into a shouting match with the doctor who had interviewed him.[14]

By the mid-1990s, Thompson was doing "cool stuff" at his own lab at the University of Chicago. His focus was not cancer but immunology, and he had arrived at a fundamental question. When someone develops an infection, the immune system responds by rallying an army of cells to attack the invaders. The result, if the response is strong enough, is a swollen mass that, like a tumor, forms from rapidly growing and dividing cells. But the new immune cells, unlike cancer cells, are only temporary workers. As the underlying infection heals, the cells, no longer needed, begin to die off. It was this disappearing act, an act of collective suicide, that had captured Thompson's imagination. How, he wondered, does the body distinguish between cells that are chosen to live and cells that are chosen to die, so that only the unwanted cells are eliminated?

It wasn't a new question. August Weismann, the celebrated German evolutionary biologist, whose lectures Otto Warburg had attended at the University of Freiburg, had marveled over the organized death of cells more than a century earlier. The phenomenon is now known as "programmed cell death," or "apoptosis," from the Greek for "falling off." Though long overlooked, apoptosis is arguably as fundamental to the biology of multicellular organisms as cell division. During the development of an embryo, apoptosis, like a sculptor chipping away at a stone to reveal a human form, helps shape our bodies. (If not for apoptosis killing off the connective tissue, our fingers and toes would be webbed.) Even as our cells grow and multiply, they never stop dying. Approximately 10 billion cells are thought to die in a person every single day. In some cases, as with immune cells after an infection, the cells die because they no longer serve a purpose. In other cases, apoptosis rids the body of cells that are damaged beyond repair.

Despite the fundamental role of apoptosis in biology, the precise mechanisms that govern the process remained unclear well into Thompson's time. In a sense, he was trying to solve a murder mystery. And he had a suspect: a group of proteins known as the BCL-2 family. From work conducted in other labs, it was clear that the BCL-2 proteins had a role in cell death, but exactly what the proteins were doing, Thompson couldn't say. Most curious of all, the BCL-2 proteins were interacting with the mitochondria, the tiny power stations of the cell.

The involvement of the mitochondria in apoptosis made the mystery all the more intriguing. In 1913, Warburg had made an important observation while examining the liver cells of a guinea pig. The cells were able to breathe oxygen, he suspected, thanks to small particles he could see inside the cells under his microscope. Warburg called the particles "grana" and refused to call them anything but "grana" for the rest of his life—long after the rest of the scientific world had begun to refer to them as mitochondria.[15]

Thompson had at once arrived at an exciting finding and a fairly significant problem: no one in the new world of molecular biology knew anything about the mitochondria. The mitochondria were part of the old world of biochemistry, the world of Warburg and his colleagues, the world that modern molecular biology was supposed to have left behind decades earlier.

If Thompson was going to make any progress in understanding how the body gets rid of unneeded cells, some unlucky soul in his lab would first have to relearn the science of a bygone era. It was exactly the sort of job one gives to the person who is least able to object. In Thompson's Chicago lab, that person was a new arrival, Matthew Vander Heiden, a soft-spoken Midwesterner who was just beginning his MD-PhD.

In some ways, Vander Heiden was Thompson's opposite. A product of a small Wisconsin town once known for its lawn mower factories, he discovered his love of research while washing equipment in a lab as part of a work-study job. And he was a better fit for a return to outdated science than Thompson could have known. Vander Heiden's wife, the biologist Brooke Bevis, said that her husband struggles to throw anything out that he can still use, be it his watch, his phone, pots and pans, or laboratory equipment. "The list goes on and on," Bevis said. "He carries his Midwestern sensibilities with him everywhere he goes."[16]

To refamiliarize himself with mitochondria, Vander Heiden turned back to his old biochemistry textbooks from college. The textbooks were full of useful facts on the subject. He read about the work of Lynn Margulis, the legendary evolutionary biologist who had studied at Chicago 40 years earlier. Margulis had revealed the remarkable origins of the mitochondria: they are descendants of ancient bacteria that moved inside another unicellular organism more than a billion years ago.

It turned out to be a symbiotic relationship, and it would give rise to the eukaryotic cells that form the building blocks of all plants and animals. The mitochondria could burn food with oxygen; the host cell could ferment. This is why our cells, as Warburg put it, have "two engines." Though the two organisms evolved together, the mitochondria held on to a small number of their own genes. Even at the cellular level, we have split personalities.

Vander Heiden's old biochemistry textbooks were less exciting when it came to the role of the mitochondria in the cell, which was thought to be fairly one-dimensional and completely figured out. The mitochondria were understood to be the power source in the basement of the cellular building. They supplied energy and heat so that the genuinely interesting aspects of molecular life could continue.

In need of additional expertise on mitochondria and metabolism, Vander Heiden teamed up with Navdeep Chandel, of Northwestern University, who was then a cellular physiology student in another University of Chicago lab and among the few young researchers of the era interested in the structure of mitochondria.

In 1996, as the two began to run experiments on BCL-2 proteins, Xiaodong Wang, a researcher then at Emory University, made a stunning discovery: The mitochondria, it turned out, weren't peripherally involved in the suicide process. They were driving it. When the mitochondria are no longer able to function, either due to cellular damage or lack of fuel, they collapse and send out a signal that triggers other proteins outside the mitochondria to begin slicing up the cell.

That a cell's mitochondria could kill it from the inside was a shock in itself. The mitochondria's relationship to the host cell suddenly looked more sinister than symbiotic. Stranger still was the manner in which a mitochondrion's death signals went out. As the cellular machinery that burns food comes to a halt, a mitochondrion's outer membrane bursts open, allowing the molecules inside, like refugees fleeing a burning city, to rush out. Among the fleeing molecules is cytochrome c, an enzyme that allows us to breathe by passing electrons to oxygen. And cytochrome c, of all molecules, is critical to activating the death signal. "Nothing in biology quite compares with this two-faced Janus," as the biochemist and author Nick Lane put it. The very same protein that gives us life had been revealed to also be an assassin.[17]

Once it became clear that the mitochondria were directing apoptosis, Vander Heiden and Chandel could piece together the role of BCL-2 proteins. Like emergency workers fixing leaks in a dam before a storm, the proteins would fill pores in the mitochondrial membrane, keeping the charges inside and outside of the membrane in balance and preventing the escape of cytochrome c. Electrons would continue to make their way to oxygen and the given mitochondrion would continue to function.

But if a mitochondrion was too damaged or unable to obtain food or oxygen, the dam would no longer hold. Sensing the mitochondrion failing, other proteins from the BCL-2 family would open additional pores in the membrane and the entire system would collapse, allowing cytochrome c to escape and alerting the cell that a brownout was underway. A single failing mitochondrion wouldn't trigger cell death, but if enough of the mitochondria began to fail, the brownout would become a blackout and the cell would self-destruct.

Even more important than the precise mechanisms of apoptosis was the broader picture coming into focus for Vander Heiden: the common understanding of the causal arrow of cellular life, it increasingly seemed, was backward. The energy-releasing reactions inside the mitochondria weren't only responding to the cell's needs. They were also dictating what a cell should do, governing even the cell's most important decision of all: whether to live or die.

Vander Heiden remembers the breakthrough on apoptosis as a "watershed moment." The feeling, he said, was, "Oh my goodness. We don't really understand metabolism." But almost as surprising as the new thinking on metabolism in Thompson's lab was that so few others were interested in it outside of the lab. "It's as fundamental as you get in terms of how biology works," Vander Heiden said. "I looked around, and no one was studying it."[18]

CRAIG THOMPSON HAD set out to understand how the body eliminates immune cells when they're no longer needed. Vander Heiden and Chandel's research on the mitochondria and apoptosis helped provide at least part of the answer. Millions of unwanted immune cells being driven to commit suicide meant something was undermining their mitochondrial power stations. Thompson wanted to determine what that something was.

He turned his attention to the messengers that instruct our cells to stay alive and grow. Owing to the work of the biologist Martin Raff, among others, it was already known that such growth factors are necessary to prevent cells from turning to apoptosis. When the signals don't arrive from other cells, the suicide process is referred to as "death by neglect." Raff once remarked that every cell in the body wakes up each morning thinking about suicide and has to be talked out of it by its nearest neighbors.

If Thompson already knew that growth factors were likely part of the story and that they had to eliminate cells by way of the mitochondria, he didn't yet know how the process unfolded. Jeff Rathmell, then a young researcher in Thompson's lab (he now runs his own lab at Vanderbilt University), began to work with Vander Heiden to find answers. In one experiment, Rathmell took a single human cell and placed it in a small dish that contained the glucose and other nutrients the cell needed to grow and thrive. It was an environment, as Thompson put it, "that would keep a yeast cell happy for the rest of its life."[19] And yet, when Rathmell examined the lonely cell, he did not find a picture of happiness. Without growth factors telling it to take up nutrients, the cell went on a hunger strike. The lack of food would be registered by its mitochondria, and within 48 hours, the cell would die by apoptosis. Rathmell repeated the test again and again with different cell types, and the results were always the same.

The body gets rid of unwanted cells, Thompson and his colleagues realized, by cutting off growth factors and starving the cells into suicide. As Thompson described the process, as cells learned to live together as collectives, tough sacrifices had to be made. What cells "really gave up as multicellular organisms," he explained, was the freedom to eat.[20]

Craig Thompson was still running an immunology lab, but the more he learned about how cells die, the more he found himself thinking about cancer. Cancer cells were the opposite of the isolated cells that refused to eat. They were cells that ate whenever they felt like it, cells that couldn't be starved into suicide, even though they had no purpose in the body. "Cancer really is saying, 'Let's override all the control and just grow whenever there's nutrient availability,'" said Lewis Cantley, director of the Meyer Cancer Center at Weill Cornell Medicine. "It's just going back to the wild."[21]

With each new finding, Thompson grew more confident that a return to metabolism was going to transform modern molecular biology and open up new avenues of cancer research. Rathmell was much less sure. For all his excitement about his research, the Warburg revival was only in its infancy. In the late 1990s, academic journals still weren't interested in publishing papers on metabolism. With the exception of Chi Van Dang and a small number of holdouts from the Warburg era, such as Peter Pedersen at Johns Hopkins, almost no one saw a future for the field. To the outside world, it still looked as though Thompson's lab was attempting to make sense of a vastly complicated construction site by studying the trucks showing up with fuel.[22]

Rathmell became particularly concerned after attending a major scientific conference where he and other postdoctoral researchers presented their work on posters. At these "poster sessions," young scientists stand by their presentations as older and more established colleagues amble by. Waiting silently by his poster, Rathmell felt as isolated as one of the cells in his experiments. "There were 400 people walking around in this relatively small space and only one person stopped at my poster," he recalled. "And that person stopped to tell me that I was wrong and wasting my time."

And so, in 1999, when Thompson decided to relocate to the University of Pennsylvania and devote his research program to metabolism, Rathmell had reason to worry. As Thompson remembered it, Rathmell told him that he might be jeopardizing the careers of all the young scientists in his lab. Thompson, confident as ever, encouraged his students to stick it out. He had a new agenda: to identify the specific genes that allow a cell to eat without permission from growth factors. Thompson thought that if he could identify these genes, it might do more than bring Warburg's metabolism research into the age of molecular biology. It might also explain why we get cancer.[23]

CHAPTER EIGHTEEN

The Metabolism Revival

CRAIG THOMPSON HOPED to bring Warburg's science into the era of molecular biology, to connect the old knowledge that cancer cells take up lots of glucose and ferment it to the new knowledge of how oncogenes signal cancer cells to divide and multiply. But before moving ahead, Thompson faced a considerable obstacle: working with the metabolic enzymes would require an expertise in the biochemistry of Warburg's era, an expertise that Thompson and most members of his lab still lacked.

Thompson called a lab meeting, and the half dozen postdocs and PhD students came to an agreement: each member of the team would be assigned a different metabolic enzyme and be required to learn how to study it in a modern lab. That meant digging up the classic papers on enzymes by Warburg and his contemporaries. It also meant spending a lot of time with Britton Chance, a legendary biochemist of Warburg's era, then in his mid-80s, who was still at the University of Pennsylvania. "We would go over to Britton's lab, and he would roll out old machines that hadn't been used in 20 years," Thompson recalled. "We would dust them off and start using them again."[1]

That Chance, who died in 2010, helped facilitate Thompson's return to Warburg's science was somewhat ironic. Chance had exchanged a number of letters with Warburg in the 1950s and 1960s and visited Warburg in Berlin in the summer of 1966. Chance's notes from the visit suggest that Warburg spent most of the time lecturing him on photosynthesis and cancer. It's not clear if Chance argued with Warburg, but it's unlikely that he found Warburg's arguments persuasive. Chance had investigated Warburg's claim that cancer cells from abdominal fluid rely almost exclusively on fermentation and arrived at the opposite conclusion. As far as Chance could tell, the mitochondrial power stations of the cancer cells were in good working order.[2]

Chance proved more helpful than Thompson could have hoped. "He knew all the tricks," Thompson said. "He became our reference source for all the things that aren't written in scientific articles." Thompson's lab had stepped into a scientific time machine. Though it was no longer in use, Chance had even held on to one of the original Warburg manometers. One afternoon the members of the lab gathered around the device to examine it, fully aware that their new investigations could be traced back to the simple U-shaped tubes Warburg had spent his life gazing at.

Even with Chance's help, Thompson had no idea if his hunch about metabolism was correct. When it was time for Vander Heiden to decide if he would continue his work in the lab or finish the medical training portion of his MD-PhD program, Thompson told him to return to medicine. If they were truly onto something important, the field would still be there waiting for him in a few years. In the meantime, metabolism was still considered so archaic that it would be hard to even publish their research.

By 2004, seven years after he and his lab had turned to metabolism, Thompson's doubts were fading. He had found precisely the type of gene he had been looking for. It wasn't an entirely new discovery. The gene, AKT, had already been identified as a cancer-causing gene

that drove cells to divide and grow. But Thompson found that AKT
had another role, a role that appeared even more fundamental: it drove
cells to swallow glucose and ferment, to switch on the Warburg effect.
Thompson discovered that he could turn on "the full Warburg effect"
simply by inserting a mutated AKT gene into a healthy cell, causing the
cell to build overactive AKT proteins.[3]

AKT does not act alone. Like a commander in a castle directing
the guards to lower the drawbridge so that a shipment of wood can be
hauled in, the protein made by AKT directs other proteins to open pas-
sageways through a cell's membranes so that glucose can come inside.
AKT also relays the news that food has arrived to mTOR, a protein
that plays a key role in governing what a cell will do next based on the
supplies at hand. Depending on the type of tissue and the food available, the glucose can be broken down for spare parts, burned to heat
and power the cell, or stored for later use.

In a healthy cell, AKT waits for the proper signals to arrive from growth
factors before letting in glucose. These messages inform the cell that food is
arriving. But when AKT is hyperactivated, as it often is in cancer, the commander of the castle no longer waits for updates. The drawbridge opens,
and the cell scavenges for more and more glucose. The Warburg effect and
fermentation soon follow. A once satisfied cell is now insatiable.

Thompson, Dang, and other researchers at the forefront of the
metabolism revival were coming to see that metabolic enzymes that
govern how cells eat and use fuel are not merely a by-product of other
more fundamental cellular processes—not merely an "epiphenomenon,"
as one prominent cancer gene researcher described the Warburg effect
in 2004. They are woven into cancer signaling networks and sit at the
origin of the disease. Cells might develop mutations that spur growth
before they develop mutations that allow them to overeat, but such cells
aren't likely to survive. "If you don't have enough cement, and you try
to put a lot of bricks together," said Dang, "you're going to collapse."[4]

According to Thompson, overeating glucose doesn't pose an equal threat to all tissues. Cancers rarely originate in fat and muscle cells, Thompson suspects, because those tissues are better equipped to manage an influx of energy. Like a castle that has a huge cellar for excess wood, the cells are able to store much of the glucose rather than burning it. The epithelial cells that line our organs are in a more precarious situation. When AKT won't stop telling an epithelial cell to eat, it will soon have more glucose than it can handle. It searches for a way to adjust. Thompson believes that the Warburg effect can be understood as one such adjustment. As he explained, "Getting more glucose than you want might be what the Warburg effect is."

A cell that is overeating due to overactive AKT proteins won't necessarily turn cancerous. Genes known as tumor suppressors are designed to put a stop to a cell's self-destructive impulses. The cell might still manage to register internal damage and turn to apoptosis— or be induced to do so by an immune response. And even if growing cells can't be eliminated, the emerging colony will still have to develop new tricks before turning into a deadly cancer. The cells will need to evolve to outwit immune attackers and grow new blood vessels to support their new appetites. Eventually, some of the cells in the new colony will have to break free and venture off on their own in search of new places where they can grow.

But if a cell that can eat all it wants isn't yet cancerous, it is an emerging threat to its neighbors. If a cell manages to survive in this gluttonous state, Thompson said, "it really doesn't ever think about dying. Now it thinks, 'I've got a lot of fuel; I could do a lot of other things.'"[5]

One specific thing an overeating cell can do is build. With fermentation now supplying plenty of energy, the mitochondria no longer have to worry about doing the same and will set aside many of its nutrients for construction projects, just as soldiers in a castle, presented with a huge amount of wood, might come to see an opportunity to build new castles. In this analogy, the soldiers had not been longing for wood. They had never even asked for it. But now that the wood is arriving, their imaginations are running wild. Never mind one castle. There is enough material to support two.

The cell's mitochondria now function like workshops within the castle, taking the wood coming in (the nutrients) and, with the help of scores of craftspeople (the enzymes), turning it into the various parts needed for new castles (the daughter cells). Biologists suspect that it is this special ability of the mitochondria to create the building blocks for new cellular components—fats, proteins, and DNA—that explains how the symbiotic relationship between the mitochondria and the host cell evolved. The host cell learned to take advantage of the energy the mitochondria could provide. But the most valuable thing it gained wasn't more power. It was the ability to reinvent itself.[6]

OVEREATING CANCER CELLS use their excess glucose in other creative ways as well. In 2009, Georgia Hatzivassiliou and Kathryn Wellen, of the University of Pennsylvania, were postdocs in Craig Thompson's lab when they made a critical finding. It was known by then that cancer isn't merely a disease of genetic mistakes, or mutations. Every cell carries the entire genome, the blueprint for the body, in its DNA. A lung cell differs from a liver cell because different parts of the DNA code are expressed. When an individual cell alters the state it's in by expressing different genes, it is known as an epigenetic change.

By 2009, it was also known that nutrients can influence which parts of the blueprint are read. The most famous example of the phenomenon is found in bees. There is no genetic difference between the larvae that grow into worker bees and the larvae that grow into queens. The queen develops ovaries and a larger abdomen because she is fed royal jelly. Scientists have now identified the specific gene that the royal jelly silences to make the developmental shift from worker to queen possible.

What was not clear, until Wellen and Hatzivassiliou's breakthrough, was whether glucose plays a direct role in changing which genes are expressed. Before new genes can be expressed, the tightly wound DNA inside of the cell needs to open up, just as a paper blueprint in the form of a rolled-up scroll will need to be spread out for reading by a construction worker. Wellen and Hatzivassiliou demonstrated that excess glucose can trigger the unwinding process. And the glucose isn't merely a signal. The glucose molecule itself will be transformed into a component of the proteins that open up portions of our DNA for gene expression. Wellen described the process as a cycle that pushes the cell in a new direction. The arrival of nutrients change which genes are expressed, and the newly expressed genes, in turn, change the metabolic state of the cell, allowing still more food to be consumed.

A decade earlier, Vander Heiden had found that metabolism governed a cell's decision to live or die. Wellen and Hatzivassiliou's research on epigenetics was still more evidence that metabolism influences virtually every decision a cell makes. A 2019 paper in *Nature* revealed that even the lactate produced via the Warburg effect can directly change which genes are expressed in a cell. As Wellen once explained it, the idea that a cell could carry out its functions without sensing its metabolic state is like trying to take trips in a car with no awareness of the gas gauge. "Your cells need to be able to assess what their nutritional resources are," Wellen said, "and then make decisions about what they're doing based on that information."[7]

It's not hard to appreciate why natural selection favored cells that make decisions according to how much food they have access to. All life-forms evolved from single-celled organisms. When food is scarce, an organism's best chance at multiplying (and passing along its genes) is to conserve resources and to survive the famine for as long as possible, so that it can reproduce when food finally arrives. Researchers believe this is why eating very few calories has been shown to extend life in so many different species. Robert Koch discovered the same phenomenon when he realized that the bacteria that cause anthrax would turn into resilient spores when no food was available. When food is abundant, the opposite strategy makes the most sense. The most successful single-celled organisms will be the ones that eat and multiply as soon as they come across food.

As Thompson explained, a cancer cell behaves like a growing single-celled organism that is binging on glucose. "When they see food in their environment, they move to capture as much as they can," he said. "And when that food exceeds their need to survive, they begin to make copies of themselves, as many as they can."

To drive this point home, Thompson once showed students a slide with images of a speck of mold growing across a slice of white bread. The speck gets bigger and bigger until it has turned into an ugly dark splotch. The title of the slide, "Everyone's First Cancer Experiment," is a starkly simple and vivid explanation of how cancer works. But the lesson did not end with this fundamental comparison between how cancer cells and microorganisms eat. As the mold grows across the bread, Thompson told the students, its supply of nutrients or water will run low. When that happens, some of the mold cells will break away and migrate to another piece of the bread. In the next slide, the students saw that the slice of white bread now had not only a big dark splotch, but also another smaller splotch on the other side of the bread.

"This is exactly what cancer does," Thompson said. "It starts to grow in one place, it accumulates until it runs out of food, and then it finds a way to get to the new home." This search for a new living space is metastasis, the spread of the disease from one place in the body to another, and it is typically why cancer kills us.[8]

BY THE END OF the first decade of the twenty-first century, metabolism researchers were talking about cancer in a way that felt not only new but startlingly so. And yet, they had also brought cancer science full circle. As Warburg put it upon discovering fermentation in cancer cells, "The most important fact that we have discovered in regard to the metabolism of carcinoma tissue is, we believe, that carcinoma tissue" behaves "like yeast."[9]

Warburg had discovered something of immense importance in the 1920s: the overeating and fermenting of glucose is as fundamental to cancer as he had always argued. But when explaining the fermentation, Warburg, until the very end, found his vision obscured by the long shadow of Pasteur, who had assumed that fermentation arose only when oxygen was in short supply and respiration wasn't possible. Warburg never grasped that Pasteur had himself missed something crucial: If yeast have all the glucose and nitrogen they want, they will ferment and grow regardless of how much oxygen is available. They ferment the glucose not out of necessity but because it allows them to process more cellular building supplies and grow more rapidly.

Even so, Warburg's explanations were not as inaccurate as his critics have sometimes argued. Though most cancers are no longer thought to have damaged mitochondrial power stations in the way Warburg described, there is a growing appreciation among cancer scientists that mitochondria play a critical role in the transformation of a normal cell into a cancer cell. According to Celeste Simon, of the University of Pennsylvania, one of the leading authorities on the respiration of cancer cells, a shrinking supply of oxygen does play a role in every cancer as the disease progresses. And as cancers grow, they will invariably end up with a limited supply of oxygen from the blood and ferment more glucose as a result. But, Simon explained, that a lack of oxygen contributes to the progression of cancer does not necessarily mean it is the underlying cause of the disease. "Warburg was a real visionary," Simon said, but "some of his principles are probably not quite going to hold up the way he thought they would."[10]

In the 1960s, Warburg continued to search for ways to treat and prevent cancer by providing cells with the vitamins, or coenzymes, needed to keep cellular breathing running smoothly. If the power stations could maintain their function, he reasoned, the shift to the backup generators (fermentation) might be slowed or stopped. But in his last years, Warburg also renewed his long-standing interest in another approach: shutting down the backup generators by starving cancer cells of the glucose they rely on. The logic was straightforward. It might be impossible to target only the cancer cells, but such cells would be especially vulnerable to glucose deprivation. As Warburg wrote in 1926, "An overpopulated city is more sensitive to stoppage of food supply than a normally populated city, even when the inhabitants can all endure hunger alike."[11]

Toward the end of his life, seemingly aware that he was now too old to conduct experimental cancer research on his own, Warburg found a new scientific partner: Manfred von Ardenne, a brilliant German physicist who had worked on the Soviet atomic bomb project after the war and who arranged special permission for Warburg to travel across the Berlin Wall. In 1965, von Ardenne announced a new cancer therapy that involved heating the body to as high as 110°F for over half an hour while giving patients DL-glyceraldehyde, a compound Warburg was then testing as a means to disrupting a cell's ability to ferment glucose.

Some 45 years later, Chi Van Dang, then at Johns Hopkins, decided to investigate this old idea of disrupting the fermentation of cancer cells. He began with a simple experiment designed to understand how cancerous and noncancerous growing cells respond to a lack of food. In the presence of glucose, Dang found, both healthy growing cells and cancerous cells took up the glucose and used it to power their growth. But when Dang removed the glucose from both groups of cells, the differences immediately became apparent. The healthy cells, sensing the lack of nutrients, slowed down their metabolism and shifted into a resting state. The cancer cells couldn't stop themselves. In becoming cancerous, they had lost the internal checkpoints and feedback loops that tell a cell there's nothing left to eat. They were like addicts, and when they couldn't get their glucose fix, they would die. "Contrary to popular belief," the University of Southern California biologist Valter Longo said, "cancer cells are dumb."

Dang's preliminary research led him and others to revisit the idea of starving cancer with modern therapies. But if the new molecular biologists turned metabolism experts were returning to ideas from generations past, they were also returning to a hard lesson that Warburg and others had learned long before: cancers are remarkably adept at finding new ways to fuel their growth. The molecular pathways, as MIT's David Sabatini explained, "can go in many, many different directions and change very, very quickly."

"You block glucose, they use glutamine," as Dang put it, referring to another primary fuel used by cancers. "You block glucose and glutamine, they might be able to use fatty acids."[12]

These challenges haven't stopped the Warburg revival from leading to new cancer treatments. A biotechnology company started by Thompson and several other prominent metabolism researchers recently created a drug that treats one type of leukemia by inhibiting a mutated form of the metabolic enzyme IDH-2. The treatment is said to be the most significant advance for this specific leukemia in decades.[13]

At the same time, the metabolism revival is prompting researchers to rethink traditional cancer therapies. Matthew Vander Heiden is now exploring why a given chemotherapy will work against one cancer but not against another, even when the two cancers have identical mutations. Vander Heiden suspects the answer has a lot to do with the types of nutrients that were available in the particular tissue where the cancer formed. "It has nothing to do with the genetic mutations," Vander Heiden said.[14]

FOR ALL THE EXCITEMENT surrounding new metabolism-focused therapies, and for all the money investors are now pouring into such drugs, the most important findings to emerge from the return to metabolism may ultimately be about cancer prevention rather than treatment. If Warburg was correct, in the broadest sense, about cancer originating in changes to how cells eat, but wrong that the process always begins with a struggle to breathe, then the most important question, when it comes to prevention, is simple: What does cause the Warburg effect?

One possibility is that the cancer cell's huge appetite for glucose is strictly a product of unlucky mutations in genes that control cellular eating. If that's the case, the Warburg revival will still have fundamentally changed our understanding of cancer, and yet it will not have told us anything new about cancer prevention.

But at the turn of the twenty-first century, something unexpected happened: like two lost legs of an expedition suddenly encountering each other after years apart, two fields of cancer science that had long been separated began to converge. With each new advance showing that cancer is tied to how our cells eat, researchers studying cancer trends were arriving at advances of their own showing that cancer is tied to being overweight. Not even Doll and Peto had seen it coming. In "The Causes of Cancer," they highlighted the importance of Tannenbaum's studies of overfed mice, but with the exception of cancers of the endometrium and gallbladder in women, they remained unconvinced that overweight people were more likely to have cancer. The evidence for a connection between obesity in humans and cancer, they wrote in 1981, was "not particularly impressive."[15]

Doll and Peto could not have anticipated how soon those words would be obsolete. In 1982, researchers at the American Cancer Society selected a population of more than 900,000 Americans and asked them to fill out surveys that included basic personal information, such as their weight, height, and smoking habits. By 1998, almost 60,000 of the participants had died of cancer, and the American Cancer Society was anxious to figure out why.

Among those digging through the data was Eugenia Calle, an American Cancer Society epidemiologist. An exercise enthusiast known to drag her colleagues to the gym, Calle wondered whether the participants in the study who had succumbed to cancer were more likely to have been overweight. The question had been asked before, but no one had ever looked for the relationship in such a large data set.

After Calle finished her number crunching, many researchers

would never think of obesity and cancer in the same way again. Her study, published in the *New England Journal of Medicine* in 2003, found that being overweight or obese increased the risk of nearly every cancer she looked at. Compared with a woman of normal weight, the women in the highest-weight category were 62 percent more likely to die from cancer. The most obese men, in turn, were 52 percent more likely to die from cancer. Calle estimated that every year some 90,000 Americans were losing their lives to cancers linked to being overweight or obese.

Even Calle was surprised by the strength of the evidence. The connection between excess weight and cancer, she said, "was the rule more than the exception."[16] Tragically, in 2009, Calle was murdered during a robbery at her Atlanta condominium. In the following years, the importance of her contribution to our understanding of cancer would grow increasingly clear. As we have grown heavier—nearly three-quarters of American adults are now overweight or obese—the association with cancer has grown still more convincing. A 2017 Center for Disease Control analysis concluded that more than 600,000 Americans had been diagnosed with body-fat-related cancers in 2014 alone.

This harrowing number does not include deaths from prostate cancer and other cancers that may also be linked to obesity. (The evidence was not quite as strong for these other cancers.) As Rebecca Siegel, scientific director of Surveillance Research at the American Cancer Society, put it, when it comes to how obesity will influence cancer rates, we currently might be seeing only "the tip of the iceberg."[17]

IT IS OFTEN SAID that we are losing the war on cancer. The numbers from the obesity studies suggest that we are losing it, in large part, to what we eat and drink each day. And yet, though it is now commonly said that obesity will soon surpass smoking as the leading cause of preventable cancers, these studies do not show that obesity causes cancer. It is entirely possible that excess weight leads to cancer, but it is also possible that another underlying phenomenon is at work that is driving both obesity and cancer.

Understanding the obesity-cancer connection has been challenging, in part, because getting to the bottom of the obesity epidemic itself has been challenging. The simple explanation—that people started to eat more food—leaves too many questions unanswered. Many societies have had an abundance of food without developing high rates of obesity or cancer. Tellingly, animals in the wild do not grow fat and sick when they are fortunate enough to get all the food they want. For animals, coming into an abundance of food is akin to winning natural selection's lottery: a well-fed population can be expected to grow in number and thrive. Likewise, prior to the nineteenth century, obesity and cancer were rare regardless of how much food a human population might have. As one British specialist noted in 1908, it seemed conceivable that eating too much led to cancer, but it was not at all clear why overeating caused pathological rather than normal growths.[18]

There are other complicating factors as well. The Warburg effect involves cells overeating glucose, but our bodies are designed to keep blood glucose levels within a tight range regardless of how much food we consume. And Thompson's own research had demonstrated that healthy cells will starve unless told to eat.

Obesity had been definitively linked to cancer at the very moment that scientists were rediscovering Warburg's metabolism research. It seemed clear that the two phenomena fit together, that the way we eat must somehow be connected to the way that cancers eat, but the precise nature of the connection was not at all obvious. Some seventy years after Warburg first observed the ravenous appetite of the cancer cell, the mystery of the cancer-diet connection remained to be solved.

CHAPTER NINETEEN

Diabetes and Cancer

IN THE FINAL YEARS OF the last century, Warburg was not the only scientist whose work experienced a revival. The return to metabolism was also a return to the research of Efraim Racker, another legendary biochemist from Warburg's era. Racker, who was Jewish, had just finished medical school in Vienna in 1938 when the Nazis arrived. He fled to England, then made his way to the United States, where he would discover how cells capture the energy released by fermentation.

Though he would eventually see that Racker was correct, Warburg initially thought that Racker's explanation of fermentation involved too many steps and dismissed it as "Racker's detour." Racker confessed to being annoyed by Warburg's insult, but he also felt "distinctly flattered" that "this great scientist" had found him worthy of his derision. As Racker appreciated, to be publicly attacked by Warburg was virtually a rite of passage for a notable twentieth-century biochemist.[1]

Like almost everyone else in the field, Racker was mystified by Warburg's personality: "His vanity matched his brilliance," Racker wrote. He once even composed a Warburg limerick:

There was a great scientist named Otto
Who lived by the following motto
"I am always right
My enemies I'll fight
(But I'll be glad to send them my photo")[2]

It was Racker, then at Cornell University, who coined the term "the Warburg effect" in 1972 to describe the cancer cell's turn to fermentation in the presence of oxygen. That same year, Lewis Cantley arrived at Cornell to begin his doctorate in physical biochemistry.

Cantley, now the director of the cancer center at Weill Cornell Medicine, grew up on a farm in West Virginia. His family was not especially well-off, but Cantley never lacked for entertainment. By age 10, he was taking apart and reassembling car engines by himself. When he asked for fireworks, his father took him to the drugstore to buy sulfur and potassium nitrate and encouraged him to figure it out. "That I survived, was pretty amazing," Cantley said.[3]

Though Racker was not Cantley's thesis adviser at Cornell, he would become a mentor and close friend. Among countless other topics, the two sometimes discussed Warburg's cancer theories and why cancer cells consume so much glucose. Cantley was curious, but it wasn't his field. He had never even bothered to take a biology class after high school.

In the early 1980s, Cantley was at Tufts University studying how phosphatidylinositol (PI), a type of fat molecule found in cell membranes, influenced basic metabolic functions. Cantley had no interest in finding oncogenes, but if you were researching cellular mechanisms in the 1980s, oncogenes had a way of finding you. In 1983, he learned that another scientist had linked SRC, the famous oncogene discovered by Bishop and Varmus, to a molecule not unlike PI.

When Cantley saw that PI, too, was woven into cancer signaling networks, the molecule started to seem all the more fascinating. PI wasn't supposed to have anything to do with cancer. It was as if Cantley had been working on a water pump in the basement of a house only to discover that the pump also somehow controlled the home's electricity.

If PI's connection to cancer was strange, bigger surprises were still to come. PI, like many of the molecules in our cells, has an on switch. When an enzyme attaches a phosphate (an atom of phosphorus bonded to 4 atoms of oxygen) to PI, it will spring to life and direct other molecules into action. That much Cantley already knew. He also knew that a phosphate can be attached to PI in different locations on the molecule and that where it ends up makes all the difference.

To find out where the phosphate was located on the form of PI linked to cancer, Cantley ran a test known as thin-layer chromatography, which can distinguish one molecule from another according to how far it moves along a slide. What Cantley saw on that slide changed his life. The PI molecule connected to cancer wasn't doing what a PI molecule with a phosphate was supposed to do. It was moving too far down the slide. The difference was barely a single millimeter, but for a biochemist, that millimeter was all the difference in the world. Cantley was stunned. "I went home and said, 'Jesus Christ!'"

Cantley had good reason to be exhilarated. That extra millimeter meant that he was looking at an entirely new form of PI. Outside of the world of biochemistry, it's difficult to appreciate the novelty and significance of the discovery. But for those who studied such molecules, it was mind-blowing. As Cantley once put it in an effort to explain why his finding had generated excitement in the field, "For physicists, it would be like finding a quark no one had ever seen before."[4]

The newly discovered form of PI, Cantley soon determined, had a phosphate attached to the third position of its molecular ring, rather than the fourth or fifth position, where it had long ago been found on other PI molecules. It became known as PIP_3. Cantley also succeeded in isolating the enzyme that attaches the phosphate to PI in the third position. It became known as PI3K. (The K stands for kinase; a type of enzyme that transfers a phosphate from one molecule to another.)

That Cantley had not only identified previously unknown molecules but also ones that appeared central to cancer was difficult for some to believe. A colleague told him that he would "eat his hat" if it turned out to be true. Yet Cantley never doubted his work. On the night of the definitive experiment in 1987, he told his wife that the molecules were "going to completely revolutionize cancer."[5]

Cantley was more right than he could have known. The importance of PI3K to cancer became all the more clear thanks to other researchers studying the gene PTEN. As a tumor suppressor, PTEN's job is to turn off the signaling networks that drive cancer. Such genes— sometimes referred to as a cell's "braking system"—cause cancer not by their presence but by their absence. Researchers knew that PTEN was often missing from cancer cells and that its disappearance was a critical step in the development of tumors, but in the late 1990s, PTEN's specific function became clearer: it is PI3K's nemesis. PI3K drives cancer by attaching phosphates to PI and setting off a network of growth signals. PTEN works by removing the phosphates and shutting down the PI3K pathway.

More evidence of PI3K's critical role in cancer arrived as researchers began to sequence the genomes of tumors. More than 20,000 human tumors have now been sequenced. Mutations in the molecules in the PI3K pathway are found in up to 80 percent of all cancers, making it the most commonly mutated pathway of all. According to Cantley, the majority of cancer deaths in the United States can be linked to either PI3K or PTEN.

Cantley had set out in the early 1980s to understand a very fundamental metabolic reaction in a well-known fat molecule. By the first years of the twenty-first century, it was increasingly clear that the molecule had led him to one of the most important molecules in cancer science. It was a significant breakthrough, and yet, for all their importance, the enzymes and fat molecules themselves aren't the full story. Before Cantley had become intrigued by how PI's signals are turned up by way of mutated cancer genes, he had been studying how its signals could be turned up by a very different type of molecule: the hormone insulin. If PI signaling was connected to cancer, it meant that insulin likely was as well, since insulin sent PI into action. But that made little sense. Insulin wasn't supposed to cause cancer. It was the protagonist of a seemingly unrelated story, the story of diabetes.

LIKE CANCER, diabetes is an ancient disease that had been relatively rare prior to the nineteenth century. Aretaeus of Cappadocia, the Greek physician who gave the disease its name, noted that it was "not very common to man." Galen, who called diabetes "diarrhea of the urine," reported only having seen two cases. By the nineteenth century, diabetes was far more common than it had been in the ancient world, but still vastly less common than it is today. From 1824 to 1898, Massachusetts General Hospital admitted almost 50,000 patients for a variety of different ailments. Only 172 patients were identified as having diabetes. Data from other hospitals during the same period reveal the same striking absence of hospitalizations for diabetes.[6]

And then something changed. In 1875, only 3.4 of every 100,000 Berlin residents died of diabetes. By 1890, that rate had tripled. Fifteen years later, it had tripled again. Berlin was not alone. Other Western cities saw similar increases. In 1897, the *Journal of the American Medical Association* stated that "it appears that the general frequency of this disease is steadily increasing." Over the next 20 years, the diabetes death rate in American cities would increase by as much as 400 percent.[7]

That the rise in diabetes deaths occurred alongside the rise in cancer deaths didn't go unnoticed. While Ernst Freund's 1885 monograph showing elevated glucose in the blood of cancer patients was the first concrete evidence of a connection between diabetes and cancer, others would soon follow. In his influential 1908 book, *The Natural History of Cancer*, the British surgeon W. Roger Williams noted a recent uptick in reports linking diabetes to malignant tumors. Most who had "specially studied this subject," Williams wrote, "maintain that the diabetic state" favors "the development of malignant disease."

The association wasn't so strong as to be indisputable. Williams made clear that his own cancer patients were rarely diabetic and that there remained "a lack of conclusive evidence" for a connection between the two diseases. That uncertainty would linger for years to come. In his 1937 book on rising cancer rates in industrialized societies, Frederick Hoffman wrote that diabetes had begun to increase "at about the same rate as cancer." But Hoffman, too, felt the evidence wasn't yet definitive.[8]

The doubts about the cancer-diabetes connection were well founded. The most that could be said was that both diseases had become more common over the same general time frame. And even to be confident about that claim, one first had to demonstrate that the increase in diabetes was a real phenomenon. That was never easy to do. Reports of the dramatic increase in diabetes were often met with the same skepticism as reports of the dramatic increase in cancer. There were more cases of diabetes, the argument went, because it had become easier to diagnose and people were living to older ages. Elliott Joslin, the most famous diabetes doctor in American history, was among the skeptics. In 1917, he mockingly wrote that the "frequency of diabetes in a community may be the index of the intelligence of its physicians."[9]

The counterarguments, in turn, were essentially the same counterarguments made with regard to the increase in cancer: physicians had known the symptoms of diabetes for thousands of years, and it was not difficult to detect the disease at its most advanced stages. Hospitals had been using tests to detect glucose in a patient's urine since the 1850s, decades before the rise in diabetes became pronounced. And old age was not a late nineteenth-century innovation; most of the gains in average life expectancy could be attributed to fewer deaths in childhood.

Though it did little to settle the debate, another interesting parallel between diabetes and cancer was sometimes noted: diabetes was not becoming more common everywhere. Among populations following their traditional diets, the disease remained extremely rare. In many cases, the absence of diabetes was discovered by the very same traveling doctors who had observed the absence of cancer in indigenous populations. After traveling to the Arctic in 1905, the surgeon Nicholas Senn declared that cancer was a "disease of civilization," yet he also noted that the "Eskimos" were "almost exempt" from diabetes. Sir Robert McCarrison's most surprising claim after several years among the Hunza was that the locals never got cancer. But he also pointed out that they never got other diseases, including diabetes.[10]

Such populations, it soon became clear, were not free of diabetes because of their genes or shorter life spans. As soon as they adopted Western lifestyles—often it happened first among the wealthier classes—diabetes would arise, sometimes so rapidly that medical experts of the time were left stunned. The same pattern could be seen in every part of the world, from Thailand and Tunisia to Native American reservations in the western United States.

The most compelling evidence that Western living habits could cause diabetes emerged from studies of migrant populations. In 1966, the physician and researcher George Campbell wrote of a "veritable explosion of diabetes" among the poor Indian laborers he studied in South Africa. In some of the villages, approximately one-third of the middle-aged Indian men had diabetes. This was striking in itself, but all the more striking because, across India, only one in every 100 citizens had the disease.

The Israeli diabetes specialist Aharon Cohen recorded a similar explosion of diabetes among Yemenite Jews in Israel. When Cohen carried out his studies in the 1950s, he had the advantage of looking at two waves of immigrants. The Yemenites who had immigrated to Palestine in the early 1930s were as likely to have diabetes as anyone else in Israel. But among the 5,000 Yemenites who came to Israel in 1949, Cohen found only three cases of diabetes. Living in Israel for a mere 15 or 20 years made a Yemenite immigrant's chances of developing diabetes almost 50 times greater.

Scientific researchers today no longer question whether diabetes became far more common starting in the second half of the nineteenth century. Rates have only continued to increase. Since 1960, the diabetes rate in the United States has increased by 800 percent, and half of American adults are now estimated to have either diabetes or prediabetes. Nor do researchers still question whether diabetes is associated with cancer. As a 2010 "consensus report" published by the American Diabetes Association and the American Cancer Society put it, "Epidemiologic evidence suggests that people with diabetes are at significantly higher risk for many forms of cancer." People with diabetes who are diagnosed with cancer are also less likely to survive than those without diabetes.[11]

MAKING SENSE OF the relationship between diabetes and cancer was not easy because diabetes itself was not well understood. By the late nineteenth century, it was clear that diabetes came in two primary forms, known today as type 1 and type 2. Type 1 diabetes tends to arise in children and is typically more severe. Type 2 diabetes, the more common form of the disease, afflicts adults, especially the overweight. Despite these important differences, both forms of diabetes have the same effect: blood coursing with glucose that our cells are unable to eat.

The first critical breakthrough in understanding the role of insulin in diabetes took place in April 1889. Two German researchers fell into a debate about whether secretions from the pancreas were needed to digest butter. To find out, they removed the pancreas from a dog and noticed something far more interesting: the dog wouldn't stop urinating on the laboratory floor. Without a pancreas, it had immediately developed diabetes.

From that point on, it was understood that the pancreas released a mysterious substance that gave cells permission to eat glucose. The substance became known as "the internal secretion," and researchers across the world set out in search of it. Some three decades later, Frederick Banting, a young Canadian surgeon, read a scientific article about the pancreas and had a new idea for how to capture the "internal secretion." The idea turned out to be entirely wrong, but Banting's efforts nevertheless succeeded. The "internal secretion" became known as "insulin," and it immediately revolutionized diabetes care. Another "dreaded" disease, the *New York Times* reported, had been "overcome by science."[12]

As insulin treatments were used more often in the following years, researchers found a new way to distinguish between the two types of diabetes. Some people with diabetes were insulin dependent (type 1 diabetics), while others (type 2 diabetics) could often survive by changing their diets or taking other medications. And yet, even after this new distinction had been made, the two conditions were still said to be essentially the same. If people with type 2 diabetes had elevated glucose, the thinking went, they couldn't possibly have enough insulin in their blood.

It all made perfect sense, and yet, it was only a guess. No one could say exactly how much insulin was in the blood of people with diabetes—or without diabetes, for that matter. Insulin is a tiny protein and there was no good way to measure it in our blood. As one newspaper article put it, trying to measure the amount of insulin in a person's blood was like trying to measure a teaspoon of one substance in "a lake 62 miles long, 62 miles wide and 30 feet deep."

The problem seemed insurmountable, at least until the 1950s, when Rosalyn Yalow took it on. Yalow, a physicist, had a history of breezing past seemingly insurmountable obstacles. After graduating from Hunter College in 1941, she found that there were few, if any, opportunities for a woman who wanted to become a physicist, particularly a Jewish woman. Yalow took a job as a secretary at Columbia University in the hope of being able to enroll in graduate courses. She was finally admitted to the physics program at the University of Illinois only because the war had left positions available. Even then, Yalow first had to agree to a stipulation: the physics department would not help her find a job after she earned her degree. It was an outrage, but an opportunity nonetheless. Before leaving New York for Illinois, Yalow took the time to rip up her stenography books.

By 1947, Yalow was working at the Veterans Administration hospital in the Bronx, studying medical uses of radioactive carbon in a room that had once been a janitor's closet. She had hired a physician, Solomon Berson, to work with her, and in the late 1950s, they arrived at a breakthrough that relied on the same principle Warburg had used to identify his respiratory ferment. Warburg saw that light and carbon monoxide compete for his ferment. Yalow and Berson took insulin from someone's blood and let it compete for antibodies with a radioactive insulin—which, unlike normal insulin, they were able to measure. The amount of radioactive insulin that would end up unbound to antibodies could then be used to calculate the amount of insulin in the sample. The test became known as a radioimmunoassay, and it gradually revolutionized modern medicine.[13]

Yalow won the Nobel Prize in 1977, but researchers were initially slow to appreciate the full implications of the new test. In the early 1960s, Yalow and Berson had the field largely to themselves, and they explored a series of critical questions about insulin and disease that had previously been unanswerable. Among their many findings, one in particular stood out: type 2 diabetes had been completely misunderstood.

Type 1 diabetes is a disease in which glucose rises to dangerously high levels in the blood because the pancreas loses the ability to make insulin. In most cases, Yalow and Berson discovered, type 2 diabetes wasn't like type 1 diabetes at all. It was almost the opposite condition. The blood of type 2 patients had *more* insulin than normal (in addition to more glucose), rather than less. The problem of type 2 diabetes, Yalow and Berson saw, was not that the pancreas could no longer produce insulin but that cells were no longer listening to insulin's instructions to eat. Such cells, in the scientific terminology, are "insulin resistant." The pancreas responds to the resistance by pumping out more and more insulin.

Insulin resistance does not set in uniformly. Cells of the muscle and liver might be slow to heed insulin's call even as the fat tissue continues to listen all too well. And because insulin tells fat cells not only to take up energy but also to store it, the extra insulin in the blood needed to overcome insulin resistance will make us put on weight. (Yalow and Berson were the first to see that obese individuals commonly had elevated insulin levels.) Over the years, the effects of the elevated insulin will accumulate, and more and more energy will be trapped inside of our fat cells. In the words of Harvard endocrinologist and metabolism researcher David S. Ludwig, "Insulin is like a Miracle-Gro for fat cells."

The extent to which elevated insulin is responsible for obesity is still debated, but that insulin tells our fat tissue to take up and hold on to nutrients was clear to diabetes doctors from the moment they began to inject the hormone into their patients. Before insulin therapy, people with type 1 diabetes remained stick thin no matter how much they ate. Once they began to inject insulin into their blood, the change was dramatic. The before and after photos look as if someone had blown new life into a deflated balloon animal. Frederick Allen, one of the first American physicians to prescribe insulin, noticed that the gaunt faces of his patients would begin to fill out in a matter of days, sometimes a matter of hours. "The patients seem to tend readily to become obese," Allen wrote.[14]

The fattening effect of insulin was hard to miss, and it became still more obvious in the second half of the twentieth century, when type 2 diabetics also began to rely on insulin to keep their blood sugar under control. The more insulin they would use, the more weight they would typically gain.

Insulin's role in fat storage is easy to appreciate in the context of an evolutionary past in which the timing of one's next meal was never certain. Insulin in the blood is a sign that glucose is being served and that the body shouldn't waste its own energy reserves. Burning stored fat when glucose is available makes no more sense than a village tapping into emergency grain reserves amid a plentiful harvest. When insulin levels are low, by contrast, it means that glucose hasn't been ingested for an extended period. The energy stored inside of fat cells will now flow back out to feed cells throughout the body.

Insulin is supposed to rise as we eat and then fall in the hours after a meal. If insulin levels remain elevated throughout the day, a condition known as hyperinsulinemia, our fat tissue will hold on to its stores of energy, and we will grow hungry even if we have eaten relatively recently. It is as if a village has all the grain reserves it needs but cannot manage to unlock the silo. "A sort of anarchy exists," wrote the influential Viennese endocrinologist Julius Bauer in 1929. The fat tissue now "lives for itself."

Bauer, yet another brilliant Viennese Jew who would flee the Nazis in 1938, noticed something else as well: when it is gobbling up nutrients and growing with no regard for the rest of our cells, our fat tissue is no longer behaving like a tissue in a body of cooperative cells. The expanding fat tissue is now acting, Bauer wrote, like a "malignant tumor."[15]

THAT A HORMONE COULD cause cancer wasn't a new idea in the late twentieth century, when Lewis Cantley began to wonder about the insulin-cancer connection. In 1907, Paul Ehrlich wrote that "specific growth-stimulating bodies or 'hormones,'" were "intimately connected" with how cells gather nutrients and grow.[16] And that sex hormones play a central role in some cancers has been widely appreciated at least since the early 1940s. Charles B. Huggins, the surgeon who would carry out the definitive studies linking testosterone and estrogen to cancer, only took an interest in the field after meeting Warburg in 1931.[17] The two remained friends for the rest of Warburg's life. Warburg nominated Huggins for a Nobel Prize in 1950, and Huggins won in 1966. The growth of a cancer, Huggins stated in his Nobel lecture, is "a function of the interaction of the tumor and its soil."

In 1958, near the end of his life, Leo Loeb, reflecting on his own research on sex hormones and cancer, wrote that normal growth-stimulating hormones may "become a very important cause of cancer." "Growth processes," Loeb added, are "presumably not only a characteristic of tumors and of cancer in particular but the essential cause of cancer."[18]

Warburg's close friend Dean Burk may have been the first scientist to fully explore the possibility that too much insulin could be fertilizing the soil of our bodies for cancer. In a 1953 paper on insulin and cancer, Burk made a commonsense point that would rarely be made again over the next 50 years: since insulin instructs cells to take up glucose, its possible role in the Warburg effect "is of special importance."

Burk's interest in insulin and the Warburg effect extended into the late 1960s. For most of this period, he had no way of knowing that cancer patients often had unusually high levels of insulin in their blood or that cancer cells had extra insulin receptors. But by 1968, Burk felt confident in writing that insulin's impact on the rate in which cells take up glucose "is one of the most fundamental and universal changes accompanying the malignant transformation of normal cells into cancer cells." Burk even anticipated one of the key concepts of the Warburg revival: excess glucose, he grasped, was primarily needed not for energy, but to supply raw materials for synthesizing DNA and proteins.[19]

Warburg almost certainly saw Burk's papers on insulin in the

1950s and 1960s. Regardless, Burk told him about his insulin research in their correspondence. Burk wrote to Warburg about insulin's effect on fermentation in cancer cells in January of 1954 and seems to have brought the subject up regularly. "I shall never give up trying to convince you of the reality and importance of our addition, which carries your dominating fact higher up the flagpole of truth!" he wrote in a 1965 letter to Warburg on insulin and cancer (emphasis in the original). Three weeks later, Burk sent Warburg another letter on his research suggesting that insulin could increase the rate of fermentation by as much as 20 percent.[20]

Burk also repeatedly reminded Warburg of something else: he was not working alone. All of his groundbreaking papers on insulin and cancer in those decades were coauthored by Mark Woods and Jehu Hunter, Burk's colleagues at the National Cancer Institute. Hunter was a Black cancer researcher at a time when there were very few such opportunities for people who were not white men. During World War II, he had fought the Nazis in Europe as a member of an all-Black unit nicknamed the "Buffalo Soldiers," a tribute to the Black cavalry regiments that patrolled the American West after the Civil War. It's "a deep, dark secret," Hunter once pointed out, "that blacks fought valiantly on the plains of France."[21]

Though it's not clear if they ever met, Hunter was more than an incidental figure in Warburg's life. Much of Warburg's postwar influence can be traced to his influential 1956 paper in *Science*, the elegantly written "On the Origin of Cancer Cells." Warburg composed the original in German, and Hunter helped Burk translate it. Warburg and Hunter are also listed as coauthors on a photosynthesis paper. In a January 1954 letter updating Warburg on his insulin work, Burk described Hunter as a "a scholar and a gentleman" who "can do every sort of job." "Captain Hunter," Burk wrote, is a "Godsend."[22]

Burk was trying to convince Warburg about the importance of insulin to cancer, but it was no use. Warburg had made up his mind about the cause of fermentation and cancer 40 years earlier. Warburg couldn't hear what Burk was telling him about insulin any more than he could consider the implications of the feeding studies of Ehrlich, Rous, and others. And if Warburg wouldn't give serious consideration to the role of insulin in the Warburg effect, it was unlikely that many others would.

By the late 1980s, the study of insulin had progressed. Diabetes researchers were coming to understand that in the first stages of type 2 diabetes, someone was likely to have too much insulin in their blood rather than too little. And Harvard metabolism researcher George Cahill, among others, had revealed insulin's central role in determining how and when the body takes up fuel and stores fat.

But the study of insulin and cancer had not progressed at all. And though Lewis Cantley had seen that a fat molecule, PI, could be activated both by signals from oncogenes and by insulin, it was difficult to draw any conclusions. There was still virtually no knowledge of what happens inside of a cell after insulin connects with a receptor on the cell's surface. The entire subject, Cantley said, was "a black box."

Cantley set himself the task of prying the "black box" open. What he found, he said, is "a little scary to think about."[23]

CHAPTER TWENTY

The Insulin Hypothesis

DEAN BURK, Warburg's closest friend after the war, had failed to convince him of the importance of insulin to cancer. It was too late to convince Warburg of a new idea; it had always been too late. But Lewis Cantley, who had grown up in West Virginia taking apart and rebuilding car engines, was a different type of scientist. He had no battles to fight with other scientists, no favored theories to champion.

Cantley had found that insulin activates particular molecules in the cell and that those same molecules, when mutated, cause cancer. His science had brought him to an inescapable question: Could too much insulin signaling overactivate the PI3K pathway in much the same way as a genetic mutation might, flooding the cell with more nutrients and growth signals than it should ever have? Diabetes was an insulin story. Could cancer be as well?

It was a tantalizing idea. And if there wasn't yet good evidence that insulin could set off the signaling cascades linked to cancer, there was already a growing body of indirect evidence that supported Burk's original hunch about the insulin-cancer connection. By the 1980s, scientists already knew that many tumors have an unusually large number of insulin receptors and that insulin, a growth factor, causes both healthy and cancerous cells to multiply.

But it all still sounded absurd. Insulin was a metabolic hormone that governed how cells eat. Like the Warburg effect, it was assumed to be, at most, peripheral to the emergence of cancer in an organism. Even Cantley had his doubts. To find out whether insulin could really be as important to cancer as his research seemed to suggest, Cantley would have to determine what happens inside a cell once insulin arrives on the scene. He began to do so in the early 1990s, and what he found was unambiguous. While other growth factors circulating in the blood can activate PI3K—the molecule Cantley had linked to cancer and the Warburg effect—none does it more effectively than insulin. Cantley calls insulin "the champion of all PI3K activators."[1]

Even as Cantley was carrying out this research, cancer epidemiologists were making striking observations of their own. In 1995, Edward Giovannucci, of Harvard, found that people diagnosed with colon cancer also tended to have high insulin. Giovannucci was soon followed by Rudolf Kaaks, who published a groundbreaking paper titled "Nutrition, Hormones, and Breast Cancer: Is Insulin the Missing Link?" Yet it wasn't just breast and colon cancer. Each year seemed to bring evidence of yet another cancer that could be connected to insulin. The list includes cancer of the pancreas, uterus, kidney, esophagus, and prostate. People who have high insulin when they are diagnosed with cancer also tend to have worse outcomes.[2]

It is sometimes said that too much insulin in our blood may be driving cancer only indirectly, by making us heavier. And yet elevated insulin is a risk factor for cancer even in the non-obese. Conversely, someone who is overweight but has normal insulin does not appear to be at greater risk for cancer. Excess fat might contribute to tumors by causing inflammation and the release of additional hormones, but even in this scenario, insulin can be blamed for causing that excess fat to accumulate in the first place.[3]

With the start of the twenty-first century, the evidence for insulin's role in cancer would only build. Cantley's work on insulin and PI3K turned out to fit perfectly the findings of Craig Thompson's lab. Thompson had found that the body eliminates unwanted cells by starving them. Insulin gives cells the opposite message. It is a growth factor that tells cells to take up nutrients. AKT, the enzyme Thompson connected to the Warburg effect, is "downstream" of PI3K, part of the same pathway, or chain of reactions. Once PI3K is activated, AKT, too, will be activated. The cell will have more food than it ever should and begin to ferment much of it via the Warburg effect.

Insulin, according to this model, isn't responsible for the first mutations that arise. Mutations appear all the time as our cells divide and age. Even healthy tissues are often rife with mutations, including mutations linked to cancer. Some cells with these mutations will be quickly eliminated; others might contribute to tiny, incipient cancers that remain harmless. Studies of the thyroid, breast, and prostate have shown that even healthy bodies often harbor these harmless cancers. "We have had a lot of men die *with* prostate cancer rather than *from* prostate cancer," said Derek LeRoith, a pioneer of insulin and cancer research at Mount Sinai Medical Center.[4]

Insulin's role is to whisper encouragement into the ear of an incipient tumor. As Michael Pollak, head of the Division of Cancer Prevention at McGill University, explained, when cells are "insulin stimulated" and there is "a lot of PI3K and AKT signaling," they become "pro-survival." In this state, they are "more likely to stay alive, even if they have genetic damage." According to Vuk Stambolic, of the Princess Margaret Cancer Centre in Toronto, even "a smidgen of a survival signal" from insulin could allow a cell an "extra division or an extra step leading to more cancer."[5]

Stambolic had always been at least vaguely aware that insulin helped cancer cells thrive. Scientists growing cancer cells in the laboratory place the cells in a culture, a mixture of food and growth factors that allows the cells to eat and flourish. For many cancer types, insulin is part of the standard recipe. Stambolic, like most researchers, had never thought much about why insulin was commonly used in cell cultures. That changed in 2006, when he learned that the vast majority of breast cancers overexpress the insulin receptor. "That just blew my mind," Stambolic said.[6]

Cancer cells often have more insulin receptors, Stambolic explained, because in the Darwinian landscape of a growing cancer, the cells that are best equipped to take advantage of elevated insulin in the blood are more likely to multiply. As Stambolic thought about the many insulin receptors on breast cancer cells, he became "obsessed" with a question. If cancers of the breast were so sensitive to insulin, could lowering insulin levels in the body serve as an effective treatment? If so, the benefits might extend far beyond breast cancer. Insulin receptors are also overexpressed in a wide array of cancers, including cancers of the prostate, uterus, colon, and lung.

As a first step, Stambolic returned to the recipe he had long used to grow cancer cells in his lab. Although he knew it included insulin, he did not know if the insulin was absolutely necessary for the cancer cells to live. To find out, he gradually weaned the cancer cells in his cultures off insulin, as though the cells were addicts in a rehab program.

The experiments were simple, not very different from those Paul Ehrlich had conducted a century earlier to determine if bacteria could grow in a cell culture without hemoglobin. Stambolic's results were startling: some types of cancer could survive without the insulin, but many others could not. The vulnerable cells had plenty of nutrients and other growth factors, but, without insulin, their demise was swift. "There's actually a dependency," Stambolic said.[7]

Insulin doesn't act alone in telling cells to eat and grow. Another central player is called insulin-like growth factor 1 (IGF-1), a closely related hormone. There are hybrid receptors on many cells that respond to both molecules. But because excess insulin leads to higher levels of IGF-1, insulin can be blamed for excessive IGF-1 signaling. Like insulin itself, IGF-1 tells cells to grow and divide and is strongly associated with cancer in population studies. (Growth hormone works, in large part, by increasing IGF-1, which in turn tells cells to eat and proliferate.)

Cancer cells often have many more receptors for IGF-1 than the cells of the surrounding tissue, just as they have more receptors for insulin, and IGF-1 signaling can also prevent cells from dying. In studies of mice, at least, raising IGF-1 levels leads to tumor growth and the Warburg effect. Lowering IGF-1, by contrast, slows tumor growth.[8]

Perhaps the most remarkable evidence of IGF-1's influence on our health was uncovered in the 1990s in a remote village of Ecuador. The village is home to a community of dwarfs who are believed to descend from Sephardic Jews who were forced to convert during the Spanish Inquisition. Researchers have traced their dwarfism, known as Laron syndrome, to an inherited genetic mutation that impairs their growth hormone receptors and thus the release of IGF-1. The Laron dwarfs are not known to live particularly healthy lifestyles, and yet, according to the scientists who study them, it makes no difference when it comes to cancer. Without IGF-1 signaling, they are virtually immune to cancer.[9]

The research on Laron dwarfs, like much of the research on insulin and cancer, has only produced indirect evidence. But more direct evidence is now emerging. In 2019, James Johnson, of the University of British Columbia, published a paper covering five years of experiments on insulin and pancreatic cancer in mice. The results were clear. Even a little less insulin led to a "significant reduction in the development of pancreatic cancer." Though effects found in mice often fail to materialize in people, Johnson believes that the findings "will be clinically relevant to humans because of the strong known associations between hyperinsulinemia and a number of different human cancers."[10]

It is perhaps no accident that both Johnson and Stambolic are studying insulin in Canada, where the hormone was first discovered. Stambolic's current lab is in the same building where Frederick Banting and his colleagues made their world-changing discovery of "the internal secretion," insulin. To inspire his new students, Stambolic takes them on tours of the historic space.

That young cancer researchers, rather than diabetes researchers, now pay tribute to Banting is fitting. Banting had moved on to cancer research in the late 1920s, with the hope of doing for cancer patients what he had done for diabetes patients. For a number of years, he put all of his energies into cancer, filling notebook after notebook with ideas. "He fished and fished for the answer to cancer," according to Banting's biographer, Michael Bliss. "Nothing took the hook." Banting died in a plane crash in 1941, unaware that his great diabetes breakthrough might itself have been the very cancer breakthrough he had sought.[11]

THERE IS STILL more evidence of the importance of insulin. In studies of mice, researchers can make tumors shrink or grow simply by enhancing or diminishing insulin signaling through genetic alterations. Then there is the metformin story. Metformin is a popular diabetes drug that helps control blood glucose levels. As it became ubiquitous in the 1990s, it wasn't thought to have any connection to cancer. But studies of cancer trends found that metformin appeared to have a remarkable side effect. Diabetes patients taking metformin were anywhere from 25 to 40 percent less likely to get cancer than patients taking other diabetes drugs. Cantley has said that metformin "may have already saved more people from cancer deaths than any drug in history."[12]

A number of different ideas have been proposed to explain how metformin might help prevent cancer. But the simplest explanation requires no special knowledge of molecular biology or biochemistry. Metformin lowers glucose levels in type 2 diabetes patients, and in so doing lowers insulin as well.

Elevated insulin appears to be especially hazardous to the epithelial cells that form a protective layer around our organs. Under normal conditions, Cantley explained, many such epithelial cells "rarely see insulin." But when hyperinsulinemia (the condition of continuously elevated insulin) sets in, the situation is very different. Someone with hyperinsulinemia might have "50 times more insulin" than normal circulating in the blood all day and night.[13]

Some researchers, including Cantley, believe that elevated insulin may also be driving a process that leads to dangerous new mutations. The mechanisms involved, like so much of our current understanding of biochemistry, can be traced directly back to Warburg's work in the 1920s and 1930s. Warburg elucidated both the first and last steps of respiration, but as the Cambridge researcher David Keilin made clear in the 1920s, electrons travel along a chain of molecules, now known as the electron transport chain, before reaching oxygen at Warburg's respiratory ferment. (Warburg never forgave Keilin for changing the name of his "respiratory ferment" to "cytochrome c oxidase.")[14]

Electron transport chains, found throughout the inner membranes of our mitochondria, are essentially tiny electrical wires containing copper and iron. (We need to eat, first and foremost, because we need electrons to supply the electricity for this system.) The wires are one of evolution's most remarkable inventions, but they also create one of life's great vulnerabilities: as electrons are passed along the chain, some will inevitably leak, leading to the formation of molecules that are commonly—and somewhat inaccurately—known as "free radicals" or "reactive oxygen species." Such molecules are unstable and hungry for electrons. They will take them from whatever nearby molecules they can, including from our DNA.

Reactive oxygen species aren't inherently bad for us. They are components of a healthy signaling system, and other molecules, the antioxidants, have evolved to sweep them up. Problems tend to arise only when the electron processing machinery malfunctions. Sometimes a damaged mitochondrion will lack the appropriate components to handle the electron flow. But other times, such as when a cell has more glucose than it should, there will be more electrons than the system can handle. More of the electrons will leak and more reactive oxygen species will form.

In the castle analogy, it is as if the guards are sending wood into the castle at all hours. To keep a chaotic situation under control, some of the desperate workers inside the castle have tossed piles of wood to the side, but the short-term solution comes at a price. The discarded wood is catching fire and everyone inside the castle is being sickened by the smoke. The excess fuel that a cell takes up and burns "actually creates a way to mutagenize itself," Thompson said.[15]

The biochemist Nick Lane compares reactive oxygen species to muggers stealing a handbag. But the analogy, as Lane points out, is not perfect: the molecules that are mugged are now short an electron and become highly reactive. As Lane writes, "It is as if having your handbag snatched deranges your mind and turns you into a mugger yourself, restless until you have snatched someone else's bag."[16]

That reactive oxygen species can cause cancer is well understood. Radiation drives mutations by splitting the water molecules inside an organism. The splitting gives rise to the same reactive oxygen molecules and the same chain reaction of electron snatching. And the reactive molecules don't have to make it all the way to the nucleus of the cell to attack our DNA. The radicals form in the mitochondria, which has its own genes that might now become damaged. There is a fire burning in the castle, and some of the most precious objects are in the very same room.

The damage to the mitochondria that ensues can have far-reaching effects, including making it harder for a cell to use oxygen—which is very much in keeping with Warburg's vision of how cancer arises.

It is a bad situation for a cell, and as it progresses, it only gets worse. When it is tightly wound, the DNA in the nucleus is hard for the electron snatchers to attack. But signals from distressed mitochondria and the excess nutrients in the cell can both lead to new gene expression, meaning the DNA will unspool and become still more vulnerable to mutations.

In this genetically unstable state, many of the mutations that arise will be irrelevant to cancer. But some of the changes will matter. KRAS is among the best-known cancer-linked genes. Once activated, it will activate PI3K and speed up glucose metabolism. As the genetic switches flip, one by one, proteins known as transcription factors will dramatically change which genes are being expressed in the nucleus and thereby reprogram a cell. In addition to driving the Warburg effect, MYC will now direct the cell to scavenge for glutamine, a source of nitrogen, and other key ingredients needed for dividing and growing. "All you have to do is make one mistake out of that more vulnerable state," Chi Van Dang explained, "and then, all of a sudden, things start cascading from there."[17]

As mutations accumulate, the cells that eat more and grow faster will triumph in the Darwinian competition. This explains why so many cancer cells end up not only with more insulin receptors on their surfaces but also with mutations in the insulin signaling pathway inside the cell. Both the receptors and the mutations make cancer cells more sensitive to insulin. Other mutations might be as likely to arise, but they are not as likely to help a cancer cell survive and flourish. The cells that are most responsive to insulin will take up more glucose and use it to build new parts, express new genes, and produce still more reactive molecules. A deadly cycle will run faster and faster.

Insulin cannot explain everything. Foodborne illnesses contributed to a rapid increase in stomach cancers in the late nineteenth century. More sophisticated diagnostic techniques and aging populations made real increases in cancer deaths appear all the more dramatic. As the twentieth century progressed, smoking, sun exposure, and cancer-associated viruses would each contribute to the growing death tolls. The interplay of genetics and bad luck will always lead to some portion of cancers.

Even among the cancers most closely tied to insulin, the story requires nuance. Insulin and IGF-1 interact with other hormones and growth factors; the PI3K pathway intersects with other signaling networks. The precise role of obesity and inflammation in the progression of cancer has yet to be fully worked out. A hundred other factors might be listed as well. The question is not whether too much insulin explains our modern epidemic of cancer, but whether it is the crucial factor we have long overlooked, the missing piece of the puzzle that best explains why prevention efforts continue to fail.

Research on insulin and cancer has long faced a significant obstacle: a siloed medical and research edifice. The people who know insulin best, the endocrinologists, are not thinking about cancer, Lewis Cantley explained. Likewise, oncologists are not typically thinking about insulin.

But then, it did not take the scientific world nearly as long to accept that other hormones and growth factors can be carcinogenic. "Everybody knows about what's called the paradigm of hormone dependency of cancer," said Michael Pollak, of McGill University. "And people think about that as androgens in prostate cancer and estrogens in breast cancer. You just have to extend that paradigm to insulin and many, many cancers." Indeed, the paradigm hardly needs to be extended at all, given that high levels of insulin will also lead to increases in sex hormones—meaning that insulin can be both directly and indirectly implicated in cancers of the breast, prostate, and other organs.

For Pollak, insulin's connection to diet is what makes it so interesting to study and so promising in the context of cancer prevention. While some cancer risk factors are largely out of our control, insulin levels are determined by what we choose to eat. And yet what makes insulin an opportunity in Pollak's mind may be precisely what prevents other scientists from taking insulin's role in cancer more seriously. Suggesting that insulin could be connected to hundreds of thousands of cancer deaths each year requires cancer specialists to overcome an often reflexively skeptical position on diet. That skepticism is understandable. Diet has been the realm of quackery and miracle cures since the earliest days of medicine. But that dietary approaches to cancer are commonly promoted by unscientific thinkers does not make the relationship between cancer and diet any less fundamental or important.[18]

Among the most important contributions of the Warburg revival is that it has changed the attitudes of many medical professionals about diet and cancer. Colin Champ, a radiologist and cancer researcher at Duke University, recounts that when he first started giving talks on cancer and diet, doctors would walk out of the room. Now the response is often the exact opposite. "Everyone is totally intrigued by the idea," he said.

A small but rapidly growing body of research now suggests that following a very low carbohydrate (ketogenic) diet—which eliminates carbohydrates and so makes less glucose available to tumors—might make traditional therapies more effective in fighting some cancers. "It makes perfect sense that the nutrients outside of the cell will influence the tumor's metabolism," said Heather Christofk, a pioneering cancer metabolism researcher at UCLA who did her graduate work in Cantley's lab. "And so it makes perfect sense that changing your diet will influence those circulating nutrients, which will, in turn, affect the tumor metabolism."[19]

Siddhartha Mukherjee, the oncologist and cancer historian who did not mention the Warburg effect in his book *The Emperor of All*

Maladies, is now investigating how a ketogenic diet (which keeps insulin levels very low) might make drugs that block insulin signaling more potent. He is also calling for more research on diet as a therapy that might be used in concert with cancer drugs. In his own lab at Columbia University, he is working with Cantley's Weill Cornell lab to test whether drugs that block PI3K signaling might be more effective against cancer when combined with a diet that lowers glucose and insulin levels. "A careful scientific examination of diet as medicine is now long overdue in oncology, and in most fields of medicine," Mukherjee wrote in 2018.

Mukherjee, to be sure, is not giving up on the promise of targeting specific mutations with specific chemicals. But he suggests that it is time for researchers to think more broadly. "Perhaps," he wrote, "we had been seduced by the technology of gene sequencing—by the sheer wizardry of being able to look at a cancer's genetic core and the irresistible desire to pierce that core with targeted drugs."[20]

The search for such "magic bullets" against cancer will continue, as it should. But no one, perhaps not even Ehrlich himself, seems to have remembered that in *The Marksman*, the nineteenth-century opera that inspired him, the longing for "magic bullets" was not something to be celebrated. The opera is a Faustian morality tale about the dangers of overreaching. The magic bullet is controlled by the devil. When Max fires the bullet, he nearly murders his beloved.

In the case of cancer science, the danger of seeking magic bullets isn't so much that we can't control which targets the bullets hit—although that is often true as well—but that the quest for magic bullets can distract us from less sophisticated, but perhaps more effective weapons. In an article in *Science* in 1956, Warburg wrote that the "struggle against cancer" was being undermined by the "continual discovery of miscellaneous cancer agents and cancer viruses." The problem, Warburg maintained, wasn't that these discoveries were false but rather that they obscured "underlying phenomena" and so stood in the way of the "necessary preventive measures." This is the paradox of the ongoing fight against cancer. We still don't know nearly enough about the disease, and yet the onslaught of new discoveries can sometimes be more dizzying than clarifying.

The renewed focus on diet in cancer research is not limited to diet's impact on insulin levels. But if elevated insulin is as hazardous as many researchers now believe, we may still not know the full scope of the threat. In the early stages of type 2 diabetes, higher insulin levels will overcome the resistance of cells and successfully clear glucose from the blood—meaning that standard blood panels, which measure glucose but not insulin levels, will not indicate a problem, even as insulin resistance is setting in. This raises the question of how many might be at risk of insulin-related cancers without knowing it. A 2019 University of North Carolina study that looked at a number of markers of insulin resistance, such as elevated triglycerides in the blood, concluded that the "prevalence of metabolic health in American adults is alarmingly low, even in normal weight individuals." In total, only 12 percent could be deemed "metabolically healthy."[21]

IF MANY OF the most dangerous cancers can be linked to hyperinsulinemia, then another, still more fundamental, question inevitably follows: what causes insulin resistance and the subsequent elevation of insulin levels? Carbohydrates raise insulin more than protein or fat and so are the most obvious suspects. Low-carbohydrate diets were commonly prescribed to treat both obesity and diabetes at the turn of the twentieth century. But blaming carbohydrates for today's epidemics of obesity, diabetes, and insulin-linked cancers leaves important questions unanswered. There are many examples of societies eating carbohydrate-rich diets without developing any of the visible signs of insulin resistance. If the answer was as simple as "carbohydrates," Asian cultures, where rice is a central part of the diet, should have had epidemics of obesity and diabetes long before their Western counterparts. The reality is almost the opposite. Up until the end of the nineteenth century, diabetes in China was almost unheard of.

For the insulin-cancer hypothesis to make sense, it would have to explain both what changed in the Western diet in the nineteenth century and how that change led to higher levels of insulin in our blood. Once again, Lewis Cantley believed he knew the answer.

CHAPTER TWENTY-ONE

Sugar

GIVEN HIS INCREASINGLY erratic behavior, the most surprising aspect of Warburg's life in the late 1950s and the 1960s might be that he was still capable of sound, sometimes even groundbreaking, research. Though he was wrong about the source of the oxygen released by photosynthesis, Warburg's explanation involved a new mechanism he had discovered, known as the bicarbonate effect, that would continue to be studied for decades.

During the same period, Warburg discovered that cancer cells typically have less of an antioxidant enzyme known as catalase. It was an important observation that helped explain why cancer cells are especially vulnerable to the highly reactive molecules formed by radiation. Warburg soon proposed a catalase-focused therapy that would leave cancer cells still more vulnerable to reactive molecules. Warburg's specific treatment was far-fetched, and his contribution to this area of research has been forgotten. But it would prove to be another example of his remarkable foresight. A 2017 review paper on catalase noted that many new drugs are now being investigated with the hope of treating cancer by targeting the balance between reactive molecules and antioxidants in cancer cells. Some 60 years after Warburg proposed the idea, the article stated that "modulating catalase expression is emerging as a novel approach to potentiate chemotherapy."[1]

Warburg even made efforts to keep up with new science on occasion. One colleague recalled meeting with Warburg in the early 1960s and telling him about his recent work on proteins. Warburg, who would have been about 80, peppered this colleague with detailed questions, continuing the conversation long after Heiss had announced that lunch was ready.

The questioning may have been as much a show of dominance as of interest. A researcher working under Hans Krebs in England recalled the day in 1965 that Krebs arrived at his lab in a pin-striped suit. Krebs tried to deny that his attire was unusual but finally showed the young researcher a telegram he had received from Warburg. As the researcher remembered it, the telegram said, "I want you to come to Berlin. I have a new theory. I do not want your opinion. I want an audience." Krebs, then in his mid-60s, flew to Berlin, spent a day listening to Warburg, then immediately got back on a plane and flew home.[2]

Warburg made his dominance known in other ways as well. Peter Ostendorf recalled that when not at his laboratory bench, Warburg could be found in his library, sitting before his typewriter at the head of the long wooden table in the center of the room. If the door was closed, it was understood that Warburg was not to be bothered. But even if the door was open, you could not interrupt Warburg, who would sometimes be gazing out the window at his institute's garden. The only thing to do was wait. "You would stand there," Ostendorf recalled, "rustling your papers or clearing your throat, hoping to get noticed."[3]

This was the man Warburg had always been. He was sitting in the same spot in his Dahlem institute he had sat in during the 1930s, when he had refused to change his ways even when his life was in danger. But when Warburg looked out of the window of his library in the 1960s, there was one difference to the scene. After the war, Warburg had erected a statue of Emil Fischer, the man who had taught Warburg organic chemistry.

BY THE TIME Warburg arrived in Fischer's Berlin lab in 1903, the older man had already made one of the greatest discoveries in the history of science: he had unlocked the secrets of carbohydrates.

Carbohydrates are formed from sugars. Glucose, a simple sugar, is the most ubiquitous and most central to life. It is the sugar made by plants during photosynthesis, and it is kept within a carefully regulated range in the body of every animal at all times. But the driving force behind Fischer's interest in sugars was not glucose. In the late nineteenth century, virtually all Germans were interested in sucrose, or "table sugar"—the sugar that sweetens so many of our foods and drinks. (Both glucose and sucrose were—and still are—colloquially referred to as "sugar." In the pages that follow, "sugar" will refer to sucrose unless otherwise indicated.)

The German sugar story can be traced to the middle of the eighteenth century, when the Berlin chemist Andreas Marggraf gazed down at a solution of pulverized beets under his microscope and noticed something intriguing. The crystalline structures he observed looked remarkably similar to the crystalline structures found in pulverized sugarcane. It wasn't a total surprise. The juice from the beets, Marggraf knew, tasted like the sweet juice from sugarcane. But knowing that the two different plants contained the identical chemical was an important breakthrough. Cane sugar was already an enormous global industry. If sugar could be extracted from beets, it had the potential to change the world economy.

Only there was a problem: Marggraf could obtain very little sugar from each beet. It took decades of technological advances before beet sugar could be extracted and refined efficiently. Napoleon, who hoped to wean his empire off of sugar exports from the British colonies, would become the first great champion of the beet sugar industry. (A satirical cartoon from 1811 depicts Napoleon haughtily squeezing a beet over his coffee.) But over the course of the nineteenth century, the Germans would come to dominate the beet sugar industry. Prussian Saxony, it turned out, had the ideal soil for growing beets.[4]

As German beet sugar production increased, so, too, did German sugar consumption. In 1800, a typical German ate about 1 pound of sugar a year. When Fischer turned to sugars in the 1880s, Germany was supplying almost a third of the world's sucrose supply, and the typical German was consuming 15 pounds a year. (The English and Americans were already eating far more sugar than the Germans.) Once a luxury item, sugar—in coffee and tea and cocoa and jam and pastries—became a daily indulgence and sometimes a substitute for meat and fats. The candy, ice cream, chocolate, and soft drink industries all stem from the mid-nineteenth century, as does the expectation of a sweet dessert at the conclusion of every lunch and dinner.[5]

But though sugar was suddenly everywhere, it was still not well understood. Chemists mistakenly thought that all sugars were merely carbon atoms attached to molecules of water—hence the name "carbohydrates." By Fischer's day, it was known that sugars were formed out of varying arrangements of carbon, hydrogen, and oxygen atoms. Chemists even knew the basic recipe of glucose: 6 atoms each of carbon and oxygen and 12 of hydrogen. But how, precisely, the atoms of carbon, hydrogen, and oxygen were arranged in each different sugar was still anyone's guess. Sugars had been lumped together as a group simply because they tasted sweet and could support fermentation. As late as the 1870s, a leading chemist referred to theories of how molecules are arranged in space as "fancy trifles" that "are completely incomprehensible to any clear-minded researcher."[6]

To make any progress, Fischer would need to isolate pure crystals of the different sugars, but neither he nor anyone else knew how to do that. The problem chemists of the era faced is familiar to anyone who has eaten pancakes: crystals are solids, but sugars tend to form syrups. Fischer's first breakthrough involved some luck. Like so many of the great German chemists of his generation, he had begun his career working on dyes. While doing so, he synthesized a compound, phenyl-hydrazine, that gave him a pale-yellow solution. It looked like urine. Fischer moved on, never imagining he would return to the seemingly worthless liquid in his glass beaker.[7]

But in the mid-1880s, Fischer discovered that phenylhydrazine was able to do one very important thing: it could react with sugars and give rise to the pure crystals he needed to move ahead with sugar chemistry. It was only a first step. And in the hands of an ordinary chemist, such crystals would have been far less significant. But Fischer—a man who turned to chemistry only after his father declared him "too stupid to be a businessman"—was far from an ordinary chemist. With the pure crystals at his disposal, Fischer could subject them to a litany of chemical tests to glean new insights about the nature of the bonds. Soon he was synthesizing new sugars from scratch. Out of a sticky goo emerged an entire class of organic molecules. As Yale chemist Frederick Ziegler once put it, Fischer's work on the structure of sugars was to organic chemistry "what Newton and Einstein were to physics." Another chemist once compared Fischer's analysis of sugars to "Shakespeare."[8]

Among the many sugars Fischer worked on was fructose, a sugar, as the name indicates, that can be found in fruits. Chemists before Fischer had already determined that fructose was similar, but not quite identical, to glucose. The two simple sugars rotated light waves in opposite directions, and fructose was slightly sweeter than glucose. Fischer could now explain why fructose and glucose were so similar and yet so different. The two molecules have the identical number of carbon, oxygen, and hydrogen atoms. The difference between glucose and fructose is only a matter of how the atoms are arranged.

Though glucose and fructose can be found apart, the two molecules can also bond to each other. When that happens, the result is sucrose, the sugar we add to our foods and drinks. Fischer could not figure out exactly how glucose and fructose bond to form sugar.

That discovery would be left to other scientists after Fischer committed suicide in 1919. Though he had cancer, it is also believed that Fischer was dying from his long-term exposure to phenylhydrazine. The molecule that revealed the true nature of sugars turned out to be a poison.

Fischer's poisoning is not the only sad irony of his story. His work on sugars changed his field, and yet the full implications of his findings would be absorbed only slowly into the bloodstream of modern medicine. When it came to nutrition and disease, sugars would remain a sticky mess for decades to come.

WHILE FISCHER AND others worked to understand how fructose and glucose bonded to form sugar, some medical thinkers of the era were taking note of something else having to do with sugar: the more refined sugar that people ate, the more likely they seemed to develop diabetes and cancer. In 1902, the British biochemist Robert Plimmer spent a year in Fischer's Berlin lab, leaving just before Warburg arrived. Two decades later, Plimmer and his wife, Violet, published *Food and Health*, a popular nutrition guide. "Not so very long ago sugar was a rare luxury kept under lock and key in the tea caddy," the Plimmers wrote. But with the rise of German beet sugar and falling prices, the consumption of sugar had "increased enormously" and was "still increasing in all civilised countries." They added, "Incidentally, cancer and diabetes, two scourges of civilisation, have increased proportionately to the sugar consumption."[9]

The possible connection between sugar and diabetes was always more apparent than any connection between sugar and cancer. Hindu physicians discovered that the urine of diabetics was sweet as far back as the sixth century CE. In the United States, the person who saw the emerging diabetes epidemic most clearly—and connected it to sugar most convincingly—was Haven Emerson, the father of Warburg nemesis Robert Emerson and the onetime commissioner of the New York

City Department of Health. Diabetes was not only on the rise, Emerson wrote in 1924, but over the previous 50 years it had increased more rapidly than any disease for which there were records.

Though he was not certain whether refined sugar from cane and beets was dangerous because of its unique impact on metabolism or because it was merely a source of excess calories, Emerson was certain that Americans were endangering their health: "Do we need to eat three times as much sugar each year as our grandparents did?"[10]

Emerson was able to show that sugar consumption in the West had risen together with diabetes beginning in the nineteenth century and also that when sugar consumption fell during World War I, diabetes rates in the following years fell as well. But no matter how closely the trends overlapped, it was still only a correlation. Elliott Joslin, the leading American authority on diabetes, looked at the same evidence as Emerson and remained unconvinced. Joslin understood that diabetics needed to avoid sugar and other carbohydrates to keep their blood sugar (glucose) under control, but that didn't mean that sugar was responsible for causing the disease in the first place.

Joslin's skepticism was largely based on a single observation. The Japanese diet, he noted in 1923, consisted "largely of rice and barley," and yet diabetes was "not only less frequent but milder in that country." The Japanese example, Joslin said, "would seem to save us from [the] error" of blaming sugar for diabetes.

It was a simple point, but Joslin himself had made a critical error. The Japanese were consuming lots of glucose in the form of rice and barley, but not lots of sugar, or sucrose—glucose bonded to fructose. It was a strange oversight, given that Joslin knew that fructose comprised half of sugar. If Joslin was glossing over the difference between sucrose and glucose, it was likely because he believed that the fructose half of sugar was harmless.[11]

Elliott Joslin would prove far more influential than Haven Emerson or anyone else who believed that refined sugar might be causing diabetes. And if the leading medical thinkers of the era were not ready to tie diabetes to sugar, it was all the less likely that they would be ready to tie cancer to sugar. The doctors most inclined to make the connection tended to be those who had traveled to faraway regions to treat indigenous populations. Sir Robert McCarrison, marveling over the absence of cancer and other ailments among the Hunza of the Himalayas, noted in 1921 that the amount of sugar coming into the country in a single year was seemingly less than what was consumed in a single day in a "moderately sized hotel" in Pittsburgh.[12]

A handful of researchers continued to point out the suspicious correlations between sugar and cancer deaths in the second half of the twentieth century. The British physiologist John Yudkin would prove the most outspoken. (Though his 1972 book, *Pure, White and Deadly*, is only about sugar and nutrition, Yudkin, who served in the Royal Army Medical Corps during World War II and whose wife escaped Nazi Germany in 1933, perhaps couldn't resist a title that sounded like a warning about Nazism.)

Yudkin's own research found that both breast and colon cancer rates in a given country would rise in accordance with how much sugar the population consumed. As Yudkin readily acknowledged, such comparisons say nothing about cause and effect, but "only provide clues as to possible causes." But the clues were gradually piling up. More arrived in 1975 when Richard Doll looked into the relationship between cancer and sugar consumption across different countries and found that sugar could be tied to a number of other cancers, including those of the prostate, ovaries, uterus, rectum, testicles, and kidneys. A few years later came a review of sugar consumption and breast cancer deaths in women over 65. The five countries with the most deaths turned out to be the very five countries that consumed the most sugar. The five countries with the fewest deaths, in turn, were the five countries that consumed the least amount of sugar.[13]

These studies, however inconclusive, might have led to a sugar scare, given that far less compelling evidence would regularly lead to panics over synthetic chemicals. At approximately the same time the evidence linking sugar to cancer was emerging, the United States cracked down on artificial sweeteners based on studies that found that rats consuming extraordinary quantities were more likely to develop bladder cancer. (The studies were later found to be irrelevant to humans.)

When it came to the natural ingredients in the Western diet, it was not the refined sugar added to our diets but meat and animal fats that would be singled out as potentially dangerous. Mice fed high-fat diets did develop cancer more readily, but the evidence said to support the relationship in humans—for instance, that migrating to the United States made Japanese women much more likely to develop breast cancer—might just as easily have been used in support of the sugar hypothesis. Americans, after all, typically ate both far more sugar and far more animal fat than the Japanese.

The animal fat hypothesis faced other problems as well: up until the early twentieth century, there were societies across the world—the Inuit of the Arctic, various Native American tribes of North America, the Maasai of Africa—where cancer was rare even though meat and fat were staples of the diet. Much later, a series of large-scale studies in the 1980s and 1990s failed to turn up evidence that dietary fat caused cancer.

Though some, like Yudkin, would continue to sound the alarm about sugar, it made little difference. The belief that animal products were the true threat had crystallized into accepted wisdom as readily as if it had reacted with Fischer's phenylhydrazine. Consumption of refined sucrose from sugarcane, beets, and corn (in the form of high-fructose corn syrup, which is essentially identical to sugar from cane and beets) would continue to increase in one country after another, decade after decade. By the first years of our current century, the average American, by conservative estimates, was swallowing more than 90 pounds of sugar a year—not including the natural fructose from fruits. (The glucose and fructose consumed together in whole fruits, though potentially still fattening, are generally considered less problematic than refined sucrose because the molecules are absorbed more slowly and lead to less dramatic spikes of glucose and insulin in the blood.)[14]

Given the evidence already available in the 1970s, it's difficult to understand why sugar was not considered at least as likely to cause cancer as animal fats. But then, Joslin and others had long ago cleared sugar of any metabolic wrongdoing based on evidence concerning only glucose. Once again, it seemed, everyone had forgotten that glucose and sugar were not the same thing, that glucose had a twin.

Otto Warburg, around 1965.

Otto Warburg and West German President Theodor Heuss by statue of
Emil Fischer, 1953.

CHAPTER TWENTY-TWO

The Evil Twin

THE END WAS NEAR, and Warburg was preparing. In 1966, he sat down before a video camera and, speaking haltingly, read a short statement that highlighted what he saw as his greatest scientific accomplishments. "Since every real discovery in science means a revolution where there are victors and vanquished, every one of our discoveries has evoked long and bitter fights," Warburg said. "All were eventually decided in our favor."[1]

That the statement reads like an obituary was no accident. Warburg had not forgotten the jarring moment in 1938 when he had been mixed up with the other Otto Warburg and found his own obituary in an English newspaper. The other Otto Warburg, a botanist, did not look much like Warburg, but from Warburg's perspective, he fit the role of the doppelgänger. In literature, the doppelgänger is typically not only a twin but an evil twin, a dark alter ego who maintains a mysterious connection to the story's protagonist. The doppelgänger is simultaneously alike and yet radically different, a mirror that exposes even as it reflects. The other Otto Warburg, driven by moral conviction, had stepped away from his science to campaign for the Jewish people. He was Warburg, only turned inside out.

THE WORD "DOPPELGÄNGER" first appeared in German author Jean Paul's late eighteenth-century novel *Siebenkäs* at the moment when the twins in Paul's story depart from one another. In the molecular doppelgänger tale of glucose and fructose, the two molecules that form sugar, the drama begins with a separation. Once sugar makes its way to the small intestine, it encounters an enzyme that separates its glucose and fructose halves.

Released from their embrace, both glucose and fructose will travel from the small intestine to the liver, by way of the bloodstream. For glucose, which can provide energy for cells throughout the body, the journey might only be beginning. But most of the fructose will be metabolized in the liver, which will take the arrival of the fructose as a signal to turn carbohydrates into fat. (Fats, like carbohydrates, are assembled out of carbon, hydrogen, and oxygen.)

To diabetes doctors busily tracking glucose levels in the blood and urine, the fructose departing to the liver seemed inconsequential. Fructose appeared so benign that by 1979, the American Diabetes Association was encouraging diabetes patients to use it as a sweetener in place of sugar. Glucose had been singled out as the evil twin.[2]

Among those who noted the new fructose recommendation was Gerald Reaven, the Stanford University endocrinologist who had become the world's leading authority on insulin resistance. Reaven grew interested in studying insulin in 1960 after reading about Rosalyn Yalow and Solomon Berson's new method of measuring insulin levels. Over the next 20 years, he assembled a remarkable body of research that connected eating carbohydrates to insulin resistance, and insulin resistance to a host of metabolic problems, including high blood pressure, elevated triglycerides (fat in the blood), and low levels of HDL cholesterol.

Reaven was primarily focused on type 2 diabetes and cardiovascular disease. But given our current understanding of the connection between insulin and the pathways activated in cancer cells, what Reaven discovered when he turned his attention to fructose might be every bit as relevant to cancer: in mice, at least, fructose did not look benign at all. Eating large quantities of fructose seemed to be a direct route to insulin resistance and, by extension, elevated insulin. As Reaven later put it to the *New York Times Magazine*, the effect of fructose was "very obvious, very dramatic."[3]

Reaven's findings on fructose have been confirmed by many other researchers over the last 40 years, in both animals and humans. In one study, Kimber Stanhope, of the University of California, Davis, fed overweight and obese adults a quarter of their calories either as fructose-sweetened drinks or as glucose-sweetened drinks. The people drinking the fructose developed clear signs of insulin resistance; the people drinking the glucose did not. After reviewing her data, as Stanhope told CBS News, she "started drinking and eating a whole lot less sugar."[4]

Such feeding studies can only tell us so much. The subjects in them eat more fructose than they would likely ever eat during the course of their normal lives. And the studies last only weeks or months, whereas the chronic diseases driven by insulin resistance develop over many years. Even so, the results make clear that fructose can send the liver into a frenzy of fat production. Recent data suggest that fructose will directly turn on the genes responsible for making fat in the liver.[5]

That fructose, a carbohydrate, appears to make us add fat more readily than the fat we eat, is counterintuitive. But the basic concept— that carbohydrates can fatten the body—has been well understood for centuries. It was known in ancient Rome that to get the best *foie gras* (fatty liver), you first needed to feed your geese dates, which happen to be an excellent source of fructose. And it is unlikely that people in the ancient world found this method of fattening geese surprising. Galen had noticed that slaves who worked in the fields would grow fatter as the grapes and figs ripened. Centuries later, the same observation would be made about plantation slaves working to extract sucrose from sugarcane.

The famed German physiologist Justus von Liebig described the phenomenon of carbohydrate-driven fattening in the early 1840s. The "herbs and roots consumed by the cow contain no butter," Liebig noted, and "no hog's lard can be found in the potato refuse given to swine." Liebig, who also studied how bees turned fructose from honey into wax (a fat), thought the fattening effect of carbohydrates was self-evident. One "can hardly entertain a doubt," he concluded, "that such food, in its various forms of starch, sugar, etc., is closely connected with the production of fat." The effect was "undeniable."[6]

THOUGH RESEARCHERS ARE still working out the specific cellular mechanisms involved, the path from a fattening liver to insulin resistance is now broadly understood. As fat accumulates in the liver, some will be stored locally, and some will be shipped off to fat tissues around the body for storage. When the fat can be safely tucked away in the fat tissue under our skin, it appears to do little metabolic harm. But as the supply of fat grows, storage capacity runs out. Now the fat will end up in places it never should. The liver itself will become marbled. The fat will make its way into the pancreas and even into our muscles.

This misplaced fat may be invisible from the outside, but it is far from benign. It drives inflammation and interferes with how our cells respond to insulin. To overcome that interference, or resistance, the pancreas has no choice but to secrete more insulin, and a dangerous cycle takes off. The additional insulin will likely make us fatter, which in turn can lead to still more misplaced fat and greater insulin resistance.

This particular model of insulin resistance, known as the fat "overload hypothesis," helps explain why some people who are overweight do not develop metabolic abnormalities and why many people at supposedly "healthy" weights do. In a sense, fat beneath the skin is protective in that it offers a safe place for storage. Scientists have genetically engineered mice that can keep expanding their fat tissue and thus their safe storage capacity for fat. Such mice are said to be equivalent to 800-pound humans, but their metabolism remains healthy. Conversely, people with lipodystrophy—a genetic disorder that leaves them with very few fat cells—become insulin resistant while remaining extremely thin.

Sugar is not the entire story of insulin resistance. Any carbohydrate that is rapidly digested—beer or bread, pasta, and cereals made with refined flour—will also spike glucose and insulin levels. If fat is eaten together with these carbohydrates, the insulin spike will lead that fat to be stored rather than burned, and that dietary fat, too, will contribute to the "overload" problem. But nothing appears to drive the first stages of the process quite like refined sugar. Drinking sugar—in soda or fruit juice—is thought to be worst of all.

Lewis Cantley, the scientist who pioneered the study of how insulin activates the pathways linked to cancer, is among the researchers who have grown alarmed about sugar. Cantley does not write popular books or articles. He has received several of the highest honors in his field, including the 2015 Canada Gairdner International Award, often a prelude to a Nobel Prize. Cantley, in other words, could hardly be more different from the diet doctors on TV or on the cover of magazines in the grocery store. But he has stopped eating sugar himself for a simple reason. His research has led him to the conclusion that today's "high consumption of sugar" is "almost certainly responsible for the increased rates of a variety of cancers in the developed world."

Cantley reached this conclusion based on the evidence connecting refined sugar to elevated insulin, and elevated insulin to cancer. But in Cantley's mouse model of colorectal cancer, he found an even more direct relationship between sugar, in the form of high-fructose corn syrup, and cancer. Mice that were genetically engineered with mutations associated with colorectal cancer and then given a daily serving of sugar equivalent to the amount in a can of soda developed cancers that grew faster and bigger than mice that did not consume the sugar. The fructose in the sugar, Cantley saw, could both turn on the Warburg effect and provide the building blocks for fat molecules that the cancers use to grow. "The evidence," Cantley said, "really suggests that if you have cancer, the sugar you're eating may be making it grow faster."[7]

While most of the fructose in sugar ends up in the liver, Richard Johnson, a fructose expert at the University of Colorado, explained that fructose can be metabolized by other tissues. He said that fructose appears to be able to directly fuel not only colon cancer but also cancers of the breast, lung, and pancreas. Though Johnson does not argue, as Warburg did, that cancer stems from a problem of cellular breathing, he believes that fructose is the "perfect food" for a growing cancer precisely because it helps cancer cells to survive in the low-oxygen environments. This can be especially important when a cancer spreads to a new location and doesn't yet have a reliable blood supply to provide oxygen. "If you want to make a cancer happy," Johnson said, "feed it fructose."

According to Johnson, the same phenomenon can be seen in animals, like the naked mole rat, that live in low-oxygen environments. In their underground homes, the mole rats have less air than most animals could ever tolerate. They manage by converting a portion of their meals into fructose, which in turn increases fermentation and makes the mole rats less dependent on respiration. For the mole rats, fructose is necessary for survival. And that we crave the taste of fructose suggests that our primate ancestors benefited from it as well. Johnson, in collaboration with the anthropologist Peter Andrews, has put forward a hypothesis that we evolved from prehistoric apes that migrated from Africa to Europe and back. In Europe, these apes initially managed to find fruit for much of the year, but around 12 million years ago, Johnson and Andrews suggest, a cooling period set in that left the apes facing long winters with very little food.

As the apes starved, the capacity to store even a little extra fat was the difference between life and death. Both genetic and fossil evidence suggest that it was during this period of starvation that a mutation arose in the gene coding for the enzyme uricase. While fructose could already be converted to fat in these apes, the uricase mutation made fructose all the more fattening. During a long period without food, survival of the fittest, Johnson wrote, became "survival of the fattest."[8]

CANTLEY AND JOHNSON are only two of a growing number of researchers who are alarmed about sugar and cancer. Michael Pollak, who runs the Division of Cancer Prevention at McGill, is another. "Glucose or fructose-based drinks are really among the most unhealthy foods that you could imagine," Pollak said. It's okay to have a little sugar, but it should be consumed like a condiment, "in the same way we have pepper."

Precisely how much sugar is too much may be different for each person, depending on one's genes and age and exercise habits and capacity to store fat safely. But the path from the refined sugar added to our diets to insulin resistance, and from insulin resistance to cancer, is now well understood and based on widely accepted science. For that reason, the science journalist and author Gary Taubes believes that sugar can be thought of as "a primary cause" of cancer and other diseases linked to insulin resistance and elevated insulin.[9]

Taubes, who studied physics at Harvard and engineering at Stanford, has spent the last 20 years researching and writing about the links between insulin, obesity, and the chronic diseases associated with the modern Western diet. He does not claim with absolute certainty that eating lots of sugar leads to these diseases, only that it is the simplest answer that fits with all of the available evidence and so, according to the principle of Occam's razor, should be considered the most likely explanation.

"Too much sugar" might be the simplest explanation for the many obesity-linked cancers, but it is not a simple explanation. It is an idea built upon more than a century of science. Scientists had to figure out, among many other things, how fructose is converted to fat; how fat in our muscles and liver and other organs interferes with insulin signaling; how the pancreas responds by pumping out more insulin; how elevated insulin activates the Warburg effect and other molecular pathways within cancer cells; and how those pathways keep fledgling cancers alive and well nourished.

Warburg's devotion to "simple" explanations was unrivaled. He liked to cite the wisdom of William Bayliss, an English physiologist who wrote in 1915 that the "truth is more likely to come out of error, if this is clear and definite, than out of confusion" and also that "it is better to hold a well-understood and intelligible opinion, even if it should turn out to be wrong, than to be content with a muddle-headed mixture of conflicting views, sometimes miscalled impartiality, and often no better than no opinion at all."[10]

Otto Warburg did not live long enough to see the most convincing evidence linking sugar to the strange metabolism of cancer cells that he discovered. Even if he had, it is unlikely he would have budged from his own oxygen-focused explanation. Warburg, alas, did not absorb another lesson from Bayliss that appears at the bottom of the very same page of his 1915 book: "It is not going too far to say that the greatness of a scientific investigator does not rest on the fact of his having never made a mistake, but rather on his readiness to admit that he has done so, whenever the contrary evidence is cogent enough."

In the case of sugar and metabolic diseases, there is contrary evidence. Skeptics have argued that if sugar is truly the driving force behind America's obesity and diabetes epidemics, the rates of these conditions should have gone down in recent years, given that sugar consumption has declined of late in response to warnings about the danger it poses. Whether this argument is "cogent enough" is worth exploring, but as Taubes has pointed out, Americans today are still eating sugar in quantities that would have been unthinkable a century ago—a time at which Haven Emerson was already pointing out that Americans were eating vastly more sugar than their grandparents ever had. As Taubes sees it, to suggest that sugar is not responsible for metabolic diseases based on current trends would be like cutting back from 20 to 17 cigarettes a day and then concluding that smoking could not be responsible for lung cancer if the rate did not fall.

The smoking parallel might also explain why the connection between sugar and cancer can be so hard to accept. Although the definitive studies linking smoking to lung cancer were only carried out in the 1950s, smoking should have been the most obvious suspect long before then. A British physician had connected inhaling snuff to cancer of the nose by 1761. By the end of the eighteenth century, it was known that chimney sweeps developed cancer of the scrotum and that cancer of the lip was more common among pipe smokers. Lung cancer, meanwhile, had become far more common after cigarette smoking had become far more common, and it was understood that cigarette smokers inhaled carcinogens more deeply into the lungs relative to pipe smokers.

As a medical student, Richard Doll himself had become interested in the connection between pipe smoking and oral cancers. And yet Doll—like most British and American doctors of the mid-twentieth century—couldn't fathom that cigarettes were behind the emerging lung cancer epidemic until his own studies finally provided overwhelming evidence. As Doll once explained, the problem was that "cigarette smoking was such a normal thing and had been for such a long time." It "was difficult to think it could be associated with any disease."

In 2016, Cantley said that we may someday "view this era of massive addiction to sugar in America in the same way that we now view the period of massive addiction to tobacco." In the meantime, almost everyone agrees that we need far more research on the health effects of sugar. Nearly 100 years ago, Elliott Joslin pointed out that if Americans were dying of infectious diseases like typhoid or scarlet fever at the rates they were dying of diabetes, there would be a rapid government response "to discover the source of the outbreak." The same point could have been made about cancer.[11]

Many different debilitating conditions appear together with insulin resistance. Our numbness to the suffering they cause might, in the end, be the most debilitating condition of all.

SUGAR HAS LONG OCCUPIED a troubling place in human history and society, even setting aside what happens when we ingest it. The sugar industry, through its reliance on the slave trade, once had a long-standing relationship with evil. In the 1930s, that relationship was renewed when the Nazis took up the cause of the German beet sugar industry, which had been decimated by the Great Depression. To the Nazis, support for beet sugar was politically expedient, a chance to win the votes of poor rural laborers. "No other German industry," wrote the historian John Perkins, "was so closely integrated as beet-sugar production with the agricultural sector and the rural life that Nazi ideologues, with their 'blood and soil' (*Blut und Boden*) outlook, sought to protect, eulogize and enhance."

In the early 1940s, the beet sugar industry was integrated into the Nazi agenda in an additional way. After sucrose was extracted from beets, it left behind a brown sludge known as *schlempe*. In 1898, the German chemist Julius Bueb had discovered that if he heated the *schlempe* in a closed chamber, he could form a cyanide gas. That gas would become known as Zyklon B. The supply that made its way into the Nazi killing chambers, via a third-party distributor, was created by a company known as Dessau Works for the Sugar and Chemical Industry.[12]

After the war started, sugar was rationed in Nazi Germany along with other foods, but some were too hooked to cut back. Hitler was a man of addictions, but his addiction to sugar may have been more

intense than any other. When Baldur von Schirach, the head of the
Nazi Students' Association, sat down for a meeting with Hitler in 1928,
he looked on in amazement. "At tea time, I couldn't believe my eyes,"
Schirach recalled. "He put so many lumps of sugar in his cup that there
was hardly any room for the tea, which he then slurped down noisily.
He also ate three or four pieces of cream pie."[13]

Sugar, of course, cannot be blamed for Nazism or for turning Hitler into a madman. But as his madness grew, so, too, did his taste for
sweets. It wasn't only his cherished Viennese pastries that he longed for.
On any given day, Hitler might consume 2 full pounds of chocolates or
2 pounds of pralines. He even added sugar to his wine. Hitler's valet,
Heinz Linge, recalled that while planning for the invasion of Norway,
Hitler kept running out of the room for more sweets. Linge asked Hitler if he was hungry. "For me, sweets are the best food for the nerves,"
Hitler replied.[14]

Hitler was aware that his binging was causing him to gain weight,
and he tried to slow down. He asked his personal chef, Constanze
Manziarly, to serve him only grated apples for dessert. But it was no
use. Manziarly recalled that Hitler would "lose control" and attack her
cakes. "I bake a lot every day, often for hours," Manziarly wrote in a
letter to her sister, "but in the evening everything is always gone."

During his final days in his bunker in Berlin, Hitler was so weak
he could barely move around. His cake had to be crumbled for him so
he could slurp down the crumbs between his rotting teeth. As Norman
Ohler asserts in *Blitzed*, an examination of substance abuse in Nazi
Germany, with Hitler's supply of medicines and stimulants now cut off,
sugar "was the final drug."[15]

Hitler was always extreme but never original. He stood out not
because he was anti-Semitic but because he was fanatically anti-Semitic.
The same was true of Hitler's hunger for sugar and dread of cancer.
Many, if not most, Germans were developing a taste for sugar and a fear
of cancer during Hitler's lifetime. Hitler was merely a grotesque carica-

ture of the Germans of his day. That his sugar addiction may have left him more susceptible to the disease that terrified him most was only one more example of how Hitler got everything exactly wrong.

Hitler was born in 1889, the year cancer was first characterized as a problem of "seed and soil." The metaphor might be the best way to understand Nazi Germany itself. Hitler was the seed, but seeds do not grow without the proper nourishment. Had the German soil not already been so well fertilized with hatred, Hitler could never have metastasized.

IN THE LATE 1960S, Warburg chose Birgit Vennesland, of the University of Chicago, to succeed him at his Institute for Cell Physiology. Vennesland was a natural for the position. Though not auditioning for the job, earlier in the decade she had demonstrated the key qualification: she had become an outspoken champion of Warburg's thinking on photosynthesis.

Vennesland saw a side of Warburg few others ever would. During a visit to his institute, she asked Warburg whether he believed he had ever made any mistakes. It was the right question for Warburg, and it's a testament to Vennesland that she was one of the few people—if not the only person—to ask it.

Warburg let the question sit for a moment before answering. "Of course, I have made mistakes—many of them," he said. "The only way to avoid making any mistakes is never to do anything at all. My biggest mistake . . ." Warburg stopped, an even longer pause this time. "My biggest mistake was to get much too much involved in controversy. Never get involved in controversy. It's a waste of time. It isn't that controversy in itself is wrong. No, it can even be stimulating. But controversy takes too much time and energy. That's what's wrong about it. I have wasted my time and my energy in controversy, when I should have been going on doing new experiments."

It was an extraordinary statement. Warburg's entire life had been spent in controversy. Warburg liked, and even framed, Max Planck's famous line that a new scientific truth triumphs only when "its opponents eventually die." But waiting for his opponents to die was not truly the Warburg way. "It is not true that time helps truth to victory," he once wrote to Dean Burk. Victory required accumulating more and more facts until "nobody can stop the stream."[16]

Vennesland wasn't done with the challenging questions. In the same conversation, she asked why Warburg held so much animosity for the great physicist James Franck. It is likely that Warburg envied Franck, who had been a prized student of Warburg's father. But Warburg offered another explanation: Franck had once said that Warburg couldn't measure light. "He was a theoretician," Warburg told Vennesland. "By himself he couldn't measure anything, and he said I—*I couldn't*—measure. . . ."

"Something curious was happening," Vennesland later wrote. "Warburg was getting a little incoherent. In the course of telling me about why he got angry, Otto Warburg got angry all over again. He got as angry as he would have been if James Franck had been sitting there in the same room, now, telling him how sorry he (Franck) was, that his (Warburg's) measurements couldn't possibly be right."

Vennesland had seen Warburg for who he was, but she had failed to appreciate that she, too, would eventually face Warburg's wrath. Though the details are unclear, Vennesland, while working with Warburg in preparation to take over the institute, appears to have committed the unforgivable sin of questioning his photosynthesis findings. Warburg responded by locking her out of the institute.[17]

In a letter that he sent to multiple colleagues in the first months of 1970, Warburg claimed that Vennesland had a "severe mental disorder" and had experienced a breakdown. As Warburg's story went, she had shouted "that she had to take the lead immediately because everything we were doing was wrong." Vennesland, Warburg continued, then "attacked my co-workers violently" and "ran screaming into the street" before being arrested.

Though it would have been understandable if Vennesland, or anyone else, had responded to Warburg by running through the streets screaming, it was almost certainly a fantasy. In 1965, Warburg told Hans Krebs that after their recent meeting in England, he had been on his way to the airport when he spotted a black stallion that had escaped its stable. Warburg said that he had stopped his journey to rescue the horse before it was hit by a car. Krebs, however, was able to confirm that the stable in question had no black stallions. "Was the incident," Krebs wondered, "merely a dream?"[18]

IN NOVEMBER 1968, Otto Warburg climbed up the ladder in his library to retrieve a book from a high shelf. He had recently celebrated his 85th birthday, and he was in good spirits. "In many respects," Warburg had written to a friend earlier that year, it was better to be old than young. "The struggle for existence"—the same Darwinian phrase he used to explain how competing cells turn cancerous—"is over, and, if one possesses luck and reason, one can still live for many years."

Whatever regrets he had expressed to Vennesland about spending his life picking fights were now long forgotten. In a letter to von Ardenne the previous year, Warburg encouraged his friend to continue his cancer studies in the face of criticism, just as he himself had: "The more resistance I found, the more I attacked and the better my weapons became."

Warburg's one known ailment, heart palpitations, had first been diagnosed while he was a medical student. (One of Warburg's professors, the famed Ludolf Krehl, made the diagnosis himself and told Warburg to focus on lab work: "No one has yet died from biochemistry," Krehl pointed out.) The palpitations returned in the last months of 1956, and Warburg feared the problem might somehow be related to a blood clot. A doctor suggested cutting back on coffee but assured Warburg that he had no reason to be concerned.[19]

At 85, Warburg might have lived a while longer. But on the third step of the ladder, his foot slipped. He fell backward, crashing to the floor of his library, where he lay with a fractured hip, helpless beneath Pasteur's cold gaze. Because the door to the library was closed—a sign that Warburg did not want to be bothered—no one immediately came to his rescue.[20]

The fall appeared to be a cinematic ending for Warburg: the Faustian hunger for knowledge and the preordained punishment encapsulated in a single act. In the film, the camera, looking down on him from above, would pull away slowly. Warburg, motionless on the floor of his Dahlem palace, would become smaller and smaller: a breathing speck, a single isolated cell.

Warburg survived the fall, but was no longer the same. In the last week of July 1970, he felt a pain in his leg and was diagnosed with the blood clot he had long feared. Late in the evening of August 1, the clot broke free and traveled up his body until it came to rest in a narrow passage between his heart and lungs.

Although Warburg did not live to see the revival of his research, he never doubted that he would eventually be proved correct about cancer. And everything else.

Otto Warburg, date unknown.

Postscript

FOR OLDER RESEARCHERS who still remembered Warburg as one of the great minds of his generation, his death landed as a blow. "Scientists all over the world felt that a king—their king—had died," wrote Ernst Jokl, the German American pioneer of sports medicine.[1]

In the immediate aftermath of Warburg's death, work at the institute came to a standstill. The 71-year-old Heiss took it upon himself to make sure that nothing would change until an appropriate new director could be identified.

"It was horrible," Peter Ostendorf, Warburg's glassblower, recalled. Ostendorf considered moving to the nearby Max Planck Institute for Molecular Genetics, but Heiss refused to let him go. For the next year, Ostendorf spent most of his time at the institute playing darts.

On one occasion, two architects arrived to discuss the future of the building. When one of them suggested it might be best to tear it down, Heiss grew enraged and chased them out. Heiss was known among the employees of the institute for his short temper, but Ostendorf had never seen him so animated.[2]

Eighteen months after Warburg's death, the Max Planck Society chose Oxford physiologist Henry Harris as Warburg's replacement and invited him to Berlin to make the offer in person. Heiss and Norman, the Great Dane, were waiting to greet Harris at the institute when he arrived. Little had changed since Warburg's death. Those technicians who were still working at all were working on experiments Warburg had outlined. "Berlin was then a city where it was easy to talk to ghosts: they called out to you as you walked the streets," Harris wrote. "But I don't think I ever had a more intimate conversation with a ghost than when I visited Warburg's Institute."[3]

Harris turned down the job offer. The Institute for Cell Physiology was finally closed in March 1972. It has been renamed the Otto Warburg House, and it is now home to the Archives of the Max Planck Society. In that capacity, it serves a function Warburg would have applauded: providing a new generation of scholars with evidence of the crimes committed by German scientists, including some of Warburg's antagonists, under the Nazis.

Although the archives hold many of Warburg's letters and papers, his scientific notebooks went missing. At one time, it was thought that Heiss had burned them, but the materials can now be found in the Berlin-Brandenburg Academy of Sciences and Humanities, which is located in the former East Berlin.

What happened to the notebooks in the interim was revealed when documents from the *Stasi*, the East German secret police, surfaced in 2008. Heiss, it turned out, had been smuggling Warburg's belongings across the Berlin Wall. It was an outrageous thumbing of the nose at the entire scientific establishment of the West, and, as such, a fitting tribute to Warburg's memory.

Jacob Heiss died in October 1984. He was buried where he belonged, next to Otto Warburg.

Acknowledgments

THIS BOOK BEGAN WITH an article I published in the *New York Times Magazine* in 2016. I want to thank Claire Gutierrez, my editor at the *Times Magazine*, for seeing the possibility in the story. Bob Weil at Liveright read that story and saw the possibility for a book. I will always be grateful for his vision and belief in the project. I will also always be grateful for Daniel Gerstle's extraordinary—and extraordinarily thorough—editing. It frightens me to think of how this book might have turned out without his guidance. Everything I publish is a credit to my wonderful agent, Dan Lazar.

The story I've told in this book is built upon the remarkable scholarship of those who came before me. Petra Gentz-Werner, a German scientist and science historian, has written a number of excellent books on Otto Warburg in German. In the course of writing my own book, I reached out to her for assistance again and again. The German historian and Otto Warburg expert Kärin Nickelsen was also extraordinarily kind and patient with my pestering questions. The journalist Gary Taubes wrote about many of the scientific topics in this book years before I did. His work and friendship have been hugely important to me. The wonderful and incredibly generous chemist Willem H. Koppenol was there for me whenever I got stuck and needed help. Although I do not know him personally, Stanford historian Robert Proctor's research on Nazi Germany and cancer politics had a major influence on this book.

I could not have written this book without the help of two amazing research and translation assistants. Japhet Johnstone got me started. Dillon Bergin got me to the finish line. Their friendship turned out to be one of the best parts of this entire process. I also want to thank Haley Bracken at Liveright for managing the production process so well and Janet Greenblatt for her great copy editing. Sara Manning Peskin, Adina Singer, Sharon Christner, Paula Nedved, and Elaine Lissner were kind enough to read drafts of my manuscript. Their feedback was invaluable. My archival research was greatly helped by Thomas Notthoff and his colleagues at the Archives of the Max Planck Society as well as by Kanisha Greaves at the Rockefeller Archive Center. Frederic Burk was kind enough to let me into his home to search through old papers and photographs. Vera Enke and the entire team at the Berlin-Brandenburg Academy of Sciences and Humanities were also very generous with their time.

It was my good fortune that so many of the world-class scientists I reached out to with questions and interview requests also happen to be world-class people. Chi Van Dang, Matthew Vander Heiden, Navdeep Chandel, Carol Prives, and Jeffrey Rathmell were exceptionally helpful. The legendary photosynthesis researcher Govindjee was both a great resource and an inspiration.

During the five years I worked on this book, I found myself in need of support from hundreds of other people, from the assistants who copied documents for me at the various archives I searched, to the scientists and scholars who sat for interviews. In most cases, I was a stranger to these individuals. Their willingness to go out of their way on my behalf meant an enormous amount to me. I am sorry that I do not have the space here to thank everyone (please see dedication), but I would like to mention the following individuals: Peter Attia, Stephanie Auteri, Monika Baark, Paul Barrett, Nir Barzilai, Franzi Becker, Elizabeth Beugg, Brooke Bevis, Kivanç Birsoy, Ewald Blocher, David Botsein, Maik Bozza, Dale Brauner, Karin Buch, Cordelia

Calvert, Lewis Cantley, Colin Champ, Travis Christofferson, Jeffrey Chuang, Sharmila Cohen, Dominic D'Agostino, Paul Davies, Ute Deichmann, Victoria Doherty-Munro, Alice Dragoon, Tim Ferriss, Eugene Fine, Barbara Fried, Jason Fung, Bruce Gladden, Sheila Glaser, Nancy Grossman, Susanne Heim, Deborah Hertz, Bettina Hitzer, Dieter Hoffmann, Thomas Hutzelman, James Johnson, Richard Johnson, Mark Johnston, Tanja Johnston, Rudolf Kaaks, Bernard Kaplan, Martin Klingenberg, Young Ko, Robert E. Kohler, Heather Kristofk, Nick Lane, Anja Laukötter, Wolfgang Lefèvre, Derek LeRoith, Valter Longo, David Ludwig, Robert Lustig, Kristie Macrakis, Tak Mak, Beate Meyer, Lia Miller, Judy Moscovitz, Peter Ostendorf, Anna Pamela, Peter Pedersen, Dana Pe'er, Michael Pollak, Sebastian Rasmussen, John Rees, Manfred Rudel, David Sabatini, Katrin Sachs, Susan Sanfrey, Rachael Schechter, Richard Schneider, Thomas Seyfried, Gerald Shulman, Rebecca Siegel, Helmut Sies, Ulrich Siggel, M. Celeste Simon, Vuk Stambolic, Beth Steidle, Kevin Struhl, Rebecca Stuhr, Peter Tarr, Craig Thompson, Susanne Uebele, Annette Vogt, Douglas Wallace, Robert Weinberg, Kathryn Wellen, Vivian Yuxin Wen, Maxine Winer, and Gary Yellen.

Three extraordinary scientists who made the time to speak with me, Richard Veech, George Klein, and Harry Rubin, died during the period I worked on this project. May their memories be a blessing.

I could not have written this book without the love and support of my wife, Jennifer, and my entire family.

Notes

INTRODUCTION

1. For a full account of Warburg's dealings with the customs office, see K. Nickelsen, "On Otto Warburg, Nazi Bureaucracy and the Difficulties of Moral Judgment," *Photosynthetica* 56, no. 1 (March 1, 2018).

2. Irving M. Klotz, "Wit and Wisdom of Albert Szent-Györgyi: A Recollection," in *Culture of Chemistry*, ed. Balazs Hargittai and István Hargittai (Boston: Springer, 2015), 123–26; Cordula Koepcke, *Lotte Warburg: "Unglaublich! Dass Ich Gelebt Habe!": Eine Biographie* (München: Iudicium, 2000), 157.

3. Peter Ostendorf, interview with author, April 26, 2017.

4. E. Schütte, "Erinnerungen an Otto Warburg," *Naturwiss*, no. 36 (1983): 444–47.

5. Birgit Vennesland, "Recollections and Small Confessions," *Annual Review of Plant Physiology* 32, no. 1 (June 1981): 1–21.

6. The biochemist and science historian Petra Gentz-Werner is the German authority on the life of Otto Warburg. For more on Warburg's meeting at the New Reich Chancellery on June 21, 1941, see Petra Werner, *Ein Genie Irrt Seltener—Otto Heinrich Warburg: Ein Lebensbild in Dokumenten* (Berlin: Akademie Verlag, 1991).

7. Heinrich Himmler et al., *Der Dienstkalender Heinrich Himmlers 1941/42*, Hamburger Beiträge zur Sozial- und Zeitgeschichte, Bd. 3 (Hamburg: Christians, 1999), 178.

8. O. Warburg and D. Burk, *The Prime Cause and Prevention of Cancer: With Two Prefaces on Prevention, Revised Lecture at the Meeting of the Nobel-Laureates on June 30, 1966 at Lindau, Lake Constance, Germany* (Würzburg: Triltsch, 1969).

9. Arthur Kornberg to Alan Mehler, February 21, 1949, *Profiles in Science*. Stanford Digital Repository.

10. Thomas Seyfried, interview with author, April 30, 2015; Siddhartha Mukherjee, *The Emperor of All Maladies: A Biography of Cancer* (New York: Scribner, 2010). Later editions of *The Emperor of All Maladies* include an interview with Siddhartha Mukherjee in which he does mention Warburg.

11. Craig Thompson, interview with author, May 14, 2015.

12. Seyfried, interview with author.

13. Arthur Newsholme, "The Statistics of Cancer," *The Practitioner* 62, n.s., 9 (April 1899): 371.

14. F. L. Hoffman, *The Mortality from Cancer throughout the World* (Newark, NJ: Prudential Press, 1915).

15. Béatrice Lauby-Secretan et al., "Body Fatness and Cancer—Viewpoint of the IARC Working Group," *New England Journal of Medicine* 375, no. 8 (August 25, 2016): 794–98.

CHAPTER ONE:
"A CHEMICAL LABORATORY OF THE MOST AMAZING KIND"

1. Fritz Baltzer, *Theodor Boveri: Life and Work of a Great Biologist, 1862–1915* (Berkeley: University of California Press, 1967), 31.

2. Baltzer, *Theodor Boveri*, 34.

3. Baltzer, *Theodor Boveri*, 40.

4. Theodor Boveri, *Concerning the Origin of Malignant Tumours*, trans. and annot. Henry Harris (Woodbury, NY: Cold Spring Harbor Laboratory Press, 2008).

5. Hans Krebs and Roswitha Schmid, *Otto Warburg: Cell Physiologist, Biochemist and Eccentric* (Oxford: Clarendon Press, 1981), 4.

6. Krebs, *Otto Warburg*, 53; Ostendorf, interview with author.

7. Baltzer, *Theodor Boveri*, 20; Otto Warburg to Lotte Warburg, 11 June 1907, in Petra Werner, *Ein Genie*, 28–29.

8. Krebs, *Otto Warburg*, 72; Werner, *Ein Genie*, 203.

9. Jost Lemmerich, *Science and Conscience: The Life of James Franck*, trans. Ann M. Hentschel (Stanford, CA: Stanford University Press, 2011), 17.

10. Interview of James Franck and Hertha Sponer Franck by Thomas S. Kuhn and Maria Goeppert Mayer on July 9, 1962, Niels Bohr Library & Archives, American Institute of Physics, College Park, MD.

11. David Nachmansohn, *German-Jewish Pioneers in Science 1900–1933: Highlights in Atomic Physics, Chemistry, and Biochemistry* (Berlin: Springer, 1979): 257; Otto Warburg to Dean Burk, 12 August 1953, private collection of Frederic Burk; Frederic L. Holmes, *Hans Krebs*, Vol. 1 (New York: Oxford University Press, 1991), 200.

12. Albert Einstein to Lotte Warburg, 22 May 1935, in Werner, *Ein Genie*, 18–19; Lotte Warburg, *Eine vollkommene Närrin durch meine ewigen Gefühle: Aus den Tagebüchern der Lotte Warburg 1925 bis 1947*, ed. Wulf Rüskamp (Bayreuth: Druckhaus Bayreuth, 1989).

13. Petra Werner, *Otto Warburg: Von der Zellphysiologie zur Krebsforschung: Biografie* (Berlin: Verlag Neues Leben, 1988), 39; Krebs, *Otto Warburg*, 2.

14. Krebs, *Otto Warburg*, 3; Emil Warburg to Otto Warburg 9 December 1912, in Werner, *Ein Genie*, 77–78. The original letter is located in the archive of the Berlin-Brandenburg Academy of Sciences and Humanities, NL Warburg, 999.

15. David Cahan, "The Institutional Revolution in German Physics, 1865–1914," *Historical Studies in the Physical Sciences* 15, no. 2 (1985): 1–65, 41.

16. Ekkehard Höxtermann and Otto Warburg, *Biographien Hervorragender Naturwissenschaftler, Techniker und Mediziner*, Bd. 91 (Leipzig: Teubner, 1989), 19.

17. Jeffrey Allan Johnson, *The Kaiser's Chemists: Science and Modernization in Imperial Germany* (Chapel Hill: University of North Carolina Press, 1990); Jeffrey Allan Johnson, "From Bio-Organic Chemistry to Molecular and Synthetic Biology: Fulfilling Emil Fischer's Dream," in *Proceedings of the International Workshop on the History of Chemistry* (2015), 2–4.

18. Johnson, "From Bio-Organic Chemistry."

19. Philip J. Pauly, *Controlling Life: Jacques Loeb and the Engineering Ideal in Biology* (New York: Oxford University Press, 1987).

20. Pauly, *Controlling Life.*

21. "Science Nears the Secret of Life," *Chicago Sunday Tribune*, November 19, 1899; "Creation of Life," *Boston Herald*, November 26, 1899.

22. Jacques Loeb, "Mechanistic Science and Metaphysical Romance," *Yale Review* 4 (1915): 768–69.

23. Carl Snyder, "Theory of Life; Dr. Jacques Loeb's Mechanistic Conception of Energy," *New York Times*, October 6, 1912; Pauly, *Controlling Life*, 102.

24. "Dr. Loeb's Incredible 'Discovery.'" *New York Times*, March 2, 1905; Albert Bigelow Paine, *Mark Twain: A Biography, Vol. 3: The Personal and Literary Life of Samuel Langhorne Clemens* (New York: Gabriel Wells, 1923), 1161–62.

25. For more on Warburg's relationship to Loeb, see Petra Werner, *Otto Warburgs Beitrag zur Atmungstheorie: Das Problem der Sauerstoffaktivierung* (Marburg/Lahn: Basilisken-Presse, 1996).

26. Pauly, *Controlling Life*, 45.

27. Krebs, *Otto Warburg*, 71. Warburg said this in a statement recorded by the Institut für den Wissenschaftlichen Film in 1966. https://doi.org/10.3203/IWF/G-108.

CHAPTER TWO: "THE GREAT UNSOLVED PROBLEM"

1. Lotte Warburg, undated diary entry, private collection of the Meyer-Viol family. For more of Lotte Warburg's journal entry, see Werner, *Ein Genie*, 79–80.

2. Krebs, *Otto Warburg*, 75; Nachmansohn, *German-Jewish Pioneers*, 267; Theodor Bücher, "Otto Warburg: A Personal Recollection," in *Biological Oxidations*, Colloquium-Mosbach (Berlin: Springer, 1983), 1–29.

3. Eric Warburg, *Times and Tides: A Log-Book* (Hamburg: privately, 1983), 130; Wieland Gevers, *Personality, Creativity and Achievement in Science* (University of Cape Town, 1978); Krebs, *Otto Warburg*, 56.

4. Krebs, *Otto Warburg*, 59.

5. Vennesland, "Recollections."

6. A more detailed account of Warburg's romantic life during this period can be found in Werner, *Ein Genie*, 18–19, 78–85.

7. Otto Meyerhof to Otto Warburg, 27 June 1912, NL Warburg, 656, archive of the Berlin-Brandenburg Academy of Sciences and Humanities.

8. Vennesland, "Recollections," 13.

9. Krebs, *Otto Warburg*, 52; L. Warburg, *Eine vollkommene Närrin*, 287.

10. Vennesland, "Recollections," 12.

11. Mark Twain, *American Claimant and Other Stories and Sketches* (Hartford, CT: American Publ. Co., 1901), 502.

12. Werner, *Ein Genie*, 147.

13. Hoffman, *Mortality from Cancer*; William Roger Williams, *The Natural History of Cancer, with Special Reference to Its Causation and Prevention* (New York: William Wood and Co., 1908); Reinhard Spree, *Der Rückzug des Todes: Der Epidemiologische Übergang in Deutschland während des 19. und 20. Jahrhunderts*, Konstanzer Universitätsreden 186 (Konstanz: Universitätsverlag Konstanz, 1992).

14. "Academy of Sciences, Paris." *Provincial Medical and Surgical Journal (1844–1852)* 8, no. 16 (1844): 237–39; P. Nicolopoulou-Stamati, ed., *Cancer as an Environmental Disease*, Environmental Science and Technology Library, v. 20 (Dordrecht: Kluwer Academic Publishers, 2004); Robert Lee, "Early Death and Long Life in History: Establishing the Scale of Premature Death in Europe and Its Cultural, Economic and Social Significance," *Historical Social Research / Historische Sozialforschung* 34, no. 4 (2009): 23–60. Life expectancy data from nineteenth-century England and Sweden are nearly identical to the German data.

15. M. Bolte, D. Kappe, and J. Schmid, *Bevölkerung: Statistik, Theorie, Geschichte und Politik des Bevölkerungsprozesses* (VS Verlag für Sozialwissenschaften, 1980); Hoffman, *Mortality from Cancer*, 14. For additional discussion of age-adjusted cancer rates, see Clifton Leaf, *The Truth in Small Doses: Why We're Losing the War on Cancer—and How to Win It* (New York: Simon & Schuster, 2013).

16. Hoffman, *Mortality from Cancer*, 28. For more on the cancer statistics debate, see Robert Proctor, *Cancer Wars: How Politics Shapes What We Know and Don't Know about Cancer* (New York: Basic Books, 1995).

17. Alan I. Marcus, *Malignant Growth: Creating the Modern Cancer Research Establishment, 1875–1915* (Tuscaloosa: University of Alabama Press, 2018), 77; Robert Proctor, *The Nazi War on Cancer* (Princeton, NJ: Princeton University Press, 1999), 248.

18. See Proctor, *Nazi War*, for more on anticancer campaigns and research in Nazi Germany; Isabel dos Santos Silva, *Cancer Epidemiology: Principles and Methods* (Lyon: IARC, 1999), 386.

19. Otto Warburg, "Prefatory Chapter," *Annual Review of Biochemistry* 33, no. 1 (June 1964): 1–15; Krebs, *Otto Warburg*, 72.

20. A discussion of Hoffman's life can be found in F. J. Sypher, *Frederick L. Hoffman: His Life and Works* (Philadelphia: Xlibris, 2002); Leaf, *Truth in Small Doses*, 31.

21. Leaf, *Truth in Small Doses*; Hoffman, *Mortality from Cancer*, 218.

22. For a discussion of Hoffman's racism, see Megan J Wolff, "The Myth of the Actuary: Life Insurance and Frederick L. Hoffman's Race Traits and Tendencies of the American Negro," *Public Health Reports* 121, no. 1 (January–February 2006): 84–

91; "Dr. Hoffman Tells of Negro Health," *New York Times*, February 6, 1926; Hoffman, *Mortality from Cancer*, 147.

23. W. H. Walshe and J. M. Warren, *The Anatomy, Physiology, Pathology, and Treatment of Cancer* (Boston: William D. Ticknor & Company, 1844), 347–52.

24. Vilhjalmur Stefansson, *Cancer: Disease of Civilization?* (New York: Hill and Wang, 1960); Gary Taubes, *Good Calories, Bad Calories: Challenging the Conventional Wisdom on Diet, Weight Control, and Disease* (New York: Alfred A. Knopf, 2007).

25. F. P. Fouche, "Freedom of Negro Races from Cancer," *British Medical Journal* 1, no. 3261 (June 30, 1923): 1116.

26. A. Hrdlička, *Physiological and Medical Observations among the Indians of Southwestern United States and Northern Mexico* (US Government Printing Office, 1908), 190; Hoffman, *Mortality from Cancer*, 151. For more on Hrdlička and his contribution to the American eugenics movement, see Samuel J. Redman, *Bone Rooms: From Scientific Racism to Human Prehistory in Museums* (Cambridge, MA: Harvard University Press, 2016).

27. Stefansson, *Cancer*, 17–23; S. K. Hutton, *Health Conditions and Disease Incidence among the Eskimos of Labrador* (Poole, UK: J. Looker, 1925); G. Malcolm Brown, L. B. Cronk, and T. J. Boag, "The Occurrence of Cancer in an Eskimo," *Cancer* 5, no. 1 (January 1, 1952): 142–43. For further reading, see Taubes, *Good Calories, Bad Calories*.

28. Albert Schweitzer, "Preface," in Alexander Berglas, *Cancer: Nature, Cause, and Cure* (Paris: Pasteur Institute, 1957).

29. "Traveler Brings New Cancer Theory," *Philadelphia Inquirer*, August 5, 1906.

30. A. R. Walker, "The Assessment and Remedying of Inadequate Diets in India, as Appreciated by Sir Robert McCarrison," *Nutrition* 18, 1 (2002): 106–9; Fouche, "Freedom of Negro Races from Cancer."

31. H. Pontzer, B. M. Wood, and D. A. Raichlen, "Hunter-Gatherers as Models in Public Health: Hunter-Gatherer Health and Lifestyle," *Obesity Reviews* 19 (2018): 24–35; El Molto and Peter Sheldrick, "Paleo-Oncology in the Dakhleh Oasis, Egypt: Case Studies and a Paleoepidemiological Perspective," *International Journal of Paleopathology* 21 (2018): 96–110; Tony Waldron, "What Was the Prevalence of Malignant Disease in the Past?," *International Journal of Osteoarchaeology* 6, no. 5 (December 1, 1996): 463–70.

32. Mukherjee, *Emperor of All Maladies*, 37–45.

33. Albert Schweitzer, "Preface"; "Noted Surgeon Travels in Darkest Africa's Wilds," *Los Angeles Times*, August 5, 1906.

34. Charles Powell White, *Lectures on the Pathology of Cancer* (Manchester: University Press, 1908), 381.

35. For Bloch's recollections of treating Klara Hitler, see Eduard Bloch, "My Patient Adolf Hitler," *Collier's Magazine*, March 15 and 22, 1941. Additional accounts of Hitler's childhood can be found in Paul Ham, *Young Hitler: The Making of the Führer* (New York: Pegasus Books, 2018). Klara Hitler's medical condition is dis-

cussed in James Stuart Olson, *Bathsheba's Breast: Women, Cancer & History* (Baltimore: The Johns Hopkins University Press, 2002).

36. August Kubizek, *The Young Hitler I Knew: The Memoirs of Hitler's Childhood Friend*, trans. Geoffrey Brooks (London: Greenhill Books, 2011), 33.

37. Bloch, "My Patient Adolf Hitler"; Eduard Bloch, "Erinnerungen an den Führer und Dessen Verewigte Mutter" (Nov. 1938) in Volker Ullrich, *Hitler: Ascent, 1889–1939*, trans. Jefferson Chase (New York: Alfred A. Knopf, 2016).

38. Michael Stolberg, *A History of Palliative Care, 1500–1970* (Cham, Switzerland: Springer, 2017), 12.

39. Kubizek, *Young Hitler*, 134.

CHAPTER THREE: MAGIC BULLETS

1. H. Ward, *New Worlds in Medicine: An Anthology* (New York: R. M. McBride, 1946), 216. For more on Ehrlich's early life, see Ernst Bäumler, *Paul Ehrlich: Scientist for Life* (New York: Holmes & Meier, 1984), 3–12.

2. Paul Ehrlich et al., *The Collected Papers of Paul Ehrlich, Vol. III: Chemotherapy* (London: Pergamon Press, 1960), 54; Frederick H. Kasten, "Paul Ehrlich: Pathfinder in Cell Biology," *Biotechnic & Histochemistry* 71, no. 1 (1996): 5; Bäumler, *Paul Ehrlich*, 81.

3. Bäumler, *Paul Ehrlich*, 81; Martha Marquardt, *Paul Ehrlich* (Berlin: Springer, 1951), 14.

4. Herman T. Blumenthal, "Leo Loeb, Experimental Pathologist and Humanitarian," *Science* 131, no. 3404 (1960): 907; V. Suntzeff, "Obituary," *Cancer Research* 20, no. 6 (July 1, 1960): 972; Leo Loeb, "Autobiographical Notes," *Perspectives in Biology and Medicine* 2, no. 1 (1958): 1–23.

5. Jan A. Witkowski, "Experimental Pathology and the Origins of Tissue Culture: Leo Loeb's Contribution," *Medical History* 27, no. 3 (July 1983): 269–88; L. Loeb, "On Transplantation of Tumors," *A Cancer Journal for Clinicians* 28, no. 6 (1978).

6. Loeb, "Autobiographical Notes," 1–23; Paul Ehrlich, *Experimental Researches on Specific Therapeutics* (London: Lewis, 1908), 43–73.

7. Ehrlich, *Experimental Researches*, 43–73.

8. Marquardt, *Paul Ehrlich*, 159; Otto Warburg, "Paul Ehrlich 1854–1915," in H. Heimpel, T. Heuss, and B. Reifenberg, *Die Grossen Deutschen: Deutsche Biographie*, Vol. 4 (Berlin: Propyläen-Verlag, 1957), 186–92.

9. David Kritchevsky, "Caloric Restriction and Cancer," *Journal of Nutritional Science and Vitaminology* 47, no. 1 (2001): 13–19; Peyton Rous, "The Influence of Diet on Transplanted and Spontaneous Mouse Tumors," *Journal of Experimental Medicine* 20, no. 5 (November 1, 1914): 433–51; E. V. Van Alstyne and S. P. Beebe, "Diet Studies in Transplantable Tumors: I. The Effect of Non-Carbohydrate Diet upon the Growth of Transplantable Sarcoma in Rats," *The Journal of Medical Research* 29, no. 2 (December 1913): 217–32.

CHAPTER FOUR: GLUCOSE, CANCER, AND THE CROWN PRINCE

1. Otto Warburg to Lotte Warburg, June 11, 1907, in Werner, *Ein Genie*, 28–29.

2. Hans A. Krebs and Fritz Lipmann, "Dahlem in the Late Nineteen Twenties," in *Energy Transformation in Biological Systems*, ed. Dietmar Richter (Berlin, Walter de Gruyter, 1974), 7–27.

3. Arthur Greenberg, *The Art of Chemistry: Myths, Medicines, and Materials* (New York: Wiley, 2003), 188.

4. Otto Warburg to Jacques Loeb, 30 October 1910, in Robert E. Kohler, "The Background to Otto Warburg's Conception of the 'Atmungsferment,'" *Journal of the History of Biology* 6, no. 2 (1973): 171–92.

5. Werner, *Otto Warburg*, 100.

6. Hans Krebs and Anne Martin, *Reminiscences and Reflections* (Oxford: Clarendon Press, 1981), 29.

7. Baltzer, *Theodor Boveri*, 20.

8. David Cahan, *An Institute for an Empire: The Physikalisch-Technische Reichsanstalt, 1871–1918* (Cambridge: Cambridge University Press, 1989), 96.

9. Hans-Georg Bartel and R. P. Huebener, *Walther Nernst: Pioneer of Physics and of Chemistry* (Singapore: World Scientific, 2007).

10. Kohler, "The Background"; Garland E. Allen, *Life Science in the Twentieth Century*, (Cambridge: Cambridge University Press, 1979).

11. Cahan, *An Institute*, 204.

12. Kärin Nickelsen, "The Construction of a Scientific Model: Otto Warburg and the Building Block Strategy," *Studies in History and Philosophy of Biological and Biomedical Sciences* 40, no. 2 (June 2009): 73–86. For more on Warburg's photosynthesis research, see Kärin Nickelsen, *Of Light and Darkness: Modelling Photosynthesis 1840–1960*, Habilitation thesis, Phil.-nat. Fakultät der Universität Bern, 2009.

13. Otto Warburg to Dean Burk, 20 May 1964, private collection of Frederic Burk.

14. Nickelsen, *Of Light and Darkness*, 103.

15. Mukherjee, *Emperor of All Maladies*, 87; Richard Willstätter and Arthur Stoll, *From My Life: The Memoirs of Richard Willstätter*, trans. Lilli S. Hornig (New York: W. A. Benjamin, 1965).

16. John C. G. Röhl, *Young Wilhelm: The Kaiser's Early Life, 1859–1888* (Cambridge: Cambridge University Press, 1998); Untitled, *Dorking and Leatherhead Advertiser*, December 10, 1887.

17. Ernst Freund, *Zur Diagnose des Carcinoms: Vorläufige Mittheilung* (Vienna: L. Bergmann, 1885).

18. "The Crown Prince's Malady," *New York Times*, December 23, 1887; "Sugar and Cancer," *New York Times*, December 24, 1887.

19. Miranda Carter, *George, Nicholas and Wilhelm: Three Royal Cousins and the Road to World War I* (New York: Alfred A. Knopf, 2010); John C. G. Röhl, *Kaiser Wilhelm II, 1859–1941: A Concise Life* (Cambridge: Cambridge University Press, 2014).

20. Röhl, *Kaiser Wilhelm II*.

CHAPTER FIVE: "SLAVES OF THE LIGHT"

1. Otto Warburg to Jacques Loeb, 18 June 1914, in Werner, *Otto Warburgs Beitrag zur Atmungstheorie*, 158–59; Jacques Loeb to Otto Warburg, 1 August 1914, NL Warburg, 606/1, archive of the Berlin-Brandenburg Academy of Sciences and Humanities.

2. Loeb, *The Organism as a Whole, from a Physicochemical Viewpoint* (New York: G. P. Putnam's Sons); Charles Rasmussen and Rick Tilman, *Jacques Loeb: His Science and Social Activism and Their Philosophical Foundations*, Memoirs of the American Philosophical Society, vol. 229 (Philadelphia: American Philosophical Society, 1998); Loeb, "Mechanistic Science."

3. "Kaiser Makes War Speech to a Great Crowd in Berlin," *New York Times*, August 1, 1914; "Kaiser Forgives Enemies, Prays for Victory," *New York Times*, August 2, 1914.

4. Tim Grady, *A Deadly Legacy: German Jews and the Great War* (New Haven, CT: Yale University Press, 2017).

5. Amos Elon, *The Pity of It All: A History of the Jews in Germany, 1743–1933* (New York: Metropolitan Books, 2002).

6. Robert Proctor, *Racial Hygiene: Medicine under the Nazis* (Cambridge, MA: Harvard University Press, 1988).

7. Heiner Fangerau, "From Mephistopheles to Isaiah: Jacques Loeb, Technical Biology and War," *Social Studies of Science* 39, no. 2 (2009): 229–56. For additional reading, see Mike Hawkins, *Social Darwinism in European and American Thought, 1860–1945: Nature as Model and Nature as Threat* (Cambridge: Cambridge University Press, 1997).

8. Werner, *Otto Warburg*, 143. Theodor Heuss and Christiane Groeben, *Anton Dohrn: A Life for Science* (Berlin: Springer, 1991), 108.

9. Krebs, *Otto Warburg*, 10; Werner, *Otto Warburg*, 142–43. Warburg's reflections on his military service were recorded by the Institut für den Wissenschaftlichen Film in 1966. https://doi.org/10.3203/IWF/G-108.

10. Willem H. Koppenol, Patricia L. Bounds, and Chi V. Dang, "Otto Warburg's Contributions to Current Concepts of Cancer Metabolism," *Nature Reviews Cancer* 11, no. 5 (May 2011): 325–37; Ernst Bumm, *Bericht über das Amtsjahr 1916/1917* (Berlin: Norddt. Buchdr. und Verl.-Anst., 1917), 16.

11. Baltzer, *Theodor Boveri*, 22.

12. Brian E. Crim, *Antisemitism in the German Military Community and the Jewish Response, 1914–1938* (Lanham, MD: Lexington Books, 2014); Grady, *A Deadly Legacy*.

13. Elisabeth Warburg to Albert Einstein, 21 March 1918, in Werner, *Ein Genie*, 120.

14. Werner, *Ein Genie*, 121–23.

15. E. Warburg to Einstein, 21 March 1918.

16. Albert Einstein, Helen Dukas, Banesh Hoffmann, and Ze'ev Rosenkranz, *Albert*

Einstein, the Human Side: Glimpses from His Archives (Princeton, NJ: Princeton University Press, 2013), 88.

17. Albert Einstein to Otto Warburg, 23 March 1918, in *Ein Genie*, 121; Krebs, *Otto Warburg*, 8–9.

18. E. Schütte, "Erinnerungen an Otto Warburg," *Naturwiss. Rdsch*, 36 (1983), 444–47.

19. Krebs, *Otto Warburg*, 10; Institut für den Wissenschaftlichen Film, https://doi .org/10.3203/IWF/G-108.

20. Kubizek, *Young Hitler*, 157.

21. Ullrich, *Hitler: Ascent*, 49.

22. Adolf Hitler, *Mein Kampf*, trans. Ralph Manheim (Boston: Houghton Mifflin, 1943).

23. Adolf Hitler to Ernst Hepp, 5 February 1915, in Ullrich, *Hitler: Ascent*, 55–56.

24. Thomas Weber, *Hitler's First War: Adolf Hitler, the Men of the List Regiment, and the First World War* (Oxford: Oxford University Press, 2010).

25. Adolf Hitler et al., *Hitler's Table Talk, 1941–1944: His Private Conversations*, trans. H. R. Trevor-Roper (New York: Enigma Books, 2000), 36; Ullrich, *Hitler: Ascent*, 50.

CHAPTER SIX: THE WARBURG EFFECT

1. Adam Hochschild, "A Hundred Years after the Armistice," *New Yorker*, November 5, 2018; Röhl, *Kaiser Wilhelm II*, 177.

2. David Welch, *Germany and Propaganda in World War I: Pacifism, Mobilization and Total War* (London: I. B. Tauris, 2014), 118; Hochschild, "A Hundred Years."

3. Krebs, *Otto Warburg*, 80.

4. Peter Gruss et al., eds., *Denkorte: Max-Planck-Gesellschaft und Kaiser-Wilhelm-Gesellschaft: Brüche und Kontinuitäten 1911–2011* (Dresden: Sandstein, 2010); Otto Warburg to Paul Warburg, 5 August 1920, in Simon Flexner Papers, 1891–1946, MSS.B. F365, American Philosophical Society Archives.

5. Koppenol et al., "Otto Warburg's Contributions"; Krebs, *Otto Warburg*, 57–58.

6. Otto Warburg to Jacques Loeb, 6 September 1922, in Werner, *Otto Warburgs Beitrag zur Atmungstheorie*, 165–66. Fritz Lipmann et al., eds., *The Roots of Modern Biochemistry: Fritz Lipmann's Squiggle and Its Consequences* (Berlin: Walter de Gruyter, 1988); Krebs, *Reminiscences*, 32.

7. George Klein, interview with author, June 10, 2015; Manfred von Ardenne in Bücher, "Otto Warburg."

8. Krebs, *Otto Warburg*, 57.

9. L. Warburg, *Eine vollkommene Närrin*.

10. Holmes, *Hans Krebs*, 145; Krebs, *Reminiscences*, 33–34.

11. Krebs, *Reminiscences*, 33–34.

12. Krebs, *Otto Warburg*, 53; Kärin Nickelsen and Govindjee, *The Maximum Quantum Yield Controversy: Otto Warburg and the "Midwest-Gang"* (Bern: Bern Studies in the History and Philosophy of Science, 2011).

13. Ostendorf, interview with author.

14. Ostendorf, interview with author.

15. E. Warburg, *Times and Tides*, 128.

16. Krebs, *Otto Warburg*, 82.

17. L. Warburg, *Eine vollkommene Närrin*, 141.

18. Werner, *Ein Genie*, 291–92, 79–80.

19. Jacques Loeb to Otto Warburg, 25 July 1922, NL Warburg, 606, archive of the Berlin-Brandenburg Academy of Sciences and Humanities.

20. Otto Warburg to Jacques Loeb, 13 June 1923, in Werner, *Otto Warburgs Beitrag zur Atmungstheorie*, 170.

21. Ron Chernow, *The Warburgs: The Twentieth-Century Odyssey of a Remarkable Jewish Family* (New York: Random House, 1993), 217.

22. Emil Warburg to Albert Einstein, 16 November 1923, in Albert Einstein et al., *The Collected Papers of Albert Einstein*, Vol. 14 (English) (Princeton, NJ: Princeton University Press, 2015), 149–50.

23. Krebs, *Otto Warburg*, 13.

24. Samuel Noah Kramer, *The Sumerians: Their History, Culture, and Character* (Chicago: University of Chicago Press, 1971), 111.

25. René J. Dubos, *Louis Pasteur, Free Lance of Science* (New York: Scribner, 1976), 135; René Vallery-Radot, *The Life of Pasteur*, trans. R. L. Devonshire (New York: Doubleday, 1914), 220; Patrice Debré, *Louis Pasteur* (Baltimore: Johns Hopkins University Press, 2000).

26. Warburg, *Prime Cause*.

27. Warburg, *Prime Cause*.

28. Birgit Vennesland, "New Methods of Cell Physiology by Otto Heinrich Warburg," *Perspectives in Biology and Medicine* 6, no. 3 (1963): 385–88.

29. Peyton Rous to Otto Warburg, 5 December 1924, Rockefeller University records, Roger C. Elliot, Series 3, Warburg, Otto, Rockefeller Archive Center.

30. Ullrich, *Hitler: Ascent*, 82, 149.

31. Ullrich, *Hitler: Ascent*, 386; Joseph Shaplin, "Hitler, Driving Force in Germany's Fascism," *New York Times*, September 21, 1930; Hitler, *Mein Kampf*.

32. Timothy Snyder, *Black Earth: The Holocaust as History and Warning* (New York: Tim Duggan Books, 2015); Adolf Hitler, *Hitler's Second Book: The Unpublished Sequel to Mein Kampf*, ed. Gerhard L. Weinberg, trans. Krista Smith (New York: Enigma Books, 2006), 7.

33. Otto Warburg to Jacques Loeb, 6 September 1922, in Werner, *Otto Warburgs Beitrag zur Atmungstheorie*, 165–66.

34. Hitler, *Hitler's Second Book*, 17.

35. Hitler, *Hitler's Second Book*, 17.

36. John Lukacs, *The Last European War: September 1939/December 1941* (New Haven, CT: Yale University Press, 2001), 42.

CHAPTER SEVEN: THE EMPEROR OF DAHLEM

1. William L. Shirer, *The Rise and Fall of the Third Reich: A History of Nazi Germany* (New York: Simon & Schuster, 1990).
2. Krebs, *Reminiscences*, 26–27.
3. Albert Einstein to Jacques Loeb, 28 December 1923, in Einstein et al., *The Collected Papers*, 189–90.
4. "Cancer Causation: Importance of Cell Physiology," *Nature* 118, no. 2964 (August 1926): 284–85; L. Warburg, *Eine vollkommene Närrin*, 28.
5. Fritz Lipmann, *Wanderings of a Biochemist* (New York: Wiley, 1971), 6.
6. R. W. Gerard, "The Minute Experiment and the Large Picture: A Lifelong Commitment," *Perspectives in Biology and Medicine* 23, no. 4 (1980): 527–40.
7. Allen, *Life Science in the Twentieth Century*; Holmes, *Hans Krebs*; Krebs, *Otto Warburg*; Nachmansohn, *German-Jewish Pioneers*, 238.
8. Otto Warburg to Alan Gregg, 13 February 1930, Rockefeller Foundation records, RG 6, SG 1, Pre-war correspondence, MS Germany, 1926–1935, Rockefeller Archive Center.
9. Friedrich Glum to L. W. Jones, 22 May 1931, Rockefeller Foundation records, RG 6, SG 1, MS Germany, 1926–1935, Rockefeller Archive Center.
10. E. Henning, "Ein märkisches Herrenhaus im 'deutschen Oxford': Zur Baugeschichte des K-W-I für Zellphysiologie in Berlin, Dahlem und seines Vorbildes in Groß Kreutz," Publication of the Archives of the Max Planck Society (2004), 95–124.
11. L. Warburg, *Eine vollkommene Närrin*, 29–30.
12. E. Henning, "Otto Heinrich Warburg—Der 'Kaiser von Dahlem,'" in *Berlinische Lebensbilder, Naturwissenschaftler*, Vol. 1 (Berlin: Colloquium Verlag, 1987), 299–316.
13. Werner, *Otto Warburg*, 211; Klein, interview with author.
14. Albert Szent-Györgyi, "Looking Back," *Perspectives in Biology and Medicine* 15, no. 1 (1971): 1–6; Werner, *Otto Warburg*, 210–11.
15. Peter Reichard, "Osvald T. Avery and the Nobel Prize in Medicine," *Journal of Biological Chemistry* 277, no. 16 (2002).
16. Werner, *Ein Genie*, 205–6.
17. L. Warburg, *Eine vollkommene Närrin*, 121, 114.
18. Proctor, *Nazi War*, 21–38; Manuel Frey, "Prävention und Propaganda," *Ärzteblatt Sachsen* (April 2005): 160–62; Anja Laukötter, "Anarchy of Cells," *Studies in Contemporary History*, online edition, 7 (2010); Werner, *Ein Genie*, 157–58.
19. Claib Price, "Cancer Is Traced to Faulty Modern Diet," *New York Times*, May 10, 1925.
20. Erwin Liek, *The Doctor's Mission*, trans. J. Ellis Barker (London: J. Murray, 1930), 126; Erwin Liek, *Krebsverbreitung, Krebsbekämpfung, Krebsverhütung* (Munich: Lehmann, 1932); Proctor, *Nazi War*, 24.
21. Corinna Treitel, *Eating Nature in Modern Germany: Food, Agriculture, and Environment, c.1870 to 2000* (Cambridge: Cambridge University Press, 2017), 136.

22. F. J. Sypher, *Frederick L. Hoffman*; "The Rockefeller Foundation," *Science* 66, no. 1700 (1927): 105–6.

23. James T. Patterson, *The Dread Disease: Cancer and Modern American Culture* (Cambridge, MA: Harvard University Press, 1987), 88.

24. Frederick L. Hoffman, *Cancer and Diet: With Facts and Observations on Related Subjects* (Baltimore: Williams & Wilkins, 1937).

25. H. T. Deelman, "The Mortality from Cancer among People of Different Races," *Cancer Control*, Report of an international symposium held under the auspices of the American Society for Control of Cancer (1927), 247.

26. Erwin Liek, *Der Kampf gegen den Krebs* (Munich: J. F. Lehmanns, 1936).

27. Ullrich, *Hitler: Ascent*.

28. Treitel, *Eating Nature*, 87.

29. Hitler, *Mein Kampf*.

30. Hitler, *Hitler's Second Book*; Robert G. L. Waite, *The Psychopathic God: Adolf Hitler* (New York: Basic Books, 1977); Ullrich, *Hitler: Ascent*.

31. Hitler, *Hitler's Table Talk*, 152.

32. Albert Krebs and William Sheridan Allen, *The Infancy of Nazism: The Memoirs of Ex-Gauleiter Albert Krebs, 1923–1933* (New York: New Viewpoints, 1976), 163–65.

CHAPTER EIGHT: "THE ETERNAL JEW"

1. Sebastian Haffner, *Defying Hitler: A Memoir*, trans. Oliver Pretzel (Lexington, MA: Plunkett Lake Press, 2014), 18.

2. Philip Ball, *Serving the Reich: The Struggle for the Soul of Physics under Hitler* (Chicago: University of Chicago Press, 2014), 62; Brandon R. Brown, *Planck: Driven by Vision, Broken by War* (Oxford: Oxford University Press, 2015); J. L. Heilbron, *The Dilemmas of an Upright Man: Max Planck and the Fortunes of German Science* (Cambridge, MA: Harvard University Press, 2000).

3. Jean Medawar and David Pyke, *Hitler's Gift: The True Story of the Scientists Expelled by the Nazi Regime* (New York: Skyhorse, 2012).

4. Heilbron, *Dilemmas*; L. Warburg, *Eine vollkommene Närrin*, 187, 235.

5. E. Warburg, *Times and Tides*, 127.

6. Harry M. Miller's account of his Germany trip, December 12–16, 1933, Rockefeller Foundation records, RG 12, M-R, 1933 August 4–December 16, Rockefeller Archive Center; Werner, *Ein Genie*, 295; Kristie Macrakis, *Surviving the Swastika: Scientific Research in Nazi Germany* (New York: Oxford University Press, 1993), 53–57.

7. Medawar and Pyke, *Hitler's Gift*; Otto Warburg to Hans Krebs, 26 April 1933, in Werner, *Ein Genie*, 283–84.

8. Macrakis, *Surviving the Swastika*, 59; Ulrike Kohl, *Die Präsidenten der Kaiser-Wilhelm-Gesellschaft im Nationalsozialismus: Max Planck, Carl Bosch und Albert Vögler Zwischen Wissenschaft und Macht* (Stuttgart: Steiner, 2002), 84.

9. L. Warburg, *Eine vollkommene Närrin*, 213.

10. Lauder W. Jones's account of his Berlin trip, May 24–25, 1933, Rockefeller Foundation records, RG 12, F-L, 1933, Rockefeller Archive Center.

11. Nickelsen, "On Otto Warburg"; Lauder W. Jones's 1933 Berlin trip, Rockefeller Archive Center; Harry M. Miller's account of his European trip, January 23 to February 3, Rockefeller Foundation records, RG 12, M-R, January 14 to June 25 1935, Rockefeller Archive Center.

12. David Nachmansohn, *German-Jewish Pioneers*, 264; Nickelsen and Govindjee, *Maximum Quantum Yield Controversy*, 33; L. Warburg, *Eine vollkommene Närrin*, 217–18.

13. Otto Warburg to Walter Kempner's mother, 29 December 1933 and 14 January 1934, Walter Kempner Papers, StGA-Kempner IV, 0303, Stefan George Archive, Stuttgart; Holmes, *Hans Krebs*, 201; "Report: Rice Diet Doctor Admitted to Whippings in Depositions," *Associated Press*, October 19, 1997.

14. Alfred E. Cohen to E. W. Bagster Collins, 6 November 1933, Emergency Committee in Aid of Displaced Foreign Scholars, Warburg, Otto 1933, New York Public Library Archives.

15. Lauder W. Jones's 1933 Berlin trip, Rockefeller Archive Center; E. Warburg, *Times and Tides*, 127.

16. Alan Gregg's account of his Berlin trip, October 25, 1933, Rockefeller Foundation records, RG 12, F-L, 1933, Rockefeller Archive Center; Vennesland, "Recollections," 12.

17. Otto Warburg to Lotte Warburg, 1 July 1936, NL Warburg, 1001, archive of the Berlin-Brandenburg Academy of Sciences and Humanities.

18. Stefan Zweig, *The World of Yesterday*, trans. Benjamin W. Huebsch and Helmut Ripperger (New York: The Viking Press, 1943); George Prochnik, "When It's Too Late to Stop Fascism, According to Stefan Zweig," *New Yorker*, February 6, 2017: newyorker.com/books/page-turner/when-its-too-late-to-stop-fascism-according-to-stefan-zweig.

19. Albert Einstein to Fritz Haber, 19 May 1933, in Daniel Charles, *Master Mind: The Rise and Fall of Fritz Haber, the Nobel Laureate Who Launched the Age of Chemical Warfare* (New York: Ecco, 2005).

20. Zweig, *The World of Yesterday*; L. Warburg, *Eine vollkommene Närrin*, 217–18.

21. Barbara Newborg and Florence Nash, *Walter Kempner and the Rice Diet: Challenging Conventional Wisdom* (Durham: Carolina Academic Press, 2011); Michael Schüring, *Minervas verstossene Kinder: Vertrieben Wissenschaftler und die Vergangenheitspolitik der Max-Planck-Gesellschaft* (Göttingen: Wallstein, 2006), 120.

22. Krebs, *Otto Warburg*; L. Warburg, *Eine vollkommene Närrin*, 300; Warren Weaver's account of his European travels, January 14–27, Rockefeller Foundation records, RG, 12, S-Z, 1938, Rockefeller Archive Center.

23. Chernow, *The Warburgs*, 369.

24. Two accounts of W. E. Tisdale's meetings with Otto Warburg in 1934 can be found in Alan Gregg's diary, Rockefeller Foundation records, RG 12, F-L, 1934, Rockefeller Archive Center.

25. L. Warburg, *Eine vollkommene Närrin*, 252–54.

26. L. Warburg, *Eine vollkommene Närrin*, 265.

27. Jeremy Noakes, "The Development of Nazi Policy towards the German-Jewish 'Mischlinge' 1933–1945," *The Leo Baeck Institute Year Book* 34, no. 1 (January 1, 1989): 291–354; Norman Stone, *Hitler* (London: Bloomsbury Reader, 2013).

28. Nathan Stoltzfus, *Resistance of the Heart: Intermarriage and the Rosenstrasse Protest in Nazi Germany* (New York: W. W. Norton, 1996), 69; Hitler, *Mein Kampf.*

29. Bryan Mark Rigg, *Hitler's Jewish Soldiers: The Untold Story of Nazi Racial Laws and Men of Jewish Descent in the German Military*, Modern War Studies (Lawrence: University Press of Kansas, 2002).

30. Werner, *Ein Genie*, 302–3; L. Warburg, *Eine vollkommene Närrin*, 244.

31. "The Last Stand," *New York Times*, January 13, 1936; Brigitte Lohff and Hinderk Conrads, *From Berlin to New York: Life and Work of the Almost Forgotten German-Jewish Biochemist Carl Neuberg* (Stuttgart: Franz Steiner, 2007).

32. L. Warburg, *Eine vollkommene Närrin*, 265–66.

33. Richard Norton Smith, *The Harvard Century: The Making of a University to a Nation* (Cambridge, MA: Harvard University Press, 1986), 124; Louis M. Lyons, "Famous Englishmen Speak at Harvard Tercentenary," *Boston Globe*, September 1, 1936.

34. L. Warburg, *Eine vollkommene Närrin*, 300.

35. L. Warburg, *Eine vollkommene Närrin*, 287.

36. Chernow, *The Warburgs*, 4.

37. L. Warburg, *Eine vollkommene Närrin*, 294–96.

38. L. Warburg, *Eine vollkommene Närrin*, 294–96.

39. Deborah Sadie Hertz, *How Jews Became Germans: The History of Conversion and Assimilation in Berlin* (New Haven, CT: Yale University Press, 2007).

40. L. Warburg, *Eine vollkommene Närrin*, 294.

41. "Obituary: Prof. Otto Warburg," *The Times*, January 12, 1938.

42. Werner, *Ein Genie*, 294; "Obituary: Prof. Otto Warburg," *The Times*, January 13, 1938.

43. "Otto Warburg: Master Biochemist 1883–1970, A Personal Portrait," James Norman Davidson Papers, GB 248 DC 024, University of Glasgow Archives; Hans Adolf Krebs, "Otto Heinrich Warburg, 1883–1970," *Biographical Memoirs of Fellows of the Royal Society*, Vol. 18 (1972), 628–99; Werner, *Ein Genie*, 294.

44. Chernow, *The Warburgs*, 4.

45. Dieter Hoffmann and Mark Walker, eds., *The German Physical Society in the Third Reich: Physicists between Autonomy and Accommodation* (Cambridge: Cambridge University Press, 2012), 297; Weaver's 1938 European trip, Rockefeller Archive Center.

46. Nachmansohn, *German Jewish Pioneers*, 256. Nachmansohn translated Warburg's words as "without giving any reasons."

47. L. Warburg, *Eine vollkommene Närrin*, 300; Nachmansohn, *German Jewish Pioneers*, 255–56.

48. Davidson, "Otto Warburg," Glasgow Archives.

49. Mark Walker, *Otto Hahn: Verantwortung und Verdrängung* (Berlin: Forschungs-programm Geschichte der Kaiser-Wilhelm-Ges. im Nationalsozialismus, 2003); Werner, *Ein Genie*, 296.

50. Otto Warburg to A. V. Hill, 5 February, 1939, A. V. Hill Papers, AVHL II 4/82, Churchill Archive Center.

51. Werner, *Ein Genie*, 296; Krebs, *Otto Warburg*, 59.

52. Christian Friedrich Daniel Schubart, "The Wandering Jew," in *The German Museum, or Monthly Repository of the Literature of Germany, the North and the Continent in General*, Vol. 3 (1801).

CHAPTER NINE: "THE HERB GARDEN" OF DACHAU

1. For more on the campaigns against cancer discussed in this chapter, see Stanford historian Robert Proctor's *The Nazi War on Cancer*, a fascinating study of Nazi politics and science.

2. Proctor, *Nazi War*, 11; Hitler, *Mein Kampf*.

3. Proctor, *Nazi War*, 134.

4. Hitler, *Hitler's Table Talk*, 374; Proctor, *Nazi War*, 137.

5. "Hitler's Throat," *Time*, November 14, 1938; Ulf Schmidt, *Karl Brandt: The Nazi Doctor* (London: Hambledon Continuum, 2007), 82–83.

6. Norman Ohler, *Blitzed: Drugs in the Third Reich*, trans. Shaun Whiteside (Boston: Houghton Mifflin Harcourt, 2017), 23–24; Christa Schroeder, *He Was My Chief*, trans. Geoffrey Brooks (London: Frontline Books, 2009); Heinz Linge, *With Hitler to the End*, trans. Geoffrey Brooks (London: Frontline Books, 2009).

7. Roger Manvell, *Doctor Goebbels: His Life and Death* (London: Heinemann, 1960), 265; Leo Katz, *Bad Acts and Guilty Minds* (Chicago: University of Chicago Press, 1987); Treitel, *Eating Nature*, 215; Proctor, *Nazi War*, 139.

8. Proctor, *Nazi War*, 254-55; Ulf Schmidt, *Karl Brandt*, 88.

9. Proctor, *Nazi War*, 138–39; Peter Longerich, *Heinrich Himmler* (Oxford: Oxford University Press, 2013), 483.

10. Proctor, *Nazi War*, 265; Heinrich Himmler, *The Private Heinrich Himmler*, ed., Katrin Himmler and Michael Wildt, trans. Thomas S. Hansen and Abby J. Hansen (New York: St. Martin's Press, 2016), 199; Stanislav Zámečník, *Das war Dachau* (Frankfurt am Main: Fischer-Taschenbuch, 2010), 121.

11. Himmler, *Private Heinrich Himmler*, 199.

12. Longerich, *Heinrich Himmler*, 483–84, 335; Proctor, *Nazi War*, 157.

13. Longerich, *Heinrich Himmler*, 334–35.

14. Paul Weindling, *Victims and Survivors of Nazi Human Experiments: Science and Suffering in the Holocaust* (London: Bloomsbury Academic, 2015), 54–55.

15. Otto Warburg, *Heavy Metal Prosthetic Groups and Enzyme Action* (Oxford: Clarendon Press, 1949), 59.

16. Bücher, "Otto Warburg," 13.

17. Guy C. Brown, *The Energy of Life* (New York: The Free Press, 2000), 29.

18. Nick Lane, *Power, Sex, Suicide: Mitochondria and the Meaning of Life* (Oxford: Oxford University Press, 2005).

19. Eric G. Ball to W. Mansfield Clark, 11 December 1937, William Mansfield Clark Papers, Correspondence, 1903–1964, Mss.B.C547, American Philosophical Society Archives.

20. Hugo Theorell, "Closing Remarks," in *Pyridine Nucleotide-Dependent Dehydrogenases*, ed. H. Sund (Berlin: Springer, 1970); George W. Schwert and Alfred D. Winer, eds., *The Mechanism of Action of Dehydrogenases: A Symposium in Honor of Hugo Theorell* (Lexington: University Press of Kentucky, 1970).

21. C. Stacy French, "Fifty years of Photosynthesis," *Annual Review of Plant Physiology* 30, no. 1 (1979): 1–27.

22. Waldemar Kaempffert, "Authorities Sure of Pellagra Cure," *New York Times*, March 20, 1938.

23. Kaempffert, "Authorities"; Waldemar Kaempffert, "Science in the News: More Light on Pellagra," *New York Times*, April 16, 1939.

24. Werner, *Ein Genie*, 313, 402; L. Warburg, *Eine vollkommene Närrin*, 170, 212.

CHAPTER TEN: THE AGE OF KOCH

1. Vallery-Radot, *The Life of Pasteur*, 99.

2. Thomas Goetz, *The Remedy: Robert Koch, Arthur Conan Doyle, and the Quest to Cure Tuberculosis* (New York: Gotham Books, 2014).

3. Goetz, *The Remedy*; René J. Dubos and Jean Dubos, *The White Plague: Tuberculosis, Man, and Society* (New Brunswick, NJ: Rutgers University Press, 1987).

4. Goetz, *The Remedy*.

5. Lawrason Brown, "Robert Koch (1843–1910), an American Tribute," *Annals of Medical History*, n.s., 7 (1935); "Dr. Koch's Career," *New York Times*, July 22, 1884.

6. "Dr. Koch: Character Sketch," *The Review of Reviews* 2 (1890): 547.

7. Warburg, *Prime Cause*.

8. Fritz Redlich, *Hitler: Diagnosis of a Destructive Prophet* (Oxford: Oxford University Press, 1998); Schmidt, *Karl Brandt*, 81.

9. Vallery-Radot, *The Life of Pasteur*, 272.

10. Hitler, *Mein Kampf*.

11. Ian Kershaw, *Hitler: A Biography* (New York: W. W. Norton, 2010), 92.

12. William John Niven, *Hitler and Film: The Führer's Hidden Passion* (New Haven, CT: Yale University Press, 2018), 117.

13. Ian Kershaw, *Hitler, 1936–45: Nemesis* (New York: W. W. Norton, 2001), 470.

14. Robert A. Lambert to Betty Drury, 3 March 1938, Emergency Committee in Aid of Displaced Foreign Scholars, Warburg, Otto 1933, New York Public Library Archives.

15. Werner, *Ein Genie*, 305.

16. Noakes, "The Development of Nazi Policy," 338; Wolf Gruner, *Jewish Forced Labor under the Nazis: Economic Needs and Racial Aims, 1938–1944* (Cambridge: Cambridge University Press, 2006).

17. Gabriele Moser, "From Deputy to 'Reichsbevollmächtiger' and Defendant at the Nuremberg Medical Trials: Dr. Kurt Blome and Cancer Research in National Socialist Germany," in *Man, Medicine, and the State*, ed. Wolfgang Uwe Eckart (Stuttgart: Steiner, 2006), 203; Ernst Telschow to Rudolf Mentzel, 12 April 1941, III. Abt., Rep. 1, Nr. 47, Archives of the Max Planck Society; Rudolf Mentzel to Kaiser Wilhelm Society, 24 April 1941, III. Abt., Rep. 1, Nr. 47, Archives of the Max Planck Society.

18. Werner, *Ein Genie*, 305.

19. Melvyn Conroy, *Nazi Eugenics: Precursors, Policy, Aftermath* (New York: Columbia University Press, 2017).

20. Memorandum Re: Notice for Professor Warburg, Dahlem, 14 June 1941, Kaiser Wilhelm Society for the Advancement of Science, 1926–1941, NL CFvS, Siemens Archive.

21. Rigg, *Hitler's Jewish Soldiers*, 38.

22. Noakes, "The Development of Nazi Policy," 319; Rigg, *Hitler's Jewish Soldiers*, 192, 182.

23. Memorandum Re: Notice for Professor Warburg, Siemens Archive.

24. Affidavit of Walter Schoeller, Transcript of the trial of Viktor Brack, May 13, 1947, Nuremberg Trials Project: A Digital Document Collection of the Harvard Law School Library, 7495: nuremberg.law.harvard.edu/transcripts/1-transcript-for-nmt-1-medical-case?seq=7642; Rigg, *Hitler's Jewish Soldiers*, 187.

25. Memorandum Re: Notice for Professor Warburg, Siemens Archive; Krebs, *Otto Warburg*, 59.

26. Viktor Brack to Heinrich Himmler, 28 March 1941, Transcript of the trial of Viktor Brack, May 13, 1947, Nuremberg Trials Project: A Digital Document Collection of the Harvard Law School Library, 7490: nuremberg.law.harvard.edu/transcripts/1-transcript-for-nmt-1-medical-case?seq=7637.

27. Werner, *Otto Warburg*, 262–63; Nickelsen, "On Otto Warburg"; Affidavit of Otto Warburg, Transcript of the trial of Viktor Brack, May 13, 1947, Nuremberg Trials Project: A Digital Document Collection of the Harvard Law School Library, 7493: nuremberg.law.harvard.edu/transcripts/1-transcript-for-nmt-1-medical-case?seq=7640.

28. Affidavit of Otto Warburg, 7493; Himmler, *Der Dienstkalender*, 17; Werner, *Ein Genie*, 311–14, Hitler's Jewish soldiers, 186–97.

29. Kershaw, *Nemesis*, 386.

30. Proctor, *Nazi War*, 3–4.

31. Kershaw, *Nemesis*, 387–88.

32. Timothy Snyder, *Black Earth*, 166.

CHAPTER ELEVEN: "I REFUSED TO INTERVENE"

1. Raul Hilberg, *The Destruction of the European Jews*, Vol. 2 (New Haven, CT: Yale University Press, 2002), 434–47.

2. Ekkehard Höxtermann and Ulrich Sucker, *Otto Warburg* (Wiesbaden: Vieweg+Teubner Verlag, 1989), 142.

3. Rudolf Mentzel to Adolf Windaus, 5 October 1942, Adolf Windaus to Rudolf Mentzel, 7 October 1942, Nachlass Adolf Windaus, 2003.9, Archives of the Göttingen State and University Library; Werner, *Ein Genie*, 310; Gabriele Moser, "A Model of Joint Research? Cancer Research and the Funding Policies of the German Research Foundation and the Reich Research Council in National Socialist Germany," *Medizinhistorisches Journal* 40, no. 2 (2005): 113–39.

4. Beate Meyer, email to author, June 10, 2020; Beate Meyer, *"Jüdische Mischlinge" Rassenpolitik und Verfolgungserfahrung 1933–1945* (Munich: Dölling und Galitz, 2015).

5. William L. Laurence, "New Avenue Seen in Cancer Studies," *New York Times*, September 26, 1948.

6. Krebs, *Otto Warburg*, 45.

7. Ilya Zbarsky, Samuel Hutchinson, and Barbara Bray, *Lenin's Embalmers* (London: Havrill 1998), 128.

8. Chernow, *The Warburgs*, 506; Nickelsen, "On Otto Warburg"; Theodor Bücher, interview with the historian Ute Deichmann, June 14, 1994; Stefan Müller, *Liebensberg: Ein Verkauftes Dorf* (BoD, 2003).

9. Bücher, "Otto Warburg"; Werner, *Otto Warburg*, 266–74.

10. Nickelsen, "On Otto Warburg"; Hilberg, *Destruction*, 442–43; Otto Warburg to Walter Kempner, 11 April 1949, Stefan George Archive, Stuttgart.

11. Nickelsen, "On Otto Warburg"; Theodor Bücher, "Otto Warburg"; Theodor Bücher to Hans Krebs, 5 July 1947, Renewing contacts with Germany, 1945–50, Bücher, T, 1947-56, Krebs Collection, University of Sheffield Archives.

12. Michael O. R. Kröher, *Der Club der Nobelpreisträger* (Munich: Knaus, 2017); Bücher, interview with Deichmann; Bücher, "Otto Warburg."

13. Kärin Nickelsen, "Ein Bisher Unbekanntes Zeitzeugnis: Otto Warburgs Tagebuchnotizen von Februar–April 1945," *NTM Zeitschrift für Geschichte der Wissenschaften, Technik und Medizin* 16, no. 1 (March 2008): 103–15; Nickelsen, "On Otto Warburg."

14. Nickelsen, "On Otto Warburg"; David Farrer, *The Warburgs: The Story of a Family* (New York: Stein and Day, 1975), 152; Macrakis, *Surviving the Swastika*, 63.

15. Proctor, *Nazi War*, 259–60.

16. Hitler, *Table Talk*, 114–19; Miriam Kleiman, "Hitler and His Dentist," US National Archives website, January 31, 2012: prologue.blogs.archives.gov/2012/01/31/hitler-and-his-dentist; Linge, *With Hitler*; Traudl Junge, *Hitler's Last Secretary: A Firsthand Account of Life with Hitler*, ed. Melissa Müller, trans. Anthea Bell (New York: Arcade, 2017).

17. Hitler, *Table Talk*, 114–19.

18. Schroeder, *He Was My Chief*; Junge, *Hitler's Last Secretary*.

19. Hitler, *Table Talk*, 332.

20. Ohler, *Blitzed*; Antony Beevor, *The Fall of Berlin 1945* (New York: Penguin Books, 2003); Linge, *With Hitler*.

21. Schroeder, *He Was My Chief*.

22. Jonathan Mayo and Emma Craigie, *Hitler's Last Day: Minute by Minute* (London: Short Books, 2016).

CHAPTER TWELVE: COMING TO AMERICA

1. Roger Adams, "Situation with Respect to Dr. Otto Warburg," Roger Adams Papers, Series:15/5/23, University of Illinois Archives.

2. Excerpts from the diary of Wilhelm Lüttgens, in Werner, *Ein Genie*, 326–34; E. Warburg, *Times and Tides*, 128.

3. Otto Warburg to Lotte Warburg, 13 January 1946, private collection of the Meyer-Viol family; Klein, interview with author; A. R. Mann, "Report on Educational Conditions in Postwar Germany," Rockefeller Foundation records, projects, RG 1.1, Rockefeller Archive Center.

4. Govindjee et al., eds., *Discoveries in Photosynthesis*, Advances in Photosynthesis and Respiration, Vol. 20 (Dordrecht: Springer, 2005); Roger Adams, "Situation with Respect to Dr. Otto Warburg."

5. Parke D. Massey, "Investigation Report on Otto Heinrich Warburg," in Adams, "Situation with Respect to Dr. Otto Warburg."

6. Adams, "Situation with Respect to Dr. Otto Warburg"; Otto Warburg to Lotte Warburg, 13 January 1946.

7. Detlev W. Bronk to Edward A. Doisy, 28 April 1948, William Mansfield Clark Papers, Correspondence, 1903–1964, Mss.B.C547, American Philosophical Society Archives; Schüring, *Minervas verstossene Kinder*, 287; Correspondence of A. V. Hill, 1945, GB 117, Archives of the Royal Society; Rigg, *Hitler's Jewish Soldiers*, 39.

8. Carola Sachse, "What Research, to What End? The Rockefeller Foundation and the Max Planck Gesellschaft in the Early Cold War," *Central European History* 42, no. 1 (2009); Declaration of Otto Heinrich Warburg for the Office of Military Government, 10 June 1948, in Helmut Maier, ed., *Gemeinschaftsforschung, Bevollmächtigte und der Wissenstransfer: Die Rolle der Kaiser-Wilhelm-Gesellschaft im System Kriegsrelevanter Forschung des Nationalsozialismus*, Bd. 17 (Göttingen: Wallstein, 2007).

9. Benno Müller-Hill, "The Blood from Auschwitz and the Silence of the Scholars," *History and Philosophy of the Life Sciences* 21, no. 3 (1999): 331–65; Interview with Max Delbruck, Caltech Oral Histories, CaltechOH:OH_Delbruck_M, Caltech Institute Archives.

10. Otto Warburg to Carl Neuberg, December 1948, Carl Neuberg Papers, Warburg, Otto 1948–56, Mss.Ms.Coll.4, American Philosophical Society Archives; Affidavit

of Otto Warburg, Viktor Brack trial; Werner, *Otto Warburg*, 307–8. Petra Gentz-Werner, interview with author, May 11, 2015.

11. Martin David Kamen, *Radiant Science, Dark Politics: A Memoir of the Nuclear Age* (Berkeley: University of California Press, 1986); Robert Emerson and Govindjee, "Robert Emerson's 1949 Stephen Hales Prize Lecture: 'Photosynthesis and the World,'" *Journal of Plant Science Research* 34, no. 2 (2018): 119–25; Nickelsen and Govindjee, *Maximum Quantum Yield Controversy*, 47–75. Nickelsen and Govindjee's excellent account of Warburg's adventures in Urbana provides additional information on Warburg's antics.

12. Robert Emerson to Hans Gaffron, 29 May 1948, in Nickelsen and Govindjee, *Maximum Quantum Yield Controversy*, 37.

13. Nickelsen and Govindjee, *Maximum Quantum Yield Controversy*, 47–75; E. Rabinowitch, "Robert Emerson, 1903–1959," *Biographical Memoirs of the National Academy of Sciences* 25 (1961): 112–31; William L. Laurence, "New Avenue Seen in Cancer Studies," *New York Times*, September 26, 1948.

14. Kamen, *Radiant Science*, 101, 304; Nickelsen and Govindjee, *Maximum Quantum Yield Controversy*, 47–75.

15. Govindjee, interview with author, April 18, 2017; James Franck to Robert Pohl, 22 March 1955, in Florian Ebner, *James Franck—Robert Wichard Pohl. Briefwechsel 1906–1964*, Preprint 8, 2013, Deutsches Museum; Rabinowitch, "Robert Emerson," 123; Oral history interview with Michael Polanyi, February 15, 1962, OH 4831, American Institute of Physics.

16. Kamen, *Radiant Science*, 101–2; Otto Warburg to Walter Kempner, 14 December 1948, Stefan George Archive, Stuttgart; Nickelsen and Govindjee, *Maximum Quantum Yield Controversy*, 47–75.

17. Dean Burk to Otto Warburg, 2 January 1958, private collection of Frederic Burk; Nickelsen and Govindjee, *Maximum Quantum Yield Controversy*, 58–60.

18. Nickelsen and Govindjee, *Maximum Quantum Yield Controversy*, 47–75; Klotz, "Wit and Wisdom."

19. For more on Warburg's search for a position in the United States, see his correspondence with Walter Kempner in the Stefan George Archive in Stuttgart; Nickelsen and Govindjee, *Maximum Quantum Yield Controversy*, 32.

CHAPTER THIRTEEN: TWO ENGINES

1. Otto Warburg to Reinhard Dohrn, December 1949, in Werner, *Otto Warburg*, 296.

2. William L. Laurence, "Vital Force Found in Plants May Increase World's Food," *New York Times*, December 31, 1949; Otto Warburg to Walter Kempner, 25 December 1951, Stefan George Archive, Stuttgart; Martin Klingenberg, email to author, April 10, 2017.

3. Otto Warburg, "On the Origins of Cancer Cells," *Science* 123, no. 3191 (1956): 309; Klein, interview with author, June 10, 2015. Klein, a Holocaust survivor

and celebrated cancer biologist at the Karolinska Institute in Stockholm, died in 2016.

4. Klein interview; Robert A. Weinberg, *Racing to the Beginning of the Road: The Search for the Origin of Cancer* (New York: Harmony Books, 1996), 12.

5. A. A. Benson, "Paving the Path," *Annual Review of Plant Biology* 53, no. 1 (2002): 1–25; Herman M. Kalckar, "50 Years of Biological Research—from Oxidative Phosphorylation to Energy Requiring Transport Regulation," *Annual Review of Biochemistry* 60, no. 1 (1991): 1–38.

6. Warburg, "On the Origins of Cancer Cells"; Warburg, *Prime Cause.*

7. H. P. Hofschneider, "Adolf Butenandt zum Gedenken," *Zeitschrift für Naturforschung A* 50, no. 4–5 (1995): 505–8; Kurt Stern, "Antimetabolites and Cancer," *American Journal of Clinical Pathology* 26, no. 5 (1956): 529; Gloria M. Hanson and Richard W. Hanson, "Sidney Weinhouse, 1909–2001: A Biographical Memoir (Washington, DC: National Academy of Sciences, 2009); "Cancer Theory Overthrown," *The Science News-Letter* 65, no. 1 (1954): 5; Dean Burk to Otto Warburg, 5 January 1954, private collection of Frederic Burk.

8. Warburg, "On the Origins of Cancer Cells."

9. Waldemar Kaempffert, "German Physiologist Is Sure That He Has Discovered the Cause of Cancer," *New York Times*, March 4, 1956.

10. S. Weinhouse et al., "On Respiratory Impairment in Cancer Cells," *Science* 124, no. 3215 (1956): 267–72.

11. Mukherjee, *Emperor of All Maladies*, 88.

12. L. Warburg, *Eine vollkommene Närrin*, 21. For more about Haber, see Daniel Charles's excellent biography, *Master Mind.*

13. Ball, *Serving the Reich*, 67–68.

14. Hitler, *Mein Kampf*, 202.

15. A. A. Liebow and L. L. Waters, "Milton Charles Winternitz February 19, 1885– October 3, 1959," *The Yale Journal of Biology and Medicine* 32, no. 3 (December 1959): 143.b1–164; Vincent T. DeVita and Edward Chu, "A History of Cancer Chemotherapy," *Cancer Research* 68, no. 21 (November 1, 2008): 8643–53; Judith Ann Schiff, "Pioneers in Chemotherapy," *Yale Alumni Magazine*, May/June, 2011.

CHAPTER FOURTEEN:
"STRANGE NEW CREATURES OF OUR OWN MAKING"

1. Farrer, *The Warburgs*, 152; Felicitas von Aretin, "Under the Watchful Gaze of Minerva: Traditions, Symbols and Dealing with the Past," in *Denkorte, Centennial Book—Essays*, Max-Planck-Gesellschaft and Kaiser-Wilhelm-Gesellschaft, 1911– 2011 (Dresden: Sandstein Verlag, 2011), 3; Krebs, *Otto Warburg*, 61.

2. Manfred Görtemaker and Christoph Johannes Maria Safferling, *Die Akte Rosenburg: Das Bundesministerium der Justiz und die NS-Zeit* (Munich: C. H. Beck, 2016); Ralph Giordano, *Die Zweite Schuld oder von der Last ein Deutscher zu Sein* (Hamburg: Rasch und Röhring, 1998).

3. Daniella Seidl, *"Zwischen Himmel und Hölle": Das Kommando "Plantage" des Konzentrationslagers Dachau* (Munich: Utz, 2008); Treitel, *Eating Nature,* 237–38.

4. G. Taubes, "Epidemiology Faces Its Limits," *Science* 269, no. 5221 (July 14, 1995): 164–69; Richard Doll and Richard Peto, *The Causes of Cancer: Quantitative Estimates of Avoidable Risks of Cancer in the United States Today* (Oxford: Oxford University Press, 1981).

5. Proctor, *Cancer Wars,* 37–39; W. C. Hueper, *Occupational Tumors and Allied Diseases* (Springfield, IL: C. C. Thomas, 1942); Adam Wishart, *One in Three: A Son's Journey into the History and Science of Cancer* (New York: Grove Press, 2008); C. Sellers, "Discovering Environmental Cancer: Wilhelm Hueper, Post-World War II Epidemiology, and the Vanishing Clinician's Eye," *American Journal of Public Health* 87, no. 11 (1997): 1824–35.

6. Proctor, *Nazi War,* 14; Proctor, *Cancer Wars,* 47.

7. Linda J. Lear, *Rachel Carson: Witness for Nature* (New York: H. Holt, 1997); Rachel Carson, *Silent Spring,* 40th anniversary ed. (Boston: Houghton Mifflin, 2002); Proctor, *Nazi War,* 286.

8. Carson, *Silent Spring,* 232.

9. L. Warburg, *Eine vollkommene Närrin,* 29–30, 78; Werner, *Otto Warburg,* 298–302; Harald Lemke, *Die Kunst des Essens: Eine Ästhetik des Kulinarischen Geschmacks* (Bielefeld: Transcript, 2007), 193–206; Udo Pollmer et al., *Vorsicht Geschmack* (Hamburg: Rowohlt, 2007).

10. E. Warburg, *Times and Tides,* 129; Krebs, *Otto Warburg,* 74; Nachmansohn, *German-Jewish Pioneers,* 263.

11. W. C. Hueper, unpublished autobiography, Hueper Papers, 1920–1981, MS C 341, National Library of Medicine; Proctor, *Nazi War,* 12–15, 79.

12. Proctor, *Cancer Wars,* 38; Hueper, unpublished autobiography, 143; Proctor, *Nazi War,* 15, 71.

13. Stephen P. Strickland, *Research and the Health of Americans* (Lexington, MA: Lexington Books, 1978), 77.

14. Howard Hiatt, "The Use of Basic Research," *New York Times,* September 8, 1976; Jane Brody, "Quick, Inexpensive Test on Bacteria Detects Chemicals Potentially Dangerous to Man," *New York Times,* April 3, 1975; Philip M. Boffey, "The Parade of Chemicals That Cause Cancer Seems Endless," *New York Times,* March 20, 1984.

CHAPTER FIFTEEN: THE PRIME CAUSE OF CANCER

1. R. Clinton Fuller, "Forty Years of Microbial Photosynthesis Research: Where It Came from and What It Led To," *Photosynthesis Research: Official Journal of the International Society of Photosynthesis Research* 62, no. 1 (1999): 1–29; Helmut Sies, interview with author, April 14, 2017.

2. Otto Warburg to Dean Burk, 3 September 1952, private collection of Frederic Burk; Govindjee, "Robert Emerson and Eugene Rabinowitch: Understanding Pho-

tosynthesis," in *No Boundaries: University of Illinois Vignettes*, ed. Lillian Hoddeson (Urbana: University of Illinois Press, 2004).

3. Warburg, *Prime Cause*.

4. "Letzte Ursache," *Der Spiegel*, July 4, 1966; Personal notes of Hans Krebs, "Career, Honours and Awards," Krebs Collection, University of Sheffield Archives.

5. C. Cook, "Oral History—Sir Richard Doll," *Journal of Public Health* 26, no. 4 (2004): 327–36; Conrad Keating, *Smoking Kills: The Revolutionary Life of Richard Doll* (Oxford: Signal, 2009).

6. Wishart, *One in Three*; Keating, *Smoking Kills*.

7. R. Doll and A. Bradford Hill, "Smoking and Carcinoma of the Lung," *British Medical Journal* 2 (1950): 739; Proctor, *Nazi War*, 196–208, 217–19.

8. George Johnson, *The Cancer Chronicles: Unlocking Medicine's Deepest Mystery* (New York: Alfred A. Knopf, 2013); Jane E. Brody, "Scientist at Work: Bruce N. Ames; Strong Views on Origins of Cancer," *New York Times*, July 5, 1994.

CHAPTER SIXTEEN: CANCER AND DIET

1. Otto Warburg to Harry Goldblatt, 3 January 1967, III. Abt., Rep. 1, Nr. 207, Archives of the Max Planck Society.

2. Harry Goldblatt to Otto Warburg, 9 April 1968, III. Abt., Rep. 1, Nr. 207, Archives of the Max Planck Society.

3. Keating, *Smoking Kills*; Gary Taubes, "Do We Really Know What Makes Us Healthy?" *New York Times Magazine*, September 16, 2007.

4. Thomas Denman, *Observations on the Cure of Cancer* (London: T. Bensley, 1810); Williams, *Natural History of Cancer*, 64.

5. Proctor, *Nazi War*, 127.

6. Hoffman, *Cancer and Diet*.

7. Lester Grant, *The Challenge of Cancer* (Bethesda: National Institutes of Health, 1950); Albert Tannenbaum, "Effects of Varying Caloric Intake upon Tumor Incidence and Tumor Growth," *Annals of the New York Academy of Sciences* 49, no. 1 (1947): 5–18.

8. Colin Champ, "The Ketogenic Diet: Making It Difficult for Cancer to Latch On?" October 6, 2018: colinchamp.com/the-ketogenic-diet-making-it-difficult-for-cancer-to-latch-on; Tannenbaum, "Effects of Varying Caloric Intake."

9. Charles, *Master Mind*.

10. Vaclav Smil, *Enriching the Earth* (Cambridge, MA: MIT Press, 2000); Vaclav Smil, "Global Population and the Nitrogen Cycle," *Scientific American*, July 1997, 76.

11. Stephen Paget, "The Distribution of Secondary Growths in Cancer of the Breast," *The Lancet*, 133, no. 3421 (1889): 571–73; A. Goldfeder, "An Overview of Fifty Years in Cancer Research: Autobiographical Essay," *Cancer Research* 36 (1): 1–9.

12. "Cancer Cells Held Cause of Its Spread," *New York Times*, January 10, 1929.

13. Ernst Freund, *Metabolic Therapy of Cancer*, trans. Laurence Wolfe (London: Daniel Godwin, 1946), 21.

14. Doll and Peto, *The Causes of Cancer*, 1233; A. Tannenbaum, "Host and Environ-

mental Factors in Cancer Research," *Acta-Unio Internationalis Contra Cancrum*, 15 (1959): 861–63.

15. William J. Blot and Robert E. Tarone, "Doll and Peto's Quantitative Estimates of Cancer Risks: Holding Generally True for 35 Years," *Journal of the National Cancer Institute* 107, no. 4 (2105); Doll and Peto, *The Causes of Cancer*, 1226.

CHAPTER SEVENTEEN: LOST AND FOUND

1. Ostendorf, interview with author; Krebs, *Otto Warburg*, 55.
2. Farrer, *The Warburgs*, 152, Werner, *Otto Warburg*, 298.
3. Dean Burk to Otto Warburg, 7 April 1965, NL Warburg, 174, archive of the Berlin-Brandenburg Academy of Sciences and Humanities. Burk is quoting a statement made by Warburg.
4. Wolfgang Lefèvre, email to author, December 10, 2019.
5. Dean Burk to Otto Warburg, 4 March 1965, NL Warburg, 174, archive of the Berlin-Brandenburg Academy of Sciences and Humanities; Peter Pedersen, interview with author, May 15, 2015.
6. "Bei Krebs Hast du nur Einmal eine Chance," *Der Spiegel*, October 2, 1972.
7. Klotz, "Wit and Wisdom of Albert Szent-Györgyi."
8. A note on nomenclature: Scientists refer to DNA sequences as "genes" and to the proteins they give rise to as "gene products." The nomenclature used to distinguish one from the other varies according to the type of organism under discussion. For the purposes of this book, such distinctions aren't necessary. The DNA code and the corresponding protein can be considered parts of a whole, and in the name of simplicity, I have represented them with the same capitalized three-letter symbols, such as AKT or SRC.
9. Chi Van Dang, email to author, July 8, 2020; Chi Van Dang, interviews with author, May 19, 2015, and April 21, 2016.
10. Chi Van Dang interviews; Chi V. Dang et al., "c-Myc Transactivation of *LDH-A*: Implications for Tumor Metabolism and Growth," *Proceedings of the National Academy of Sciences* 94, no. 13 (1997): 6658–63.
11. Otto Warburg to Dean Burk, 27 November 1957, NL Warburg, 174, archive of the Berlin-Brandenburg Academy of Sciences and Humanities.
12. D. Hanahan and R. A. Weinberg, "The Hallmarks of Cancer," *Cell* 100, no. 1 (2000): 57–70; Robert Weinberg, interviews with author, April 29, 2015, and December 8, 2016; Robert Weinberg, email to author, November 29, 2016. In 2011, Hanahan and Weinberg published "Hallmarks of Cancer: The Next Generation," an updated version of their paper. The updated paper discusses Warburg in a section titled "An Emerging Hallmark: Reprogramming Energy Metabolism." D. Hanahan and R. A. Weinberg, "The Hallmarks of Cancer: The Next Generation," *Cell* 144, no. 5 (2011): 646–74.
13. Harry Goldblatt to Otto Warburg, 19 April 1968, III. Abt., Rep. 1, Nr. 207, Archives of the Max Planck Society.

14. Laura Stephenson Carter and Craig Thompson, "'75: Why Not?" *Dartmouth Medicine Magazine,* Fall 2006; Thompson, interview with author.

15. L. Ernster and G. Schatz, "Mitochondria: A Historical Review," *Journal of Cell Biology* 91, no. 3 (1981): 227s–55s; Klein, interview with author.

16. Brooke Bevis, email to author, November 28, 2017.

17. Lane, *Power, Sex, Suicide.*

18. Navdeep Chandel, interviews with author, September 27, 2016, and November 9, 2017; Matthew Vander Heiden, interviews with author, April 29, 2015, and November 6, 2017.

19. Craig Thompson, "Why Don't We All Get Cancer?" Major Trends in Modern Cancer Research, conference held at Memorial Sloan Kettering, November 10, 2010.

20. Thompson, "Why Don't We All Get Cancer?"

21. Lewis Cantley, interviews with author, May 14, 2015, and November 18, 2016.

22. Matthew Vander Heiden, interview with author.

23. Jeff Rathmell, interview with author, January 17, 2017; Ushma S. Neill, "A Conversation with Craig Thompson," *Journal of Clinical Investigation* 125, no. 6 (2015): 2181–83; Thompson, interview with author.

CHAPTER EIGHTEEN: THE METABOLISM REVIVAL

1. Thompson, interview with author.

2. Britton Chance's notes on his 1966 trip to Berlin, Britton Chance Papers, Mss. Ms.Coll.160, American Philosophical Society Archives.

3. C. B. Thompson et al., "Akt Stimulates Aerobic Glycolysis in Cancer Cells," *Cancer Research* 64 (2004): 3892–99; Thompson, interview with author.

4. Ken Garber, "Energy Boost: The Warburg Effect Returns in a New Theory of Cancer," *Journal of the National Cancer Institute* 96, no. 24 (2004): 1805–6; Dang, interview with author.

5. Thompson, interview with author; Thompson, "Why Don't We All Get Cancer?"

6. M. G. Vander Heiden, L. C. Cantley, and C. B. Thompson, "Understanding the Warburg Effect: The Metabolic Requirements of Cell Proliferation," *Science* 324, no. 5930 (2009): 1029–33.

7. K. E. Wellen, G. Hatzivassiliou, et al., "ATP-Citrate Lyase Links Cellular Metabolism to Histone Acetylation," *Science* 324, no. 5930 (2009): 1076–80; D. Zhang et al., "Metabolic Regulation of Gene Expression by Histone Lactylation," *Nature* 574 (2019): 575–80; Kathryn Wellen's presentation at Women & Science: Nutrition and Gene Regulation, The Wistar Institute, February 13, 2019; Kathryn Wellen, email to author, July 7, 2020.

8. Craig Thompson, "Discussion with Memorial Sloan Kettering's President," Major Trends in Modern Cancer Research, conference held at Memorial Sloan Kettering, November 7, 2012; Thompson, "Why Don't We All Get Cancer?"

9. Otto Warburg, *The Metabolism of Tumors: Investigation from the Kaiser Wilhelm Institute for Biology, Berlin-Dahlem* (London: Constable, 1930).

10. Celeste Simon, interview with author, February 4, 2019.

11. Warburg, *The Metabolism of Tumors*.

12. Valter Longo, interview with author, December 23, 2016; David Sabatini, interview with author, May 5, 2015; Dang, interview with author.

13. Jerome Groopman, "The Transformation," *New Yorker*, September 15, 2014. Thompson started Agios Pharmaceuticals together with Lewis Cantley and Tak Mak.

14. Vander Heiden, interview with author.

15. Doll and Peto, *The Causes of Cancer*, 1233.

16. Eugenia E. Calle et al., "Overweight, Obesity, and Mortality from Cancer in a Prospectively Studied Cohort of U.S. Adults," *New England Journal of Medicine* 348, no. 17 (2003): 1625–38; Associated Press, "Study Hailed as Convincing in Tying Fat to Cancers," *New York Times*, April 24, 2003.

17. C. Brooke Steele et al., "Vital Signs: Trends in Incidence of Cancers Associated with Overweight and Obesity—United States, 2005–2014," *Morbidity and Mortality Weekly Report* 66, no. 39 (2017): 1052–58; Rebecca Siegel, email to author, July 6, 2020.

18. Williams, *The Natural History of Cancer*.

CHAPTER NINETEEN: DIABETES AND CANCER

1. E. Racker, "Bioenergetics and the Problem of Tumor Growth," *American Scientist* 60, no. 1 (1972): 56–63.

2. Efraim Racker, "Otto Warburg at a Turning Point in 1932," *Trends in Biochemical Sciences* 7, no. 12 (1982): 448–49.

3. Cantley, interview with author.

4. Cantley, interview with author; Beth Saulnier, "Bench to Bedside," *Cornell Alumni Magazine*, May/June, 2014.

5. Cantley, interview with author; Lewis Cantley, "From Kinase to Cancer," *The Scientist*, November 30, 2007.

6. Kathryn M. King and Greg Rubin, "A History of Diabetes: From Antiquity to Discovering Insulin," *British Journal of Nursing* 12, no. 18 (2003); Robert Tattersall, *Diabetes: The Biography* (Oxford: Oxford University Press, 2009); Taubes, *Good Calories, Bad Calories*.

7. Elliott Proctor Joslin, *The Treatment of Diabetes Mellitus* (Philadelphia: Lea & Febiger, 1917); "The Frequency of Diabetes Mellitus, and Its Relation to Diseases of the Pancreas," *Journal of the American Medical Association* 29, no. 25 (1897): 1277; Haven Emerson and Louise Larimore, "Diabetes Mellitus, Contribution to Its Epidemiology Based Chiefly on Mortality Statistics," *Archives of Internal Medicine* 34, no. 5 (1924): 585–630.

8. Williams, *Natural History of Cancer*, 259, 334; Hoffman, *Cancer and Diet*.

9. Joslin, *Treatment of Diabetes Mellitus*.

10. Robert McCarrison, "Faulty Food in Relation to Gastrointestinal Disorders," *Jour-*

nal of the American Medical Association 78 (1922):1–8; Nicholas Senn, *In the Heart of the Arctics* (Chicago: W. B. Conkey, 1907).

11. T. L. Cleave and George Duncan Campbell, *Diabetes, Coronary Thrombosis, and the Saccharine Disease* (Bristol: J. Wright, 1969), 26; A. M. Cohen, "Prevalence of Diabetes among Different Ethnic Jewish Groups in Israel," *Metabolism* 10 (1961): 50–58; Taubes, *Good Calories, Bad Calories*; E. Giovannucci et al., "Diabetes and Cancer: A Consensus Report," *Diabetes Care* 33, no. 7 (2010): 1674–85.

12. B. A. Houssary, "The Discovery of Pancreatic Diabetes: The Role of Oscar Minkowski," *Diabetes* 1, no. 2 (March 1, 1952): 112–16; Joseph Collins, "Diabetes, Dreaded Disease, Yields to New Gland Cure," *New York Times*, May 6, 1923.

13. Elizabeth Stone, "A Mme. Curie from the Bronx," *New York Times*, April 9, 1978; Eugene Straus, *Rosalyn Yalow, Nobel Laureate* (New York: Basic Books, 1998).

14. Anahad O'Connor, "The Keto Diet Is Popular, but Is It Good for You?" *New York Times*, August 26, 2019; F. M. Allen. and J. W. Sherrill, "The Use of Insulin in Diabetic Treatment," *Journal of Metabolic Research* 2 (1922): 960.

15. Julius Bauer, "Obesity: Its Pathogenesis, Etiology and Treatment," *Archives of Internal Medicine* 67, no. 5 (1941): 968.

16. Ehrlich, *Experimental Researches*, 67–68.

17. Robert E. Forster, "Charles Brenton Huggins (22 September 1901–12 January 1997)," *Proceedings of the American Philosophical Society* 143, no. 2 (1999): 326–31.

18. Forster, "Charles Brenton Huggins," 326–31; Leo Loeb, "Autobiographical Notes."

19. Mark Woods, Kent Wight, Jehu Hunter, and Dean Burk, "Effects of Insulin on Melanoma and Brain Metabolism," *Biochimica et Biophysica Acta* 12, no. 1–2 (1953): 329–46; Mark Woods, Dean Burk, and Jehu Hunter, "Factors Affecting Anaerobic Glycolysis in Mouse and Rat Liver and in Morris Rat Hepatomas," *Journal of the National Cancer Institute* 41, no. 2 (August 1968): 267–86.

20. Dean Burk to Otto Warburg, 5 January 1954, 5 January 1965, 29 January 1965, private collection of Frederic Burk.

21. Ken Olsen, "Medal for a Gallant Man: The Vernon Baker Story," *The Spokesman-Review*, November 29, 1996.

22. Burk to Warburg, 5 January 1954.

23. Cantley, interview with author.

CHAPTER TWENTY: THE INSULIN HYPOTHESIS

1. Cantley, interview with author; L. C. Cantley et al., "The Phosphoinositide 3-Kinase Pathway," *Science* 296, no. 5573 (2002): 1655–57.

2. E. Giovannucci, "Insulin and Colon Cancer," *Cancer Causes & Control* 6, no. 2 (March 1995): 164–79; R. Kaaks, "Nutrition, Hormones, and Breast Cancer: Is Insulin the Missing Link?" *Cancer Causes & Control* 7, no. 6 (November 1996): 605–25; R. Dankner et al., "Effect of Elevated Basal Insulin on Cancer Incidence and Mortality in Cancer Incident Patients," *Diabetes Care* 35, no. 7 (2012): 1538–43.

3. Tetsuro Tsujimoto et al., "Association between Hyperinsulinemia and Increased Risk of Cancer Death in Nonobese and Obese People," *International Journal of Cancer* 141, no. 1 (July 1, 2017): 102–11; L. L. Moore et al., "Metabolic Health Reduces Risk of Obesity-Related Cancer in Framingham Study Adults," *Cancer Epidemiology Biomarkers & Prevention* 23, no. 10 (October 1, 2014): 2057–65.

4. Mina J. Bissell and William C. Hines, "Why Don't We Get More Cancer? A Proposed Role of the Microenvironment in Restraining Cancer Progression," *Nature Medicine* 17, no. 3 (2011): 320–29; Derek LeRoith, interview with author, December 6, 2016.

5. Michael Pollak, interview with author, September 25, 2017; Vuk Stambolic, interview with author; September 26, 2017.

6. Anna Marie Mulligan et al., "Insulin Receptor Is an Independent Predictor of a Favorable Outcome in Early Stage Breast Cancer," *Breast Cancer Research and Treatment* 106, no. 1 (2007): 39–47; Stambolic, interview with author.

7. Stambolic, interview with author.

8. Emily Jane Gallagher and Derek LeRoith, "The Proliferating Role of Insulin and Insulin-Like Growth Factors in Cancer," *Trends in Endocrinology & Metabolism* 21, no. 10 (2010): 610–18; Bas ter Braak et al., "Insulin-Like Growth Factor 1 Receptor Activation Promotes Mammary Gland Tumor Development by Increasing Glycolysis and Promoting Biomass Production," *Breast Cancer Research* 19, no. 1 (2017): 14.

9. Nicholas Wade, "Ecuadorean Villagers May Hold Secret to Longevity," *New York Times*, February 16, 2011.

10. James Johnson et al., "Endogenous Hyperinsulinemia Contributes to Pancreatic Cancer Development," *Cell Metabolism* 30, no. 3 (2019): 403–4; James Johnson, email to author, July 25, 2020.

11. Michael Bliss, *Banting: A Biography* (Toronto: University of Toronto Press, 1993), 212.

12. Cantley, interview with author; Y. Fierz et al., "Insulin-Sensitizing Therapy Attenuates Type 2 Diabetes-Mediated Mammary Tumor Progression," *Diabetes* 59, no. 3 (March 1, 2010): 686–93; Gary Taubes, "Cancer Prevention with a Diabetes Pill?" *Science* 335, no. 6064 (2012): 29.

13. Cantley, interview with author.

14. Otto Warburg, *Heavy Metal Prosthetic Groups and Enzyme Action*, trans. Alexander Lawson (Oxford: Clarendon Press, 1949), 65.

15. Thompson, "Why Don't We All Get Cancer?"

16. Nick Lane, *Oxygen: The Molecule That Made the World* (Oxford: Oxford University Press, 2003), 116.

17. Dang, interview with author.

18. Cantley, interview with author; Y. Poloz and V. Stambolic, "Obesity and Cancer, a Case for Insulin Signaling," *Cell Death & Disease* 6, no. 12 (2015); Pollak, interview with author.

19. Colin Champ, interview with author, December 22, 2016; Heather Christofk, interview with author, March 15, 2019.

20. Siddhartha Mukherjee, "It's Time to Study Whether Eating Particular Diets Can Help Heal Us," *New York Times Magazine*, December 5, 2018; Siddhartha Mukherjee, "The Search for Cancer Treatment beyond Mutant-Hunting," *New York Times Magazine*, June 13, 2018.

21. Joana Araújo, Jianwen Cai, and June Stevens, "Prevalence of Optimal Metabolic Health in American Adults: National Health and Nutrition Examination Survey 2009–2016," *Metabolic Syndrome and Related Disorders* 17, no. 1 (2019): 46–52.

CHAPTER TWENTY-ONE: SUGAR

1. O. Warburg et al., "The Partial Anaerobiosis of Cancer Cells and the Effects of Roentgen Irradiation on Cancer Cells," *Die Naturwissenschaften* 46, no. 2 (1959): 25–29; "Cancer: Poison to Cells," *Der Spiegel*, July 30, 1958; Christophe Glorieux and Pedro Buc Calderon, "Catalase, a Remarkable Enzyme: Targeting the Oldest Antioxidant Enzyme to Find a New Cancer Treatment Approach," *Biological Chemistry* 398, no. 10 (2017): 1095–1108.

2. Nachmansohn, *German-Jewish Pioneers*; Richard Veech, email to author, June 8, 2015. Veech was a biochemist who trained under Hans Krebs and made a number of important discoveries about ketone metabolism. He died in 2020.

3. Ostendorf, interview with author.

4. D. A. Cooke and R. K. Scott, *The Sugar Beet Crop* (Dordrecht: Springer, 1993).

5. Karl-Peter Ellerbrock and Hans-Jürgen Teuteberg, "Pioneering Spadework in the History of the German Food Industry during the Nineteenth and Early Twentieth Centuries," in *The Food Industries of Europe in the Nineteenth and Twentieth Centuries*, ed. Derek J. Oddy and Alain Drouard (Burlington, VT: Ashgate, 2013); Gary Taubes, "What If Sugar Is Worse Than Just Empty Calories?" *BMJ*, January 3, 2018; Gary Taubes, *The Case against Sugar* (New York: Alfred A. Knopf, 2016).

6. J. H. van't Hoff, *Imagination in Science*, trans. G. F. Springer (Berlin: Springer, 1967).

7. Catherine M. Jackson, "Emil Fischer and the 'Art of Chemical Experimentation,'" *History of Science* 55, no. 1 (2017): 86–120.

8. Graeme K. Hunter, *Vital Forces: The Discovery of the Molecular Basis of Life* (San Diego: Academic Press, 2000); F. E. Ziegler, "Emil Fischer and the Structure of Grape Sugar and Its Isomers": ursula.chem.yale.edu/~chem220/STUDYAIDS/history/Fischer/fischer.html; J. Michael McBride, speaking to his class at Yale: openmedia.yale.edu/projects/iphone/departments/chem/chem125b/transcript37.html

9. Robert Plimmer and Violet Plimmer, *Food and Health* (London: Longmans, Green, 1925).

10. Emerson and Larimore, "Diabetes Mellitus"; Lulu Hunt Peter, "Diet and Health: Overeating and Diabetes," *Los Angeles Times*, May 26, 1927.

11. Joslin, *Treatment of Diabetes Mellitus*, 3rd ed., 1923.

12. McCarrison, "Faulty Food."

13. John Yudkin, *Pure, White, and Deadly* (New York: Penguin Books, 2013).

14. Yudkin, *Pure, White, and Deadly*; Philip Buell, "Changing Incidence of Breast Cancer in Japanese-American Women," *Journal of the National Cancer Institute* 51, no. 5 (November 1973): 1479–83; Taubes, *Good Calories, Bad Calories*; Gary Taubes, "Is Sugar Toxic?" *New York Times Magazine*, April 13, 2011. American sugar consumption can also be estimated according to the amount of sugar made available by the food industry each year. At the peak of sugar availability in the United States in 1999, the industry made 150 pounds of sugar available for each American—a 34-pound increase from 1970. For more on sugar availability, see https://www.ers.usda.gov/topics/crops/sugar-sweeteners/background/.

CHAPTER TWENTY-TWO: THE EVIL TWIN

1. Krebs, *Otto Warburg*, 67.

2. John L. Sievenpiper, "Fructose: Back to the Future?" *American Journal of Clinical Nutrition* 106, no. 2 (2017): 439–42.

3. Taubes, "Is Sugar Toxic?"

4. Kimber L. Stanhope et al., "Consuming Fructose-Sweetened, Not Glucose-Sweetened, Beverages Increases Visceral Adiposity and Lipids and Decreases Insulin Sensitivity in Overweight/Obese Humans," *Journal of Clinical Investigation* 119, no. 5 (May 1, 2009): 1322–34; Kimber Stanhope on 60 Minutes, April 1, 2012: cbsnews.com/news/is-sugar-toxic-01-04-2012/2.

5. Thomas Jensen et al., "Fructose and Sugar: A Major Mediator of Non-Alcoholic Fatty Liver Disease," *Journal of Hepatology* 68, no. 5 (May 2018): 1063–75; Johanna K. DiStefano, "Fructose-Mediated Effects on Gene Expression and Epigenetic Mechanisms Associated with NAFLD Pathogenesis," *Cellular and Molecular Life Sciences* 77, no. 11 (2020): 2079–90.

6. Justus von Liebig, *Animal Chemistry, or Organic Chemistry in Its Applications to Physiology and Pathology* (Cambridge: John Owen, 1842), 80–81.

7. Lewis C. Cantley, "Seeking Out the Sweet Spot in Cancer Therapeutics: An Interview with Lewis Cantley," *Disease Models & Mechanisms* 9, no. 9 (2016): 911–16; Cantley, interview with author.

8. Richard Johnson, email to author, April 21, 2020; Richard J. Johnson et al., "Redefining Metabolic Syndrome as a Fat Storage Condition Based on Studies of Comparative Physiology: Fat Storage Condition," *Obesity* 21, no. 4 (2013): 659–64.

9. Pollak, interview with author; Taubes, *The Case against Sugar*.

10. Krebs, *Otto Warburg*; William Maddock Bayliss, *Principles of General Physiology* (London: Longmans, 1915), x–xi.

11. Wishart, *One in Three*; Cantley, "Seeking Out the Sweet Spot"; Elliott Joslin, "The Prevention of Diabetes Mellitus," *Journal of the American Medical Association* 76, no. 2 (1921): 79–84.

12. John Perkins, "Nazi Autarchic Aspirations and the Beet-Sugar Industry, 1933–9," *European History Quarterly* 20, no. 4 (October 1990): 497–518; Peter Hayes, *From Cooperation to Complicity: Degussa in the Third Reich* (Cambridge: Cambridge University Press, 2005).

13. Ullrich, *Hitler: Ascent*, 407–8.

14. Proctor, *Nazi War*, 137; Laura Shapiro, *What She Ate: Six Remarkable Women and the Food That Tells Their Stories* (New York: Viking, 2017), 155; Linge, *With Hitler*.

15. Schroeder, *He Was My Chief*; Frank Thadeusz, "Der F. Hat Gut Gegessen," *Der Spiegel*, November 19, 2017; Ohler, *Blitzed*.

16. Eric E. Conn et al., "Remembering Birgit Vennesland (1913–2001), a Great Biochemist," *Photosynthesis Research* 83, no. 1 (January 2005); Vennesland, "Recollections"; Otto Warburg to Dean Burk, 27 December 1955, private collection of Frederic Burk.

17. Vennesland, "Recollections"; Govindjee et al., *Discoveries in Photosynthesis*, 1222.

18. Otto Warburg to Charles Huggins, 20 June 1970, NL Warburg, 1097, archive of the Berlin-Brandenburg Academy of Sciences and Humanities; Krebs, *Otto Warburg*, 90–91.

19. Walter Kempner, "The Durham Connection to Germany," *North Carolina Medical Journal* 45, no. 1 (1984): 25–26; Otto Warburg to Walter Kempner, 15 December 1956, 22 November 1956, Stefan George Archive, Stuttgart.

20. Ostendorf, interview with author.

POSTSCRIPT

1. Ernst Jokl, "König der Biochemiker," *MPG Spiegel* 6 (1983): 20–22.

2. Ostendorf, interview with author.

3. Henry Harris, *The Balance of Improbabilities: A Scientific Life* (Oxford: Oxford University Press, 1987), 211–212.

Suggestions for Further Reading ·

Allen, Garland E. *Life Science in the Twentieth Century.* New York: Wiley, 1975.

Ball, Philip. *Serving the Reich: The Struggle for the Soul of Physics under Hitler.* Chicago: University of Chicago Press, 2014.

Carson, Rachel. *Silent Spring.* 40th anniversary ed. Boston: Houghton Mifflin, 2002.

Chandel, Navdeep. *Navigating Metabolism.* New York: Cold Spring Harbor Laboratory Press, 2015.

Christofferson, Travis. *Tripping over the Truth: How the Metabolic Theory of Cancer Is Overturning One of Medicine's Most Entrenched Paradigms.* White River Junction: Chelsea Green Publishing, 2017.

Deichmann, Ute. *Biologists under Hitler.* Translated by Thomas Dunlap. Cambridge, MA: Harvard University Press, 1996.

Goetz, Thomas. *The Remedy: Robert Koch, Arthur Conan Doyle, and the Quest to Cure Tuberculosis.* New York: Gotham Books, 2014.

Grady, Tim. *A Deadly Legacy: German Jews and the Great War.* New Haven, CT: Yale University Press, 2017.

Haffner, Sebastian. *Defying Hitler: A Memoir.* Translated by Oliver Pretzel. Lexington, MA: Plunkett Lake Press, 2014.

Heim, Susanne, Carola Sachse, and Mark Walker, eds. *The Kaiser Wilhelm Society under National Socialism.* New York: Cambridge University Press, 2009.

Höxtermann, Ekkehard. *Otto Warburg.* Leipzig: Teubner, 1989.

Kershaw, Ian. *Hitler: A Biography.* New York: W. W. Norton, 2010.

Krebs, Hans, and Roswitha Schmid. *Otto Warburg: Cell Physiologist, Biochemist and Eccentric.* Oxford: Clarendon Press, 1981.

Lane, Nick. *Power, Sex, Suicide: Mitochondria and the Meaning of Life.* Oxford: Oxford University Press, 2005.

Macrakis, Kristie. *Surviving the Swastika: Scientific Research in Nazi Germany.* New York: Oxford University Press, 1993.

Meyer, Beate. *"Jüdische Mischlinge" Rassenpolitik und Verfolgungserfahrung 1933–1945.* München: Dölling und Galitz, 2015.

Mukherjee, Siddhartha. *The Emperor of All Maladies: A Biography of Cancer.* New York: Scribner, 2010.

Nickelsen, Kärin. "Of Light and Darkness: Modelling Photosynthesis 1840–1960." Habilitation thesis, Phil.-nat. Fakultät der Universität Bern, 2009.

Nickelsen, Kärin, and Govindjee. *The Maximum Quantum Yield Controversy: Otto Warburg and the "Midwest-Gang."* Bern: Bern Studies in the History and Philosophy of Science, 2011.

Ohler, Norman. *Blitzed: Drugs in the Third Reich.* Translated by Shaun Whiteside. Boston: Houghton Mifflin Harcourt, 2017.

Proctor, Robert. *The Nazi War on Cancer.* Princeton, NJ: Princeton University Press, 1999.

Rigg, Bryan Mark. *Hitler's Jewish Soldiers.* Lawrence: University Press of Kansas, 2002.

Seyfried, Thomas. *Cancer as a Metabolic Disease: On the Origin, Management and Prevention of Cancer.* Hoboken: John Wiley and Sons Publishing, 2012..

Snyder, Timothy. *Black Earth: The Holocaust as History and Warning.* New York: Tim Duggan Books, 2015.

Taubes, Gary. *Good Calories, Bad Calories.* New York: Alfred A. Knopf, 2007.

Treitel, Corinna. *Eating Nature in Modern Germany: Food, Agriculture, and Environment, c.1870 to 2000.* Cambridge: Cambridge University Press, 2017.

Ullrich, Volker. *Hitler: Ascent, 1889–1939.* Translated by Jefferson Chase. New York: Alfred A. Knopf, 2016.

Warburg, Lotte, and Wulf Rüskamp. *Eine vollkommene Närrin durch meine ewigen Gefühle: Aus den Tagebüchern der Lotte Warburg 1925 bis 1947.* Bayreuth: Druckhaus Bayreuth, 1989.

Weber, Thomas. *Hitler's First War: Adolf Hitler, the Men of the List Regiment, and the First World War.* Oxford: Oxford University Press, 2010.

Werner, Petra. *Ein Genie Irrt Seltener—Otto Heinrich Warburg: Ein Lebensbild in Dokumenten.* Berlin: Akademie Verlag, 1991.

Werner, Petra. *Otto Warburg: Von der Zellphysiologie zur Krebsforschung.* Berlin: Neues Leben, 1988.

Illustration Credits

xxiii Archives of the Max Planck Society: AMPG, VI. Abt., Rep. 1, Otto Warburg, Nr. IV/14

24 Photograph by Georg Pahl, The German Federal Archives

60 Archives of the Max Planck Society: AMPG, VI. Abt., Rep. 1, Otto Warburg, Nr. I.2/1

68 Photograph by Heinrich Hoffmann, National Archives and Records Administration

100 *Top*: Archives of the Max Planck Society: AMPG, VI. Abt., Rep. 1, KWI für Zellphysiologie, Nr. I/10

100 *Bottom*: ullstein bild

112 Photograph by Heinrich Hoffmann, National Archives and Records Administration

136 ullstein bild

143 Photograph by Heinrich Hoffmann, National Archives and Records Administration

192 Photo published in the *Champaign-Urbana News-Gazette*, January 9, 1949

216 Private Collection of Frederic Burk

246 Private Collection of Frederic Burk

312 ullstein bild - Jung

313 ullstein bild

329 ullstein bild

Index

Note: Page numbers in italics refer to illustrations.

absorption spectrum, 94
Adams, Roger, 183, 184
aerobic glycolysis (Warburg effect), xviii
Africa, absence of cancer in, 25–26, 27, 28, 29, 106, 210
aging, and cancer, xix, 21–22, 148, 296
agriculture:
  development of, 82
  organic, 209
  production in, 88–89, 235, 236, 237
  and soil depletion, 235
AKT (oncogene), 261–63, 290
Aktion T4 (Nazi killing program), 159
alcohol, fermentation of grains for, 82
aldolase (zymohexase), 167–68
Allen, Frederick, 283
American Cancer Society, 270–71, 280
American Diabetes Association, 280, 315
Ames, Bruce, 219, 228–29
Ames test, 219
amino acids, 10, 48, 72, 219
ammonia, 236–37
Andrews, Peter, 319
androgens, 297
animal fat hypothesis, 310

anthrax, 153, 265
antioxidants, 294, 302–3
anti-Semitism, see Jews
apoptosis, 253–58
Arabia, absence of cancer in, 25
Arctic, absence of cancer in, 26–28, 106, 279
Ardenne, Manfred von, 268, 327
Aretaeus of Cappadocia, 277
Arrowsmith (Lewis), 11
artificial sweeteners, 310
Aryanization, via German Blood Certificate, 159–62, 166–67, 185
asbestos, 228
Asia, absence of cancer in, 28, 106, 210, 279, 309
athreptic immunity, 40–42
Atlantik Hotel, Hamburg, 110
Auler, Hans, 163

Baltzer, Fritz, 4
Banting, Frederick, 281, 293
Bauer, Julius, 284
Bayliss, William, 321
BCL-2 proteins, 253, 256
bees, epigenetic change in, 264–65, 317

beet sugar industry, 304–5, 307, 310, 323

Benson, Andrew, 220–22

benzoyl peroxide, 104, 137

beriberi, 146

Berlin, cultural and intellectual life in, 90

Berlin-Brandenburg Academy of Sciences and Humanities, 332

Berliner, Arnold, 101

Berlin Olympic Games, 125, 128

Berlin Physiological Society, 153

Berlin Wall, 268, 332

Berson, Solomon, 282–83, 315

"between ferment," 147

Bevis, Brooke, 254

bicarbonate effect, 302

biochemistry:
    building on "outdated theories" in, 252–54, 255–56, 260–62, 294
    Warburg's innovations in, 82–83, 95, 150, 253, 294

Biokhimiya, 183

biological warfare, 203

Bircher-Benner, Max, 176

Bishop, J. Michael, 245–46, 247, 248, 274

Bismarck, Otto von, 154

bladder cancer, 211

Blaschke, Hugo, 174–75, 176

Bliss, Michael, 293

Blitzed (Ohler), 324

Bloch, Eduard, 30–32

blood pressure, 315

blood sugar, *see* glucose

body fat, *see* obesity

body weight, 240

Boston Globe, 128

Bouhler, Philipp, 159–62, 171

Boveri, Theodor, 3–6, 15, 35, 46, 58, 61–62, 91, 246

Brack, Viktor, 159, 161–63, 164, 187

Brandt, Karl, 140

Braun, Eva, 177

breast cancer, xx, 289, 291, 297, 309, 310, 319

Brown, Guy, 145

Bücher, Theodor, 170–72

Bueb, Julius, 323

Burk, Dean, *192*
    on glucose and Warburg effect, 285, 287
    on insulin-cancer link, 285–87, 288
    at National Cancer Institute, 77, 193, 286
    as tireless Warburg champion, 77, 192, 193
    Warburg's communications with, 222, 245, 326
    and Warburg's publications, 230
    and Warburg-Weinhouse disagreement, 201

Butenandt, Adolf, 133, 200

Cahill, George, 287

Calle, Eugenia, 270–71

Cameron, Gladys, 198–99

Campbell, George, 279

cancer:
    and aging, xix, 21–22, 148, 296
    and asbestos, 228
    avoidable, 228
    and bad luck, 247, 297
    causes sought in, xvi-xviii, 23, 30, 91–92, 103, 137, 155, 174, 218, 223, 225–27, 232, 269, 285, 296–97; *see also specific causes*

cells transformed into, xviii; *see also* cancer cells

as common disease, xix, xxi

cure for, 247

death rates, xix, 21–23, 25, 30, 101–2, 104, 105, 218, 224, 277, 297

as defining illness of our time, xix

and diabetes, 277–87

diagnosis of, xviii, xix, 22–23, 31, 167, 174, 224–25, 296

and diet, *see* diet; food

as disease of bad information, 5

as "disease of civilization," xx, xxi, 25–30, 102–3, 104, 107, 109, 140, 210, 218, 219, 279

and energy, 245

environmental links to, xx, 103, 138, 208–10, 212–14, 228–29, 246–47

fear of, 208, 218, 324

feeding studies, 86–87, 287, 316

and fructose, 315–16, 318–21

and glucose, *see* glucose

and growth processes, 50, 285

harmless, 290

in history, xvii, xix, 28–29

"hunger cures" for, 233

increases in, xix-xxi, 21–23, 101–2, 105, 211, 218, 224–25

in indigenous populations, 25–29, 106, 279

and inflammation, 297

and insulin, 284–87, 288–300, 318

and KRAS, 296

and lifestyle, xx, 107, 210

as metaphor, 108, 140, 163, 164, 325

metastasis, 266

in migrant populations, 210

and mutations, xvii, 219, 246, 276, 288, 290, 295–96

Nazi war against, 137–38, 150, 163, 174, 185, 217, 351

and nutrition, 40, 42, 230–35, 237–40, 289

and obesity, xx-xxi, 233–35, 270–72, 289, 297, 300

and oncogenes, xvii, 245–46, 247, 249, 260, 261–63

One in Eight (film), 102, 213

paradigm of hormone dependency, 204

and PI3K, 276, 288, 289, 290, 296, 297, 299

population studies of, 25–29, 104, 106, 210, 279, 309

prevention of, 223, 230, 247, 267, 269–70, 298, 300

and PTEN, 276

and radiation, 198, 202, 214, 295, 302

research on, *see* cancer research

and respiration, xvi, 6, 84–85, 146–47, 150, 155, 199, 222–23, 267

risk factors of, 210

and sex hormones, 285, 297

shaped by chromosomes, 5, 6

signaling networks, 262, 294, 297

and smoking, 218, 223–28, 229, 232, 246, 271, 296, 322

and sugar, 307, 309–11, 318–22

and sun exposure, 296

traveling exhibit about, 102, 106

treatment of, 37–38, 42, 202–3, 205–6, 223, 230–31, 267, 269

and "tumorlike" growths, 5, 285

types and forms of, 22–23; *see also* *specific cancers*

cancer (*continued*):
  viruses in, 86, 218, 245, 297
  and Warburg effect, *see* Warburg effect
  Cancer and Diet (Hoffman), 105–6,
    107, 233
  "cancer bacillus," 91–92
cancer cells:
  in abdominal fluid, 196–97, 201,
    222, 261
  adaptability of, 269
  and catalase, 302–3
  and cell death, 258
  compared to weeds, 238
  fermentation of glucose by, xv–xvi,
    xviii, 50, 82, 84, 85–86, 260, 262,
    267, 274
  insulin receptors in, 291, 296
  metabolism of, xv, xvii–xviii, xx, 245,
    250, 268
  overeating by, xxi, 80, 258, 263–64
  starving, 268–69, 290
cancer research:
  Cantley's work in, 274–77, 289,
    318–19
  cyclical interest in, xvi–xix, 240, 250,
    253, 254, 261
  diet as focus in, 300
  genetic approach to, xviii, xx, 15,
    245–46, 297
  inducing cancer in animals, 39
  metabolic approach to, xviii, 15, 52,
    206, 223, 238, 240, 245–46, 250
  Nazi "war against cancer," 137–38,
    150, 163, 174, 185, 217, 351
  new discoveries in, 300
  on obesity, 270–72, 297
  occupational carcinogenesis, 212
  on sea urchins, *see* sea urchins

siloed fields of, xxi, 297
  Thompson's work in, 258, 260–63,
    264, 266, 269, 295
  Warburg's work in, xv–xix, 23–24,
    49, 50, 101, 106, 143–51, 160–
    61, 162, 166, 167, 171, 186–87,
    189, 195, 196–97, 198–200, 208,
    213–14, 244, 250, 266–67, 295,
    299–300, 321; *see also* Warburg
    effect
  Weinhouse's findings vs. Warburg's,
    200–202, 206, 221
Cantley, Lewis:
  cancer research of, 274–77, 289,
    318–19
  on insulin-cancer connection, 277,
    284, 287, 288–90, 294, 297, 301,
    318
  sugar studies of, 318–19, 322
  at Weill-Cornell, 258, 274, 299
carbohydrates:
  derivation of name, 305
  and diabetes, 300–301
  fat from, 315–18
  and insulin resistance, 315, 317–18
  and ketogenic diet, 298
  low-carbohydrate diet, 42, 298,
    300–301
  and sugar, 304–5; *see also* fructose;
    glucose; sucrose; sugar
  and tumor growth, 42, 237
carbon dioxide, 82, 220
carbon monoxide, 93–94
carcinogens, *see* cancer, causes sought in
Carlsberg Laboratory, Copenhagen, 197
Carson, Rachel, *Silent Spring*, xvi,
    212–14, 217, 218
Cassel, Simon von, 128–29

catalase, 302–3
"Causes of Cancer, The" (Doll and
  Peto), 228, 229, 231, 232, 233,
  239, 270
cells:
  and apoptosis, 253–58
  "braking systems" of, 276
  chemical bonds to, 34, 35–36
  as chemical laboratories, 11
  "death by neglect," 257
  death of, 253–58
  differentiated, 199
  entire genome carried in DNA of, 264
  epigenetic changes in, 264–65
  epithelial, 52, 263, 294
  eukaryotic, 255
  fermentation by, xv-xvi, xviii, 82, 85,
    222
  fuel for, xvii, xviii, 40, 42, 44, 80
  growth by division, 4
  growth factors of, 257–58, 259, 267,
    297
  insulin resistant, 283
  life or death of, 256
  metabolism of, xv, xxi, 223, 245, 250
  in multicellular organisms, 258
  mutations in, xvii, 219, 247, 251,
    262, 290, 294, 295, 296
  and natural selection, 265
  organized death ("programmed cell
    death") of, 253, 257
  overeating, 262–63, 267, 270
  reprogramming, 296
  respiration of, 15, 16, 50, 81, 93–95,
    101, 145–46, 150, 214, 267, 319
  transformed to cancer, xviii; *see also*
    cancer cells
  undifferentiated, 199

Center for Disease Control, 271
certainty, dangers of, xxii
Chamberlain, Houston Stewart, 58–59
Champ, Colin, 298
Chance, Britton, 260–61
Chancellery of the Führer, 159, 160,
  161, 171
Chandel, Navdeep, 255, 256, 257
Charles, Daniel, 235–36
chemicals:
  artificial, xvi, 23, 92, 137, 138, 198,
    210–14, 218–19, 227, 228–29,
    310
  and DuPont, 211
  synthetic dyes, 36, 39, 62, 144, 211
chemotherapies:
  and catalase expression, 303
  coining of term, 34
  Ehrlich's development of, 34, 35–38,
    51, 167, 203
  how they work, 205–6
  and "magic bullets," 36, 205
  specificity sought in, 37–38, 144,
    203
  Vander Heiden's study of, 269
Chernow, Ron, 129
Chicago Sunday Tribune, 12
Chile, nitrogen in, 236
Chi Van Dang, 247–50, 252, 258, 262,
  268–69, 296
cholera, germ as cause of, 154, 213,
  226
cholesterol, 315
Christian, Walter, 167
Christofk, Heather, 298
chromosomes:
  and cancer, 5, 6, 246
  copies of, 4–5

chromosomes (*continued*):
  new life formed from, 5
  studies of, 4, 6, 35
Civil Service Law, 114–17, 125
Clean Air Act, 212
Clean Water Act, 212
coenzymes, 146–48, 149–50, 151, 171,
  231, 267
Cohen, Aharon, 279–80
colorectal cancer, xx, 289, 309, 319
Committee on the Treatment of War
  Gas Casualties, 205
concentration camps:
  absence of cancer in, 174
  Auschwitz, 186
  Dachau, 141–43, 209
  extermination programs in, 159,
    169, 176, 204, 205, 208, 323
  gardens in, 141–42
  Jews sent to, 116, 141, 168, 170
  medical experiments conducted in,
    143, 186
  political prisoners in, 114, 123, 141
  Sobibor, 169
consumption (TB), germ as cause of,
  153–54, 157, 213
Copernicus, 13
corn: high-fructose corn syrup, 310, 318
Correns, Carl, 62
Crick, Francis, xvii, 244
cyanide, 48, 93, 323
cytochrome c, 255–56
"cytochrome c oxidase" (respiratory
  ferment), 294

Dachau concentration camp, 141–43, 209
Dahlem, Germany:
  Kaiser Wilhelm Institute in, ix, 187,
    195–96

Warburg's postwar life in, 208, 245,
  303
Dang, Chi Van, 247–50, 252, 258,
  262, 268–69, 296
Dang, Mary, 247
Darwin, Charles, 59, 88
Darwinian competition, 296, 327
Davidson, Norman, 133
DDT, 218
Deelman, H. J., 107
Delbrück, Max, 186
Denman, Thomas, *Observations on the
  Cure of Cancer,* 232
Der Spiegel, 223, 245
Der Stürmer, 140
Dessau Works for Sugar and Chemical
  Industry, 323
diabetes:
  and cancer, 277–87
  and diet, 279, 297–98, 300–301
  and insulin, 277–81, 283–84, 287,
    316
  and overweight, 280
  and population studies, 279–80
  rise in deaths from, 277
  and sugar, 307–9, 321
  type 1 diabetes, 280, 281, 282–83
  type 2 diabetes, 280, 281, 282, 283,
    284, 287, 293, 316
  and Western lifestyle, 279
diet:
  animal fats and meat, 310
  and cancer, 86, 103–4, 105–7, 138–
    41, 175, 210, 228, 230–40, 246,
    270–72, 289, 298–300, 309
  and diabetes, 279, 297–98, 300–301
  disease cured by, 149
  and growth, 238
  and insulin, 297–301

ketogenic, 299
low-calorie, 233
low-carbohydrate, 42, 298, 300
as medicine, 299
nutritional deficits in, 104
overeating, 232–35; *see also* obesity
and quackery, 298
sugar in, 309
vegetarian, 108, 109, 110, 139–40,
174–75, 209
disease:
causes of, 153–55, 156, 213, 218,
226
foodborne, 296
germ theory of, 30, 152–57, 213, 226
spread by contaminated air, 152
and sugar consumption, 307
see also specific diseases
DL-glyceraldehyde, 268
DNA:
and cancer-causing genes, 245,
295–96
damaged, 5
in every cell, 264–65
genes and gene products of, 360
replication of, 206, 247
structure of, xvii, 244
DNA code, 360
Dobzhansky, Theodosius, 50
Doctor Faustus (Marlowe), 1, 15, 69,
179, 241
Dohrn, Anton, 4
Dohrn, Antonietta, 158
Doll, Richard, 223–29, 322
and Hill, on smoking and cancer,
225–28
and Peto, "The Causes of Cancer,"
228, 229, 231, 232, 233, 239, 270
on sugar and cancer, 309

doppelgänger, 314–15
DuPont laboratories, 211
dwarfs, studies of, 292

Eastman Kodak Company, 149
eating, unrestrained, xviii; *see also*
overnutrition
eggs, development of, 6
Ehrlich, Paul, 33–38, 153–54, 287,
291
and athreptic immunity, 40–41
and chemical dyes, 36, 39, 62, 104,
144, 211
chemotherapy developed by, 34,
35–38, 51, 167, 203
death of, 62
and hormones, 284–87
and immune system, 40, 41
influence on Warburg of, 37, 152
living in two worlds, 41
magic bullets sought by, 36–38, 40,
41, 144, 154, 203, 211, 299
"nucleus of truth" for, 40
syphilis treatment developed by, 41
on transplanted tumors, 39–41, 86
Eichmann, Adolf, 157, 165
Einstein, Albert, 90, 199, 200, 306
and Civil Service Law, 116–17
discoveries of, 49, 191, 222
and Germany's descent into Nazism,
79, 116–17, 121
on personal possessions, 99, 101
and Planck, 114
and Warburg's departure from
military service, 63–64
and Warburg's family, 7, 8, 63
and Warburg's father, 7, 9, 79, 191
on Warburg's relationships with
colleagues, 91

Eitel, Wilhelm, 186
electron transport chains, 294, 295, 296
Elizabeth of Austria, Empress, 51
Elon, Amos, 57
Elsinore castle, and *Hamlet,* 197–98
Elvehjem, Conrad, 149
Emergency Association of German
    Science, 73
Emerson, Haven, 307–8, 322
Emerson, Ralph Waldo, 188
Emerson, Robert, 187–93, *192,* 197,
    201, 222, 307
Emperor of All Maladies, The
    (Mukherjee), xvii, 29, 298–99
Endangered Species Act, 212
endocrinologists, 297
endometrial cancer, 270
energy:
    cell's use of, 50, 86
    generation of, xvi
    transfer from light to matter, 49,
        187, 190–91
Enlightenment, 57, 58
environmental movement, 208–10,
    212–14, 218, 228–29, 232
Environmental Protection Agency, 218
enzymes:
    and cellular breathing, 145–46
    coenzymes, 146–48, 149–50, 151,
        231, 267
    cytochrome c, 255–56
    in fermentation, 167
    "housekeeping," xvii
    lactate dehydrogenase (LDH), 249–50
    and metabolism, 144, 262, 269
    "respiratory ferment," 45, 93, 145–
        46, 151, 206
    "Warburg's yellow enzyme," 146–47

epidemiology, 226–29, 231
epigenetic change, 264–65, 317
epithelial cells, 52, 263, 294
esophageal cancer, 289
estrogen, and cancer, 285, 297
"Eternal Jew, The" (mythical folkloric
    figure), 134–35
eugenics, 103
eukaryotic cells, 255
evolution, 50, 145
evolution in reverse, 199

fat, body, *see* obesity
fat "overload hypothesis," 317–18
fat production, 316–18
fats, structure of, 315
Faust (Goethe), 101
Faust, metaphorical figure of, 15, 134,
    299, 328
"ferment," Warburg's use of word, 45
fermentation, 82–87
    in agriculture, 82
    as backup process, 83
    carbon dioxide emitted in, 82
    by cells, xv-xvi, xviii, 82, 85, 222
    energy of, 84, 263
    and fructose, 319
    of glucose, xv-xvi, 50, 82, 84, 85–86,
        167, 251, 260, 267
    LDH in, 249
    as "life without air," 83
    and MYC, 250
    and nicotinamide, 148
    and overeating, 239–40, 267
    respiration vs., 83, 84–85, 144, 155,
        196–97, 199, 200–202, 267
    scientific studies in, 143, 249
    sugar's support of, 305

Fibiger, Johannes, 91, 92
"Fight Against Cancer, The" (traveling
    exhibit), 102
Fischer, Albert, 238
Fischer, Emil, 9–11, 20, 43
    on amino acids, 10, 72
    and carbohydrates, 304, 305
    death of, 91, 307
    and sugars, 304–7, 310
    on unlimited possibilities in science,
      10–11, 15
    Warburg's studies under, 9–10, 72
foie gras, 316
food:
    additives in, 104, 137, 141, 208,
      210, 219, 231
    availability of, 86–87, 265–66
    burning without fire, 94
    in combinations, 231–32
    consumption of, 44, 240
    cooked, 109, 175, 231
    labeling of, 215
    molecules from, 44–45
    natural, 175–76, 209, 231
    organically grown, 209
    and photosynthesis, 62, 72, 195, 235
    preservatives in, 104, 106, 208, 210
    production of, 88–89, 142, 195
    and respiration, 44
    vitamins in, 146–47
    "whole," 210
    see also diet
Food and Health (Plimmer and
    Plimmer), 307
Fouché, F. P., 26
Franck, James, 7–8, 190, 193–94, 326
Franklin, Rosalind, xvii, 244
Franz Ferdinand, Archduke, 54

free radicals, 294, 295
Free University of Berlin, 245
Freon, 211
Freund, Ernst, 51–52, 87, 238–39
    Metabolic Therapy of Cancer, 238, 277
Friedrich, Crown Prince, 238
Friedrich III, Kaiser, 51, 52–53
fructose, 306–7, 308, 310–11, 315–16,
    318–21
Fuller, Clint, 220–22

Gairdner International Award, Canada,
    318
Galen, 44, 277
gallbladder cancer, 270
Gardening and Ploughing without Poison
    (Seifert), 209
gas:
    cyanide gas, 323
    in Holocaust gas chambers, 159,
      176, 204, 205, 323
    mustard gas, 204–6
    nitrogen gas, 236
    in warfare, 61, 73, 114, 203, 204,
      205, 211, 235
    Zyklon B, 177, 204, 205, 323
gene expression, 265, 296
"gene products," use of term, 360
genes:
    allowing cells to eat, 259
    DNA molecules in, 244
    mutations in, 246, 270, 288
    oncogenes (cancer-causing genes),
      xvii, 15, 245–46, 260, 261–63
    tumor suppressors, 263, 276
    use of term, 360
gene sequencing, 299
genetics, 244, 297

Gentz-Werner, Petra, 187
Georgius of Helmstadt, 15
German Armed Forces, High
    Command, 166
German Blood Certificates, 159–62,
    166, 185
German Physical Society, 117
German Society for Cancer Research,
    105
Germany:
    anti-Semitism in, *see* Jews
    cancer deaths in, 21–23, 102
    cancer research in, 137–38, 160,
        166, 245
    Civil Service Law in, 114–17, 125
    diet in, 103–4, 209, 235
    economy of, 90, 113
    fear of cancer in, 208, 324
    "first guilt" and "second guilt" of,
        208
    food production in, 88–89, 142
    homosexuality as crime in, x, 76, 170
    industrialization of, 58, 78, 102, 107
    inflation in, 78
    international reputation of, 125
    Kristallnacht in, 133–34
    land (*Lebensraum*) sought by, 88–89,
        163
    nationalism in, 38, 58, 65
    Nazis in, *see* Nazi Party
    nitrogen in, 235–37
    Nuremberg Laws, 124–26, 157, 162
    Post-World War II years in, 181
    post-World War I years in, 71–73,
        78–79
    reparations payments by, 78, 79
    Romanticism in, 58, 102–3, 106,
        135, 214
    science as preeminent in, 20–21, 33,
        92
    Soviet Union invaded by, xv, 163–64
    synthetic dyes in, 36, 62, 211
    "the Wandering Jew" (folklore) in,
        134–35
    and Treaty of Versailles, 72, 78, 79
    uhlan (cavalry) of, 59, *60*, 61, 72,
        120
    Volkssturm in, 172–73
    and World War I, 53, 54, 56–57,
        59–62, 114, 203, 211
    and World War II, 53, 89, 172–73,
        174, 185
germs, fear of (germaphobia), 156
germ theory of disease, 30, 152–57,
    213, 226
Gilman, Alfred, 205
Giordano, Ralph, 208
Giovannucci, Edward, 289
glucose:
    basic recipe of, 305
    as blood sugar, 308, 311, 315
    cancer cell consumption of, xv, xviii,
        42, 85, 199, 251, 262, 272
    carbohydrates broken down to, 42,
        237
    and diabetes, 280–81, 282–85
    in DNA and protein synthesis, 285
    elevated levels in blood, 52, 272
    and epigenetic change, 265, 266
    fermentation of, xv–xvi, xviii, 50, 82,
        84, 85–86, 251, 260, 262, 267,
        274
    and fructose molecules, 306–7, 308,
        311, 315
    and insulin, 284
    metabolism of, 296

in photosynthesis, 49
and Warburg effect, 263, 285, 287
glucose deprivation, 268
Glum, Friedrich, 96, 129
glutamine, 269, 296
glycolysis, xviii, 126
Gobineau, Joseph Arthur de, 58
Goebbels, Joseph, 108, 140, 163
Goethe, Johann Wolfgang von, *Faust*, 101
goiters, 104
Goldberger, Joseph, 149
Goldblatt, Harry, 198–99, 252
Goldfeder, Anna, 237
Goodman, Louis, 205
Göring, Hermann, 161, 162–63, 168
Graham, Evarts, 225, 227
Great Depression, 323
Greenland, Sander, 232
guano deposits, 236
Gutmann, Hugo, 67

Haas, Erwin, 119, 120
Haber, Fritz, 7, 121
ammonia developed by, 236–37
death of, 204
and gas warfare program, 114, 203, 204, 205, 211, 235
resignation from Kaiser Wilhelm Society, 114, 203–4
Haffner, Sebastian, 114
"Hallmarks of Cancer, The" (Weinberg), 251
Hammarsten, Einar, 99
Harris, Henry, 331–32
Hatzivassiliou, Georgia, 264
Hegel, Georg Wilhelm Friedrich, 88
Heisenberg, Werner, 128

Heiss, Jacob, *216*
and customs officials xi, xii
death of, 332
lifestyle of Warburg and, 97–99, 167, 169, 172, 173, 208, 214, 215, 243, 245
move to U.S., 188, 189, 190
as Warburg's assistant, 75–77, 193, 194, 197, 221, 303
and Warburg's death, 331–32
hemoglobin, 40, 93
Heroes of Civilization (radio), 127–28
Herzl, Theodor, 168
Hess, Rudolf, 140, 209
Heubel, Emil, 33–34, 35
Hill, Austin Bradford, 225–28
Hill, A. V., 93–94
Himmler, Heinrich, xiv-xv, 161, 162, 164
and Dachau, 141–43
death of, 174
health issues of, 140
interest in diet, 140–42, 176, 209
and Nazi war against cancer, 174
Hindenburg, 145
Hindenburg, Paul von, 113, 116
Hippocrates, 110
Hitler, Adolf, *68, 112, 143*
anti-Semitism of, 79, 88, 114–15, 122, 124–25, 156, 157, 176, 324
and Aryanization, 160, 162
birth of, 155, 325
and cancer, xxi, 107, 108, 138, 140, 155, 162, 324–25
diet as focus of, 138–40, 142, 174–75, 209
fear of death, 108
final days of, 174–77
fits of rage in, 31, 64, 115

Hitler, Adolf (*continued*):
  and German nationalism, 65
  as Germany's "first guilt," 208
  and germ theory of disease, 155, 213
  health issues of, 176
  and his mother's illness and death,
    30–32, 64, 67, 108
  homelessness of, 64–65
  hypochondria of, 108–11, 138–39,
    155–56
  instability of, 115, 138, 324
  Iron Cross awarded to, 66
  and Koch, 154, 155–57
  Lebensraum sought by, 88–89, 163,
    237
  madness of, 324
  Mein Kampf, 66, 90, 108, 138, 155,
    156, 204
  Nobel Prize forbidden to Germans
    by, 171
  and Nuremberg Laws, 124, 125, 162
  and Operation Barbarossa, 163–64
  and Planck, 113–15
  public speaking by, 87
  rise to power, 57, 87–88, 111, 113–
    14, 127, 217
  as smoker, 227–28
  sugar addiction of, 323–25
  vague beliefs of, 87–89
  vulgarity of, 119, 121
  and Warburg, xiv, 116, 163–64, 183
  and World War I, 65–67, 89, 204
  and World War II, 172
Hitler, Klara (mother), illness and death
  of, 30–32, 64, 67, 108
Hoffman, Frederick, xix, 104–7
  Cancer and Diet, 105–6, 107, 233
  on diabetes, 278

*The Mortality from Cancer throughout
  the World*, 24–25, 104
  racism of, 25
Holocaust:
  denial of, 208
  as "Final Solution," 165–66
  and Kristallnacht, 133–34
  start of, 164
  see also concentration camps
Hopkins, Sir Frederick Gowland, 198
Hopkins Seaside Laboratory, Stanford
  University, 54
hormones, 284–87, 289
  growth hormone, 292
  IGF-1, 291–92, 297
  paradigm of hormone dependency of
    cancer, 297
Hrdlička, Aleš, 26, 28
Hueper, Wilhelm C., 210–13
  on artificial chemicals, 211–13, 227
  influence of, 212–13, 214, 217
  Nazi sympathies of, 217, 218
  *Occupational Tumors and Allied
    Diseases*, 212
  racism of, 217–18
Huggins, Charles B., 285
Hunter, Jehu, 286
Hutton, Samuel, 27
hydrogen, activation of, 145–46
Hygiene Museum, Dresden, 102
hyperinsulinemia, 284, 292, 294, 300

IDH-2 enzyme, 269
IGF-1 (insulin-like growth factor 1),
  291–92, 297
immortality, possibility of, 39
immune system:
  athreptic immunity, 40–42

in cancers of the blood, 205
and mustard gas attacks, 204–5
and organized death of cells, 252–53, 257
and transplantation failures, 39, 40
immunology, 37, 252, 257, 258
Imperial Physical and Technical Institute of Germany, 47, 49
inflammation, 297, 317
Institute for the History of German Jews, 166
insulin:
and cancer, 284–87, 288–300, 318
and diabetes, 277–81, 283–84, 287, 316
and diet, 297–301
discovery of, 281, 293
and epithelial cells, 294
excess, 292
and fat "overload hypothesis," 317–18
and glucose, 284
as growth factor, 289, 290, 291–92, 297
hyperinsulinemia, 284, 292, 294, 300
and IGF-1, 291–92, 297
and metabolism, 287
and obesity, 283–84, 289
and Warburg effect, 285, 287, 290, 321
insulin resistance, 283, 315, 317–18, 320, 323
insurance companies, life expectancy data of, 234–35
International Congress of Physiology, Zurich, 132

Japanese Americans, internment of, 188
J.D. (Polish victim of non-Hodgkin's lymphoma), 205
Jews:
anti-Semitism, 57–58, 62, 78–79, 88, 101, 108, 114–17, 118, 122–27, 128, 137, 140, 151, 156–57, 170, 176, 217, 224, 227, 324
assimilation of, 58, 157, 160, 165
attempts to prove "Aryanhood," 160–61
and Civil Service Law, 114–16, 117, 125
in concentration camps, *see* concentration camps
conversion to Christianity required of, 57, 114, 292
deportations of, 157, 171, 174
and Enlightenment thinking, 57, 58
and "Final Solution," 165–66
fleeing Germany, 117, 120, 122, 149, 151, 174, 204
forced out of jobs, 114–16, 119–20, 126, 157, 158, 174, 237
and Holocaust, 164, 165–66
and *Kristallnacht*, 133–34
legal cleansing of Jewish blood via German Blood Certificates, 159–62, 166, 185
Mischlinge, 125–26, 151, 157–58, 160, 162, 165–67, 170, 185, 208
as "non-Aryans," x, 58, 124–25
and Nuremberg Laws, 124–26, 157, 162
and Passover, 82
and Protocols of the Elders of Zion, 79
restrictions on, 11, 57, 114, 124–26, 132–33, 157–58, 217

Jews (*continued*):
and Romanticism, 58, 106, 135
"*Schutzjuden*" (protected Jews), 129
Sephardic, 292
sterilizations of, 157, 161, 165
traveling exhibit about, 134
"the Wandering Jew" (folklore),
134–35
and World War I, 56, 58, 62, 64
and Zionism, 56, 131, 168, 315
Johnson, James, 292–93
Johnson, Richard, 319–20
Jokl, Ernst, 331
Joslin, Elliott, 278, 308, 311, 322
Journal of Medical Research, 42

Kaaks, Rudolf, "Nutrition, Hormones,
and Breast Cancer," 289
Kaiser Wilhelm Institute for
Anthropology, 130, 186, 244
Kaiser Wilhelm Institute for
Biochemistry, 127, 133, 186
Kaiser Wilhelm Institute for Biology,
72–75, 96, 203
Kaiser Wilhelm Institute for Cell
Physiology:
bomb damage to, 168, 172
building of, 97, *100*
funding of, xiii, xiv, 96–97, 98, 101,
116, 121–22, 123, 127, 157, 185
Nazi customs official visits to, ix–xiii
relocation of, xiv, 168–69, 173
renamed Max Planck Institute for
Cell Physiology, 207, 331–32
returned to Warburg by US military,
195–96
Vennesland as Warburg's successor in,
325–27

Warburg as director of, ix-xiii, 46,
50, 73, 158, 165, 216
Warburg dismissed by Nazis from,
xiv, 158, 161, 186
and Warburg's death, 331–32
as war institute, 172–73
Kaiser Wilhelm Institute for Medical
Research, 132
Kaiser Wilhelm Institute for Physical
Chemistry, 203
Kaiser Wilhelm Society, 45–46
chemical weapons made by, 61
and Civil Service Law, 113–16, 117
founding of, 46, 51
Jewish scientists in, 118–19, 174,
186–87
and Nuremberg Laws, 127
Planck as president of, xiii, 113, 129
renamed Max Planck Society, 207
Rockefeller Institute grant to, 96,
101
Telschow as chairman of, 162
Warburg's colleagues and employees
in, 88, 91, 132, 133, 170–73, 174,
185–87, 194, 207–8, 325–27
Warburg's findings discussed in,
150–51
Kalckar, Herman, 197–98
Kaman, Martin, 191
Keilin, David, 93
Kempner, Walter, 119–20, 191–92
Kennedy, John F., 212
Kershaw, Ian, 164
ketogenic diet, 299
kidneys, cancer of, 289, 309
kinase, 276
Klein, George, 196, 197
Klingenberg, Martin, 196

Koch, Robert, 152–57, 176, 213, 223, 226, 265
Koch's postulates, 154–55
Kollath, Werner, 209–10
Kornberg, Arthur, xvi
KRAS (cancer-linked gene), 296
Krebs (German word for cancer), 110
Krebs, Albert, 109–11
Krebs, Hans, 252
  departure to England, 117
  on Warburg-Heiss relationship, 77
  on Warburg's attachment to science, 19, 95, 223
  on Warburg's attempt at Aryanization, 159–60, 161, 167
  on Warburg's influence, 75, 223
  on Warburg's personality, 17, 19, 75
  on Warburg's postwar life, 207–8, 215, 327
  on Warburg's reluctance to leave Germany, 122
  working with Warburg, 74–75, 90, 95, 303
Krebs cycle, 117
Krehl, Ludolf, 43, 327
Kristallnacht, 133–34
Kubizek, August, 31, 64, 107
Kubowitz, Fritz, 170, 172, 173
Kuhn, Richard, 132

lactate dehydrogenase (LDH), 249–50
lactic acid, 82, 84, 85
Lane, Nick, 256, 295
La Presse, 152
Laron syndrome, 292
Laue, Max von, 114
Lavoisier, Antoine, 44, 99

Law for the Protection of German Blood and German Honor, 124
Law for the Restoration of the Professional Civil Service (Civil Service Law), 114–17, 125
LDH (lactate dehydrogenase), 249–50
lead poisoning, study of, 34–35
League of Nations, 116
Leavitt, George B., 27
Lefèvre, Wolfgang, 244–45
LeRoith, Derek, 290
leukemia, 205, 269
Lewis, Sinclair, *Arrowsmith,* 11
Leyen, Ernst von, 33
Liebenberg:
  postwar Soviet takeover of, 182, 186
  Warburg lifestyle in, 172–74
  Warburg's institute relocated to, 168–69, 170, 171
Liebig, Justus von, 317
Liek, Erwin, 103–4, 106–7, 109, 175, 210
life:
  beginning of, 14
  mechanism of, 95
life expectancy, xix, 21–23, 234–35, 279
light, absorption of, 94–95, 148
Linge, Heinz, 177, 324
Lipmann, Fritz, 92
lipodystrophy, 318
liver:
  fat accumulations in, 317
  fructose in, 319
Loeb, Anne, 12, 14
Loeb, Jacques, 11–14, 38
  ambitions of, 12, 13, 32
  anti-Semitic attacks on, 59

Loeb, Jacques (*continued*)
  death of, 13
  experiments on sea creatures, 11, 12,
    18, 78, 79
  influence of, 14, 15, 43
  and instability of German life, 78, 79
  research on insects, 55–56
  and Rockefeller Institute, 45
  and Stanford University, 54–55
  The Mechanistic Conception of Life, 14
  and Warburg, 13–14, 88
  and Warburg's colleagues, 91
  on war propaganda, 55–56
Loeb, Leo, 38–39, 285
Longo, Valter, 268
Lösener, Bernhard, 126
Ludwig, David S., 283
Ludwig Institute for Cancer Research,
  247
lung cancer, 224–27, 291, 319, 322
lymphoma, 205

"magic bullets":
  Ehrlich's search for, 36–38, 40, 41,
    144, 154, 203, 211, 299
  ongoing search for, 247, 299
  targeted chemotherapies as, 36,
    37–38, 144, 203, 205
  use of term, 36, 203, 299
malaria:
  and DDT, 218
  and methylene blue, 37
malignant neoplasm, 24
Manhattan Project, 233
manometers, 80–82, 95, 117, *192,*
  196, 245, 261
Manziarly, Constanze, 324
Marggraf, Andreas, 304

Margulis, Lynn, 254
Marine Biological Station, Woods Hole,
  193–94
Marksman, The (opera), 36, 299
Marlowe, Christopher, *Doctor Faustus,*
  1, 15, 69, 179, 241
Massachusetts General Hospital, 277
Max Planck Institute for Cell
  Physiology, 207, 331–32
  renamed Otto Warburg House, 332
  see also Kaiser Wilhelm Institute for
    Cell Physiology
Max Planck Society, 207, 331–32
  Archives of, 332
  see also Kaiser Wilhelm Society
McCarrison, Sir Robert, 28, 279, 309
Mechanistic Conception of Life, The
  (Loeb), 14
Mein Kampf (Hitler), 66, 90, 108, 138,
  155, 156, 204
Memorial Sloan Kettering Cancer
  Center, 252
Mengele, Joseph, 186
Mentzel, Robert, 158, 166, 217
Metabolic Therapy of Cancer (Freund),
  238, 277
metabolism:
  and apoptosis, 256
  of cancer cells, xv, xvii–xviii, xxi, 245,
    250, 268
  of entire body, 238
  influences on cells, 256, 265
  and insulin, 287
  outdated research of, xvii, 261
  and PI (fat molecule), 274
  revival of research in, 262–64,
    269–71
  role of enzymes in, 144, 262

synthesis of, 240

Thompson's work in, 256, 258–59, 260, 261

Warburg's studies of, xvii-xix, 15, 117, 206, 272, 321

metastasis, 266

metformin, 293

methylene blue, 37

Meyer, Beate, 166

Meyerhof, Otto, 92, 122

  escape to France, 133

  on fermentation, 84

  job vulnerability of, 78–79, 127, 132

  on respiration of sea urchins, 47

  and Warburg's mental state, 18

microbes:

  and germ theory of disease, 30, 152–55, 226

  nutrients for growth of, 40

  as possible cause of cancer, 91

Mischlinge:

  Hitler's review of applications for, 162

  and postwar Germany, 208

  Warburg's status as, 125–26, 127, 151, 157–58, 160, 165–67, 170, 185

  and World War II, 185

mitochondria:

  apoptosis driven by, 253–57

  energy production by, xv, 264

  and gene expression, 296

  in old biochemistry textbooks, 254–58

  origins of, 254, 295

  as Warburg's "grana," 253–54

modernity, cancer linked to, xx, 30, 102–3, 104, 210–14

molecular biology:

  and metabolism revival, 260–64

  new era of, xvii, 244, 247, 248, 249, 254, 259

Morell, Theodor, 139

Moreschi, Carlo, 41–42

Mortality from Cancer throughout the World, The (Hoffman), 24–25, 104

mTOR proteins, 262

muesli, 176

Mukherjee, Siddhartha:

  on chemotherapy, 203

  and ketogenic diet, 299

  The Emperor of All Maladies, xvii, 29, 298–99

Müller, Franz H., 227

Müller, Karl Alexander von, 87

mustard gas, 204–6

Mutaflor pills, 139

mutation:

  arising by chance, 247

  and cancer, xvii, 219, 246, 276, 288, 290, 295–96

  of cells, xvii, 219, 247, 251, 262, 290, 294, 295, 296

  and chemotherapy, 299

  genetic, xvii, xx, 264, 269, 270, 292, 320

  and insulin, 294, 296

  in oncogenes, 246, 247

  in pathways, 276

  and radiation, 295

MYC (oncogene), 249–50, 296

Nachmansohn, David, 8, 95, 119, 216–17, 251–52

NAD; NADP, 148

Naples Zoological Station, 4, 158
  sea urchin research in, 5, 6, 13, 14, 15, 43
  tubularia studies in, 11
  and World War I, 61
Napoleon Bonaparte, 304
National Cancer Institute, 77, 193, 212, 213, 286
National Institutes of Health, 217
Native Americans, absence of cancer among, 26, 28
Natural History of Cancer, The (Williams), 278
Nature, 265
nature:
  "back to nature" movement, 214
  as highest ideal, 88, 89, 102, 109, 175, 177
  laws of, 89
  toxic chemicals in, 228–29
  unity of, 84
Nazi Party:
  and Aktion T4 (systematic killing program), 159
  anti-Semitism of, x, 116, 122, 123, 124, 126–27, 128, 137, 170, 227
  anti-smoking campaigns of, 227
  on "asocial" behavior, 170, 171–72
  and beet sugar industry, 323
  concentration camps of, *see* concentration camps
  "declaration of Aryan descent" forms required by, ix-xiii, 118
  disabled persons to be eliminated by, 159
  fall from power, 90
  gas chambers of, 205, 323
  genealogical registry of, 130
  homosexuality punished by, x
  and *Kristallnacht,* 133–34
  medical experiments on prisoners of, 143, 186
  pollutants to be eradicated by, xxi, 138, 209
  postwar return to German life, 208
  postwar trials of, 187, 208
  propaganda spread by, 128
  racism of, 227
  rise to power, 113, 118
  SS death squads, 164
  substance abuse in, 324
  and "the Eternal Jew" (metaphor), 135
  vulgarity of, 119, 123
  and war against cancer, 137–38, 150, 163, 174, 185, 217, 351
Nazi War on Cancer, The (Proctor), 163, 217, 351
Neely, Matthew, 105
Negelein, Erwin, 74, 173
Nernst, Walther, 7, 21, 47
Neuberg, Carl, 127, 186–87
New England Journal of Medicine, 271
New Reich Chancellery, xiv, 161
Newton, Isaac, 306
New York Times:
  on cancer therapies, 52, 150
  on diabetes, 281
  on Kaiser Wilhelm Society, 127
  on Koch, 154
  on Loeb, 13
  on pellagra victims, 149
  on synthetic chemicals, 219
  on Warburg, 189, 195, 201–2
New York Times Magazine, 316
nicotinamide, 148, 149–50, 159

nicotinic acid (niacin), 149–50

Nietzsche, Friedrich Wilhelm, 88

Ninkasi (goddess of beer), 82

Nishimura, Shimpe, 188

nitrogen, 235–37

Nonnevitz, Rügen, Warburg's vacation home in, 98–99

Norman (Great Dane), 243, 331

nose, cancer of, 322

Nuremberg Laws, 124–26, 157, 162

Nuremberg trials, 187

nutrients, roles of, 264, 265

nutrition, 40, 42, 230–35, 237–40, 289;
  see also diet; food

"Nutrition, Hormones, and Breast Cancer" (Kaaks), 289

obesity:
  and cancer, xx–xxi, 233–35, 270–72, 297
  and carbohydrates, 300–301
  and insulin, 283–84, 289, 300
  insurance company data on, 234–35
  and nutrition studies, 234, 237
  and overeating, 233–35, 237
  and sugar, 321

Observations on the Cure of Cancer (Denman), 232

Occam's razor, 320

Occupational Tumors and Allied Diseases (Hueper), 212

Ohler, Norman, *Blitzed,* 324

oncogenes, 245–46, 247, 249, 260, 261–63
  AKT, 261–63, 290
  cancer caused by, xvii, 245–46, 260
  identification of, 245

and LDH, 249, 250

as mutations, 246, 247

MYC, 249–50, 296

and PI, 287

SRC, 247, 274

and Warburg effect, 262

oncology, 247

One in Eight (film), 102, 213

"On the Origin of Cancer Cells" (Warburg), 286

Operation Barbarossa (German invasion of Soviet Union), xv, 163–64

Operation Otto, 164

Ostendorf, Peter, 76, 243–44, 303, 331

Otto Warburg House, 332

ovarian cancer, xx, 309

overnutrition, xx–xxi, 232–35, 237, 239

oxygen:
  burning glucose with, 85–86, 87, 274
  and photosynthesis, 220
  reaction between iron and, 144
  reaction of food with, 93
  reaction of hydrogen with, 145
  reactive oxygen species, 294–95
  and respiration, 199, 200–201
  and respiratory ferment, 145

pancreas:
  fat accumulation in, 317
  insulin secretions from, 280–81, 283, 317

pancreatic cancer, xx, 289, 292, 319

paraffin breast implants, 211

Pasteur, Louis, 176, 230
  on elimination of infection, 40
  on fermentation, 83, 84, 85–86, 267

Pasteur, Louis *(continued)*
and germ theory of disease, 30, 156, 213, 226
influence on Warburg, 23–24, 33, 37, 85, 152, 155
Pasteur effect, 83
Paul, Jean, *Siebenkäs,* 315
Pavlov, Ivan, 56
Peary, Robert, 27–28
Pederson, Peter, 245, 258
pellagra, 149–50
Perkins, John, 323
Perkins, William, Henry, 36
pesticides, 137, 177, 211, 214, 218
PET (positron-emission tomography) scans, xviii
Peto, Richard, 210, 228–29
and Doll, "The Causes of Cancer," 228, 229, 231, 232, 233, 239, 270
phenylhydrazine, 306–7, 310
phosphate (phosphorus-oxygen bond), 275
phosphatidylinositol (PI), 274–77, 287
photosynthesis:
disputes about findings in, 17, 49–50, 187–88, 190–91, 192–93, 220–22, 326
and food production, 62, 72, 195, 235
precise measurements of, 21, 190–91, 193
as respiration in reverse, 50
transfer of energy from light to matter in, 49, 187, 190–91
Warburg's work in, 15, 21, 49–50, 62, 72, 195, 208, 244, 286, 325
PI3K, 276, 288, 289, 290, 296, 297, 299
PIP$_3$, 275
plague, causes of, 155

Planck, Max, 7, 20, 49, 130, 133, 326
and Civil Service Law, 114–16
and Hitler, 113–15
as Kaiser Wilhelm Society president, xiii, 113, 129
Plimmer, Robert and Violet, *Food and Health,* 307
Pollak, Michael, 290, 297–98, 320
polypeptides, 72
population growth, 235, 236–37
Prime Cause and Prevention of Cancer, The (Warburg) "Lindau Lecture," 230
Princip, Gavrilo, 54
Proctor, Robert, 138, 218
The Nazi War on Cancer, 163, 217, 351
prostate cancer, xxi, 271, 289, 291, 297, 309
proteins:
AKT, 262
BCL-2 family of, 253, 256
"degenerate ferments," 93
metabolic, xvii, 244
mTOR, 262
study of, 10
transcription factors of, 249, 296
Protocols of the Elders of Zion, 79
Prudential Insurance Company, 24
Prussian Academy of Science, 116
PTEN, 276
Pure, White and Deadly (Yudkin), 309

Rabinowitch, Eugene, 189
racism:
anti-Semitism, *see* Jews
and eugenics, 103
philosophical justification of, 88, 102, 106

"race defilement," 126
racial hygiene, 209
racial mixing, 58, 108, 124, 218
scientific, 58, 65
Racker, Efraim, 273–74
radiation, 198, 202, 214, 295, 302
radioimmunassay, 282
Raff, Martin, 257
Rascher, Sigmund, 142–43
Rathmell, Jeff, 257–59
reactive oxygen species, 294–95
Reaven, Gerald, 315–16
Reich Citizenship Law, 124
Reich Committee for the Fight against
    Cancer, 166
Reich Education Ministry, xiv, 158,
    170, 217
Reich Ministry of the Interior, 160
research:
    causation vs. correlation in, 225
    manometer use in, 80–81, 95, 117,
        196, 261
    model systems in, 48, 94
    new discoveries being made in, 299
    observational studies, 231
    PET scans, xviii
    radioimmunassay in, 282
    randomized controlled trials,
        225–26
    signaling pathways in, 249
    theories based on experimental
        findings, 85
    thin-layer chromatography, 275
    tissue slice technique in, 81
    see also cancer research
respiration:
    and cancer, xvi, 6, 84–85, 146–47,
        150, 155, 199, 222–23, 267
    and carbon monoxide, 93–94

as cellular breathing, xvi, 6, 15, 16,
    21, 50, 81, 93–95, 101, 145–47,
    150, 214, 319
energy of, 84
fermentation vs., 83, 84–85, 143, 144,
    155, 196–97, 199, 200–202, 267
and food, 44
and nicotinamide, 148, 149
role of membrane in, 45
role of protein in, 93
roles of metals in, 47–48, 50, 93
respiratory ferment:
    and absorption spectrum, 94
    and "between ferment," 147
    in cellular breathing, 93, 145
    as "cytochrome c oxidase," 294
    and iron, 47–48, 93, 146, 177
    role of, 45
    Warburg's discovery of, 94–95, 99,
        143, 282
    Warburg's Nobel Prize for, 99, 101
    and Wieland's research, 144–45
rickets, 104, 146
Rigg, Bryan Mark, 160
Rockefeller Foundation, 116
    and Warburg's arrogance, 101, 185
    Warburg's institute funded by, xiv,
        96, 98, 101, 121–22, 123, 127,
        157, 185
    and Warburg's isolation, 157, 185
Rockefeller Institute, 45, 86
Röhl, John, 53
Romanticism:
    and anti-Semitism, 58, 106, 135
    on corruption in modernity, 102–3
    and nationalism, 58
    nature as highest ideal in, 102, 214
    Warburg's disparagement of, 144,
        198

Rous, Peyton, 86–87, 233, 287
Rügen, Warburg's vacation home in, 98, 169, 173, 181
Rust, Bernhard, 217

Sabatini, David, 269
San Francisco, cancer death rate in, 104
Sauerbruch, Ferdinand, 102, 158
Schirach, Baldur von, 324
schlempe, 323
Schoeller, Paula, 159
Schoeller, Walter, 147, 158–59, 171
Schrödinger, Anny, 150
Schroeder, Christa, 177
Schubart, Christian Friedrich Daniel, 135
Schweitzer, Albert, 27, 29
Science, 189, 201, 202, 286, 299
science:
    and Civil Service Law, 114–17
    crop yields increased via, 88–89
    foundational, 21
    German preeminence in, 20–21, 33, 92
    unlimited possibilities in, 10–11, 12, 13, 14, 15, 20
sea urchins:
    as basis of cancer science, 15, 50
    chromosomes of, 5, 6
    development of, 5, 6, 12, 41
    Loeb's studies of, 12
    research questions about, 3, 6
    respiration of, 47–48, 80
    Warburg's studies of, 6, 13, 14, 15, 30, 41, 43, 47–48, 50, 80, 95
Seidl, Daniella, 209
Seifert, Alwin, *Gardening and Ploughing without Poison,* 209

Senn, Nicholas, 27–28, 29
sex hormones, 285, 297
Seyfried, Thomas, xvii, xix
Siebenkäs (Paul), 315
Siegel, Rebecca, 271
Siemens Company, 74
Silent Spring (Carson), xvi, 212–14, 217, 218
Simon, Celeste, 267
single-celled organisms, 265–66
Smil, Václav, 236–37
smoking, and cancer, 218, 223–28, 229, 232, 246, 271, 297, 322
Snow, John, 226
snuff, inhaling, 322
Soviet Union:
    German invasion of (Operation Barbarossa), xv, 163–64
    German scientists recruited by, 181–82, 183, 184
    postwar German sector ruled by, 181–83, 186
    Red Army in Germany, 173
Spanish Inquisition, 292
SRC (oncogene), 247, 274
Stalin, Joseph, 163, 164, 173, 181, 183
Stambolic, Vuk, 290–91, 293
Stanhope, Kimber, 316
Stasi, 332
Stefansson, Vilhjalmur, 27
Steinhoff, Hans, 156–57
stock market collapse, U.S., 113
stomach cancer, 296
Streicher, Julius, 140
sucrose, 304, 305, 306–7, 308, 310–11, 317
sugar, 304–11
    addiction to, 322–25

and artificial sweeteners, 310
from beets, 304–5, 307, 310, 323
and cancer, 307, 309–11, 318–22
and carbohydrates, 305
and diabetes, 307–9, 321
and fermentation, 305
Hitler's addiction to, 323–25
and obesity, 321
refined, 307, 308, 310–11, 318, 320
and slave trade, 323
structure of, 306
use of term, 304
see also fructose; glucose; sucrose
sugarcane, 310, 317
sun exposure, 297
Sylt, Germany, Warburg country home
    in, 244
synthetic dye industry, 36, 39
syphilis, treatment of, 41
Szent-Györgyi, Albert, 98

Tanchou, Stanislas, 25–26
Tannenbaum, Albert, 233–35, 239,
    270
Taubes, Gary, 320, 321–22
Taylor, Maxwell D., 196
Teflon, 211
Telschow, Ernst, 158, 162, 185–86
Tesch (customs official), ix-xiii
testicular cancer, 309
testosterone, and cancer, 285
Theorell, Hugo, 147, 148
Thompson, Craig, xviii, 257–59
    and AKT gene, 261–63
    cancer research of, 258, 260–63,
        264, 266, 269, 295
    on cell death (apoptosis), 252–53,
        254, 256, 257–58, 272, 290

and epigenetic change, 264
    and growth factor, 257–58, 259
    immunology research of, 252–53,
        257
    and metabolism, 256, 258–59, 260,
        261
    and mitochondria, 254, 257, 263–64
    and Warburg revival, 252, 258,
        260–63, 269
Thunberg, Torsten, 131
thyroid cancer, xx
Times, The (London), 131
Tisdale, W. E., 122–23
tobacco, addiction to, 322
Train, Russell, 218
transcription factors, 249, 296
Traube, Wilhelm, 171
Treaty of Versailles, 72, 78, 79
Treitel, Corinna, 209
triglycerides, 300, 315
tuberculosis (TB), germ as cause of,
    153–54, 157, 213
tubularia, studies of, 11
tumors:
    and diet, 237, 298
    growth of, 6, 42, 237, 285
    and insulin, 293
    sequencing the genomes of, 276
    spontaneous rise of, 86
    transplantation of, 38–39, 42, 86
tumor suppressors, 263, 276
Twain, Mark, 13

UCSF, 248–49
Ullrich, Volker, 67
University of Chicago, 252
University of Heidelberg, 13, 15, 16,
    33, 43, 45, 47, 91

University of Illinois, Urbana, 187–93, *192*, 197
University of Pennsylvania, 259, 260, 267
University of Würzburg, 3
uricase (enzyme), 320
US Military Government, FIAT (Field Information Agency, Technical), 184
uterine cancer, xx, 289, 291, 309

Vander Heiden, Matthew, 254–57, 261, 265, 269
Varmus, Harold, 245–46, 247, 248–49, 274
vegetarian diet, 108, 109, 110, 139–40, 174–75, 209
Vennesland, Birgit, 17, 325–27
Verschuer, Otmar von, 186
vitamin B$_2$ (riboflavin), 146, 231
vitamin B$_3$ (niacin), 149–50
vitamins:
    and coenzymes, 151, 267
    discovery of, 146–47
    protective effects of, 228–31, 237, 267
    Warburg's gift to Soviet Union, 183

Wagner, Richard, 59, 108
Warburg, Dr. Betty (cousin), 169
Warburg, Elisabeth (mother), 7, 62, 63, 77–78
Warburg, Emil (father), 57, 186, 190
    and anti-Semitism, 79
    criticism of, 9
    death of, 99
    dedication to work, 8
    and Einstein, 7, 9, 79, 191

high standards of, 8–9, 43
influence on Otto, 8–9, 20, 50, 144, 326
and manometer, 81
and Otto's research, 49–50, 62, 72
professional influence of, 47, 49
Warburg, Eric (cousin), 17, 76–77, 122, 182, 207
Warburg, Felix (cousin), 79
Warburg, Gerta (aunt), 169
Warburg, Käthe (sister), 215
Warburg, Lotte (sister):
    death of, 215
    diary of, 8, 125, 150–51, 214
    and family, 8, 130
    leaving Germany, 120, 123–24, 129
    on Otto and women, 16, 18, 96
    on Otto's fear of death, 214
    Otto's letters to, 16, 120–21, 128
    on Otto's lifestyle, 98, 129, 214
    on Otto's personal traits, 77, 101
    on Otto's reluctance to leave Germany, 123–24, 125, 128, 130
    and Otto's work, 8, 18, 50, 116, 150–51
    in Paris, 129
    and Planck, 115–16, 130
Warburg, Max (cousin), 79, 120
Warburg, Otto, *xxiii, 24, 136, 312, 329*
    aging of, 223, 267–68, 303, 314, 321
    ambition of, 15, 24, 32, 37, 46, 74, 144, 150
    army service of, 59, *60*, 61, 62–64, 72, 98, 120, 160, 166
    arrogance of, x-xiv, xvi, xxii, 6–7, 13, 64, 80, 88, 99, 101, 119, 120, 144, 181, 185, 222, 251

and bicarbonate effect, 302
cancer research of, xv-xix, 23–24, 49,
    50, 101, 106, 143–51, 160–61,
    162, 166, 167, 171, 186–87, 189,
    195, 196–97, 198–200, 208,
    213–14, 244, 250, 266–67, 295,
    299–300, 321
cellular breathing experiment of,
    94–95
childhood of, 8–9
coenzymes discovered by, 146–48,
    149–50, 171, 231
death of, 327–28, 331–32
dismissal from his institute, xiv, 158
and doppelgänger, 314–15
in exile, 169
family of, x, 7, 8, 79, 129, 130
fear of death, 214–15
on fermentation, *see* fermentation
German Blood Certificate requested
    for, 159–62, 166
and "grana" (mitochondria), 253–54
and Heiss, 75–77; *see also* Heiss,
    Jacob
and his father, *see* Warburg, Emil
and his mother, 7, 62, 63, 77–78
and Hitler, xiv, 116, 163–64, 183
homes of, 7, 8, 97–98, 214, 244
homosexuality accusation against,
    170
and horses, 6, 98, 99, 183, 207,
    246
influence of, x, xvi-xviii, 23–24, 81–
    82, 106, 107, 127–28, 148–49,
    166, 202, 213–14
influences on, 8–11, 13–15, 21, 33,
    37, 41, 43, 63–64, 72, 83, 85–86,
    99, 144, 152, 155, 213, 326

as institute director, *see* Kaiser
    Wilhelm Institute for Cell
    Physiology; Kaiser Wilhelm
    Society
Iron Cross awarded to, 61, 66
isolation of, 92–93, 116, 151, 157,
    174, 182
Jewishness of, 59, 114, 116–18, 121,
    124, 125–28, 130, 132, 134, 151,
    157–62, 166–67, 170
at Kaiser Wilhelm Institute, *see*
    Kaiser Wilhelm Institute for
    Cell Physiology; Kaiser Wilhelm
    Society
letters of denunciation against,
    169–70, 191
lifestyle of, xiv, 97–99, 117, 167,
    169, 172–74, 208, 214, 215, 221,
    243–44, 245
"Lindau Lecture" by, 222–23, 230
metabolic approach of, 206, 251,
    259, 272, 321
move to U.S., 188–93
at National Cancer Institute, 193
Nazi harassment of, 157, 166
Nazi officials provoked by, ix-xiii,
    117–19, 127, 133, 159
Nazi restrictions on, 171
Nobel Prize awarded to, xiv, 7, 49,
    74, 99, *100*, 101
"On the Origin of Cancer Cells,"
    286
outdated ideas of, xvii, 223, 244–45,
    247, 251–52, 258
papers of, 80, 98, 173, 183, 186,
    200, 244, 260, 299, 332
paranoia of, 19, 91, 170, 191–92,
    215–17, 250–52

Warburg, Otto (*continued*)
and Pasteur effect, 83
personal traits of, x-xiv, 6, 17–19, 45,
    50, 74, 75, 77, 80, 91, 101, 121,
    132, 158, 170–71, 181, 184, 185,
    187, 190, 211, 222, 251, 273–74,
    303, 326–27
photosynthesis work of, *see*
    photosynthesis
physical appearance of, xii, 16
political connections of, xiv-xv,
    158–59, 171
political usefulness to Nazi Germany,
    128–29, 150, 157, 166
postwar job search of, 185, 187
post–World War II life of, 182–87,
    207–8, 215, 327
premature obituary for, 131–32,
    314–15
*The Prime Cause and Prevention of
    Cancer,* 230
professional resentment toward,
    250–52
professional training of, 9–11, 14,
    16, 33, 43
public speaking by, 18–19, 96, 197,
    222–23
rediscovery of, xix, xx-xxi, 252, 258,
    260–64, 269–71
relationships with women, 16,
    17–18, 96
reluctance to leave Germany, 120–
    24, 128–30, 184, 194, 250
reputation of, xvi, 15, 80, 96, 99, 127,
    166, 185, 197, 201–2, 251–52
respiratory ferment discovered by, 45,
    93, 94–95, 99, 101, 143–46, 151,
    282, 294
return to Germany, 195–97

sailing, 208, 243–44
"schizoid Warburg character" of,
    131–32
scientific breakthroughs of, 148–49,
    150–51
sea urchin research of, 6, 13, 14, 15,
    30, 41, 47–48, 50, 80, 95
simple explanations preferred by,
    321
as unable to admit error, 50, 121,
    200, 201–2, 222, 321, 325–26
at University of Illinois, 187–93, *192*
videotape of, 314
at Woods Hole, 193–94
work as highest priority of, xii, 7, 8,
    18, 19, 20–21, 73–75
Warburg, Paul (cousin), 73, 79
Warburg, Otto (cousin; botanist), 131,
    314
Warburg effect:
as aerobic glycolysis, xviii
and AKT enzyme, 290
cause of, 269–71
coining of term, 274
defined, xviii
as described by other researchers,
    262–63, 265
on fermentation of cancer cells, 84,
    251, 274, 287, 290, 319
and fructose, 319
and glucose, 263, 285, 287
and IGF-1, 291–92
insulin's role in, 285, 287, 290, 321
and MYC, 296
and obesity, 270–71, 272
and PI3K molecule, 289, 290, 296
result of, 270
revival of interest in, *see* Warburg
    revival

Warburg family:
  background of, 7, 8
  bankers in, x, 79
  as "non-Aryan," x, 116, 129
Warburg manometer, 81–82, 95, 117, *192*, 196, 245, 261
Warburg revival, xviii, xix, xx-xxi, 252, 258, 260–64, 269–71, 285, 298
"Warburg's yellow enzyme," 146–47
Watson, James, xvii, 244
Weaver, Warren, 132
Weber, Thomas, 66
Weill Cornell Medicine, 258, 274, 299
Weinberg, Robert, "The Hallmarks of Cancer," 251
Weinhouse, Sidney, 200–202, 206, 221
Weismann, August, 253
Wellen, Kathryn, 264–65
White, Charles Powell, 29–30
Wieland, Heinrich, 131, 144–45
Wilderness Act, 212
Wilhelm, Prince (Sweden), *100*
Wilhelm I, Kaiser, 52, 154, 168
Wilhelm II, Kaiser, xiv, 51, 53, 54, 56, 62, 66, 71, 76
Williams, W. Roger, 233
  The Natural History of Cancer, 278
Willstätter, Richard, 120
Windaus, Adolf, 166
Winternitz, Milton, 204–5
Wood, Grant, *American Gothic,* 161
Wood, Harlan, 220–22
Woods, Mark, 286
World Health Organization, 210
World War I, 59–62
  beginning of, 54, 236
  German Jews in, 56–58, 59, 61–62, 64, 65
  German uhlans (cavalry regiments) in, 59, *60*, 61, 72, 120
  Hitler in, 65–67, 204
  poison gas used in, 114, 203, 204–5, 211, 235
  propaganda in, 55–56
  setting the stage for World War II, 53, 89
  sugar consumption in, 308
  and Treaty of Versailles, 72, 78, 79
  Warburg's military service in, 59, 61, 62–64
  war propaganda, 55–56
World War II, 156, 172–73, 174, 185, 224
  Buffalo Soldiers in, 286
  World War I setting the stage for, 53, 89
World Zionist Organization, 131
Wynder, Ernst, 227

Xiaodong Wang, 255

Yalow, Rosalyn, 281–83, 315
yeast, carcinoma tissue compared to, 266–67
Yudkin, John, 310
  Pure, White and Deadly, 309

Zhukov, Georgy, 183
Ziegler, Frederick, 306
Zionism, 56, 131, 168, 315
Zweig, Stefan, 121
Zyklon B, 177, 204, 205, 323
zymohexase (aldolase), 167–68